10.99

advanced

LEVEL

study aids

biology

advanced

LEVEL study aids

biology

John Churchman and Kath Pedder

JOHN MURRAY

Titles in this series:

Advanced Level Study Aids Biology   0 7195 7630 X
Advanced Level Study Aids Chemistry 0 7195 7631 8
Advanced Level Study Aids Physics   0 7195 7629 6

First published in 2000
by John Murray (Publishers) Ltd
50 Albemarle Street
London W1S 4BD

Reprinted 2001, 2002

Layouts by Amanda Easter
Artwork by Oxford Designers & Illustrators
Cover design by John Townson/Creation

Typeset in 10/12 Garamond by Wearset, Boldon,
Tyne and Wear
Printed and bound in Great Britain by St Edmundsbury Press,
Bury St Edmunds

A catalogue entry for this title is available from the British
Library

ISBN 0 7195 7630 X

Contents

Acknowledgements

The authors wish to thank their respective wife Jean, and husband Ian, for all their support and encouragement, and gratefully acknowledge the help and assistance given by Katie Mackenzie Stuart and Michelle Sadler of John Murray (Publishers) Ltd.

Exam questions have been reproduced with kind permission from the following examination boards:

AQA: The Associated Examining Board (AEB)
         Northern Examinations and Assessment Board (NEAB)
Edexcel, including London Examinations
OCR: University of Cambridge Local Examinations Syndicate (UCLES)
         Oxford
Welsh Joint Education Committee (WJEC)

The answers provided in this book are the sole responsibility of the authors, and no responsibility whatsoever is accepted by the examination boards for the accuracy or method of working given.

Thanks are due to the following for permission to reproduce copyright photographs: **p.40** Biophoto Associates; **p.68** Dr. Kari Lounatmaa/Science Photo Library; **p.70** Biophoto Associates; **p.212** © Gene Cox; **p.232** Peter Parks/www.osf.uk.com; **p.250** © Gene Cox

The publishers have made every effort to contact copyright holders. If any have been overlooked they will be pleased to make the necessary arrangements at the earliest opportunity.

Introduction

This book is a Study Aid rather than simply a revision guide to be used at the end of an A level course. A very important part of Advanced Level Study Aids Biology is the questions and model answers accompanying each of the 45 topics, which cover much of the core material included by the examination boards. The text is written in such a way as to help the student understand the biological concepts. Important scientific words are in **bold** type, and words and phrases of which the student should take special note are in *italics*. Ideally, each section should be revised as it is covered in the student's A level course, so that understanding of key concepts can be improved and then tested by attempting the relevant questions.

The questions and answers are an extension of the learning but are chosen not only to complement the text but also to illustrate the types of questions the different examination boards are currently setting. The simpler introductory questions are designed to help the student gain confidence to tackle the later more complex ones.

The mark scheme for each examination board question has been provided, with hints on tackling essay questions and a break down of the essay mark scheme. Many students lose valuable marks by vague or incomplete answers. This book stresses the main biological principles and their application to a particular topic. It also gives examples of the amount of detail required in answers. Examination technique can be improved by referring to the guidance notes in *italics* as these emphasise the skills and knowledge being tested.

Answers to questions have been presented in a distinct way so that students can clearly see how marks would be allocated for each question.

- Where questions have several appropriate answers that can be given to gain full marks, separate points are given in the form of a bullet list.
- Where all of the answers provided have to be given to gain full marks, each point is given on a new line to separate the relevant answers required.
- If statements given in the answers are linked and do not make sense if written independently, statements have been separated by a semi-colon.
- Alternative words or phrases for the same answer are all given, and separated with a solidus.

The section dealing with synoptic questions is especially valuable since the new Synoptic Papers will be unfamiliar to many students. The questions are designed to test the student's ability to link different areas of biology, and therefore their overall understanding. Examples are given of applying statistics to biological data, drawing graphs, interpreting diagrams, providing explanations from data, comprehension of scientific text and synoptic essays.

Biological molecules, cells and cell systems

1 Cells and tissues

All cells have an outer **plasma membrane** enclosing the cell **cytoplasm** (a watery liquid – the cytosol) in which are the organelles and the DNA, usually within a **nucleus**.

Prokaryotes

The cyanobacteria (photosynthetic blue–green algae) and bacteria both consist of prokaryotic cells.

- They lack a true nucleus bound by a nuclear membrane and, instead, have a single long strand of DNA in the shape of a ring.
- Cell division is by binary fission rather than by mitosis or meiosis.
- If chlorophyll is present, it is not in chloroplasts.
- The mesosome is the site of respiration and therefore has the same function as the mitochondria of eukaryotic cells.
- Cyanobacteria are unicellular or filamentous.

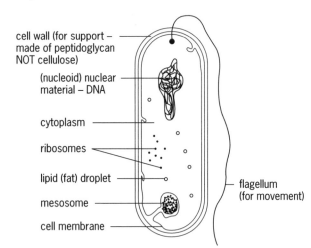

Figure 1.1 Simplified diagram of a bacterium, for example *Escherichia coli*

Eukaryotes

Apart from the bacteria and cyanobacteria, all other living organisms have eukaryotic cells with:

- a nucleus bounded by a nuclear membrane (or envelope)
- a number of **organelles** with highly specialised functions within the cytoplasm.

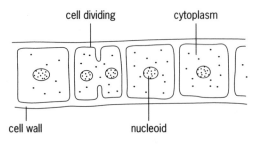

Figure 1.2 Simplified diagram of a filamentous cyanobacterium, for example *Oscillatoria*

Organelles are surrounded by their own membrane so that the enzymes and raw materials needed for reactions are always at hand. Thus the speed of reactions is increased. Also, those substances which would be harmful can be isolated from the rest of the cell.

Plant parenchyma, a tissue composed of unspecialised thin-walled cells with a packing function, is made of typical eukaryotic cells (Figure 1.3).

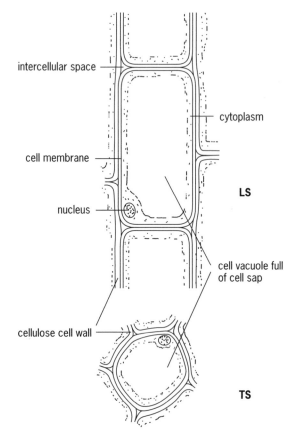

Figure 1.3 Plant parenchyma cells as seen under high power

Animals have a layer of covering cells called epithelium which line the organs. Simple epithelium (Figure 1.4) is found lining the cheek in mammals.

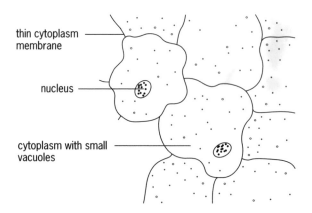

Figure 1.4 Epithelial cells from the lining of the alveoli in mammalian lungs as seen under high power

The organelles

Nucleus

The nucleus is ovoid or spherical. As it is more acidic than the surrounding cytoplasm it takes up different biological stains, for example methylene blue. This makes it easier to see under the microscope. Within the nucleus are one or more **nucleoli** (*singular nucleolus*) which contain RNA (see **6** The genetic code and protein synthesis). The nucleus contains the chromosomes which are concerned with heredity.

Endoplasmic reticulum (ER)

ER consists of a three-dimensional network of saclike, flattened structures called **cisternae**. They provide a large surface area for synthesis. Large surface areas are needed for the efficient diffusion of substances across membranes. Large surface areas are also needed as 'workbenches' for biochemical reactions. In some ER the membranes are covered with **ribosomes** which consist of proteins and ribosomal DNA. This type of ER is called *rough* ER (Figure 1.5). Elsewhere the ribosomes are missing and it is called *smooth* ER.

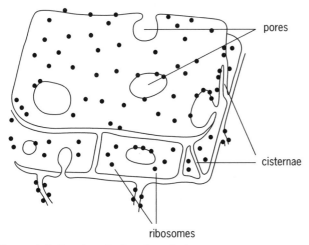

Figure 1.5 Rough endoplasmic reticulum

The ribosomes make proteins which are passed through the membranes to be stored in the sacs of the rough ER. They are then either used or passed to the outside of the cell. Rough ER is abundant in enzyme-secreting cells such as those of the pancreas. The smooth ER found in the testes secretes steroids.

Mitochondria

Mitochondria are just visible under the light microscope (about 0.2–0.5 µm). They are the site of aerobic respiration and are therefore found in large numbers in cells such as the liver and muscle, where a lot of energy is needed.

Mitochondria have a double membrane, with the inner membrane folded into a number of projections called **cristae**. These provide a large surface area. Situated on the cristae are the **oxysomes** which are where ATP is produced during respiration (Figure 1.6). The space inside the inner membrane is called the **matrix**. It contains enzymes and DNA. The DNA codes for the synthesis of proteins.

Mitochondria replicate when a cell divides.

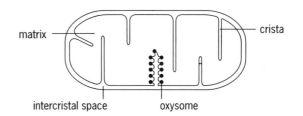

Figure 1.6 Diagrammatic cross-section of a mitochondrion

Golgi apparatus

A stack of **cisternae** are temporarily linked to, but not connected with, the endoplasmic reticulum. Secretory vesicles are formed and move to the periphery of the cell where they fuse with the cell membrane and release excretory or secretory substances to the outside of the cell – a process known as **exocytosis**.

In the Golgi apparatus simpler molecules may be assembled into more complex ones, for example, carbohydrates and proteins may be packaged into glycoprotein.

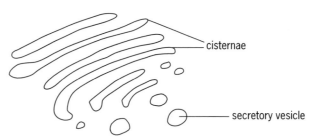

Figure 1.7 Diagram of Golgi apparatus

Summary of the differences between prokaryotic and eukaryotic cells

Prokaryotic cells	Eukaryotic cells
possess a nucleoid with a single strand of DNA	possess a membrane-bound nucleus
no ER, chloroplasts or mitochondria	possess chloroplasts, ER and mitochondria
respiration takes place on a plasma membrane – the mesosome	respiration takes place at the mitochondria
cells divide by binary fission	cell division is by mitosis and meiosis

Tissues

Tissues are large groups of cells which together perform a specific function, for example, xylem and phloem (see **28** and **29** Transport systems in flowering plants 1 and 2) and squamous (flattened and plate-like) epithelium (Figure 1.4).

Tissues may be drawn under the high power of the light microscope (about 400× magnification) when individual cells are shown in detail, or using low power (about 100× magnification) when tissue plans are drawn (see Figure 28.4, **28** Transport systems in flowering plants 1).

CELLS AND TISSUES

1 a Label the diagram of a sensory neurone.

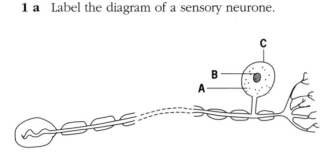

(3)

b i Name 3 structures found in a typical plant cell, which are not present in a typical animal cell. (3)
ii Give the function of each of these structures. (3)

2 The table below shows structures found in some typical cells. Place a tick or a cross to show whether the structures would be present or absent in the three types of cells.

Structures	Plant cell	Animal cell	Prokaryote
cell wall			
nuclear membrane			
mitochondria			
chloroplasts			

(4)

3 a Distinguish between tissues and organs, giving an example of each in an animal. (4)

b Name the structures **A** to **D**. (2)
c Calculate the actual diameter of the structure labelled **F**. Show your working. (2)
d Describe the functional relationship between structures **A**, **D** and **E**. (3)

OCR, Biology Foundation Paper 4801, March 1999

4 Plant material may be ground up to break open cells. A buffered ice-cold solution is then added and the mixture is centrifuged differentially to isolate the cell components. Each time the material at the base of the tube is collected and the centrifuge process is repeated at a higher speed with the supernatant. The first centrifugation resulted in nuclei and cell fragments being deposited at the base of the tube.

The following cell organelles were obtained at different centrifuge speeds:

mitochondria, chloroplasts and ribosomes.

a Fill in the names of these organelles in the correct positions on the scheme below.

very slow	slow spin	medium spin	fast spin
↓	↓	↓	↓

cell fragments
and nuclei

(2)

b Suggest why an ice-cold solution is used to homogenise the tissues. (1)
c Why is a buffered solution used? (1)
d The organelles are separated at different speeds of centrifugation. What causes this? (1)

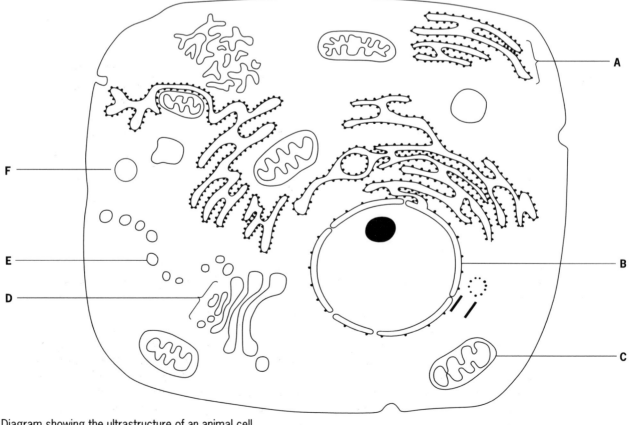

Diagram showing the ultrastructure of an animal cell

5 Use a tick or cross in the box to show whether the statements are correct or not. (3)

Statement	Bacteria	Plants	Viruses
able to divide by mitosis			
cannot reproduce independently			

6 Describe the functions of the following cell organelles:
   a microtubules (2)
   b ribosomes (2)
   c nucleolus (1)
   d Golgi apparatus (2)

7 a Describe how a prokaryotic cell would differ from a eukaryotic cell in the location of its DNA. (2)
   b i Name two cell types in mammals where you would expect large numbers of mitochondria to be present. (2)
   ii Name the part of the mitochondrion in which the enzymes for the Krebs cycle reactions would be found. (1)

8 a From your knowledge of electron micrographs, compare the ultrastructure of plant and animal cells. (10)
   b Describe how a stage micrometer and a graticule could be used to find the width of a guard cell. (4)
   c i State two advantages of using electron microscopy to study the structure of cells. (2)
   ii Describe how material is prepared for the electron microscope. (4)

9 Explain:
   a Why some cells have their membranes folded into microvilli. (2)
   b Why the inner membrane of mitochondria is folded into cristae. (2)

10 a Relate the structure of eukaryotic cells to their function. (10)
   b List the advantages of eukaryotic organisation. (5)

1 a **A** – cytoplasm
  B – nucleus
  C – cell membrane
  b **i, ii** Three of:
  - The cellulose cell wall which supports and gives shape to the cell.
  - The central vacuole which helps to provide turgor pressure for support.
  - The chloroplast which is responsible for photosynthesis.
  - The inner cell membrane which controls entry and exit of substances to the vacuole.

2

Structures	Plant cell	Animal cell	Prokaryote
cell wall	✓	✗	✓
nuclear membrane	✓	✓	✗
mitochondria	✓	✓	✗
chloroplasts	✓	✗	✗

3 a
  - A tissue is a collection of specialised cells/similar cells/limited number of cell types.
    Examples include adipose tissue, blood, etc.
  - An organ is two or more tissue types, working together to perform a particular function.
    Examples include heart, eye, etc.

  b **A** – rough endoplasmic reticulum
  B – nuclear envelope/nuclear membrane
  C – mitochondrion
  D – Golgi (body/apparatus)

  c *It is useful to remember that the following units are related by a factor of 1000.*

  1 metre = 1000 mm = 1 000 000 µm = 1 000 000 000 nm

  Remember that a high magnification is needed to see organelles because they are very small.
  Always show the steps in your calculation. These earn you some marks even if you make an arithmetical error.
  First change the length you have measured into micrometres.

  8 mm = 8 × 1000 µm = 8000 µm

  Divide by the magnification to obtain the actual length.

  8000/9000 = 0.89 µm

  Alternative answers:
  0.000089 cm/0.00089 mm/889 nm/8.9×10^{-5} cm/ 8.9×10^{-4} mm.

  d A – ribosomes manufacture protein.
  D – Golgi apparatus packages/modifies/processes proteins.
  E – secretory vesicle removes proteins from the cell/carries them to the cell membrane.

4 a *The heavier particles will sediment at lower speeds. This question requires you to place the organelles in order of decreasing size.*

very slow ↓	slow spin ↓	medium spin ↓	fast spin ↓
cell fragments and nuclei	chloroplasts	mitochondria	ribosomes

  b The low temperature will slow down the deterioration of the cell organelles by slowing the rate of enzyme action.
  c A buffered solution maintains a constant pH.
  d It is separation by difference in mass (weight) since heavier particles will sediment to the bottom of the tube at lower centrifugation speeds than lighter particles.

5 *(Hint – take care with negative statements)*

Statement	Bacteria	Plants	Viruses
able to divide by mitosis	✗	✓	✗
cannot reproduce independently	✗	✗	✓

6 a Two of:
  - Microtubules help to provide **an internal structure** for the cell and also form the nuclear spindle in cell division.
  - They may allow **transport of materials** inside the cells.
  - Microtubules are found in cilia and flagellae which help in **cell movement**.

  b Ribosomes are the sites of **polypeptide/protein synthesis**.
  It is a site for the attachment of messenger RNA molecules, which dictate the order of the amino acid molecules which are brought to the ribosome by their tRNA molecules to form a specific polypeptide.
  c The nucleolus produces ribosomal RNA/ribosomes.
  d The Golgi apparatus is responsible for **modifying protein** materials from the ER.
  Substances such as enzymes and conjugated proteins such as mucus are then **packaged** and transported as vesicles across the cell.

7 a DNA is found organised into chromosomes within the **nucleus** in eukaryotic cells.
  In prokaryotic cells the DNA is circular in form and not bounded by a nuclear membrane.
  b **i** *Note: Mitochondria produce ATP, a useful short term store of energy. Metabolically active cells will have a high energy requirement and will therefore contain many mitochondria.*
  Two of:
  - kidney tubule cells
  - muscle fibres
  - the terminals of axons.

  ii The matrix.

8 a *It is often a good idea to illustrate your answers, in this case with a simple diagram of a typical plant and animal cell as seen in electron micrographs.*

You must compare or contrast one type of cell with the other. Describing each separately will gain few marks. Do not describe the functions.

Ten of:
- Both are surrounded by a fluid mosaic membrane (describe).
- Only the plant cell has an outer wall made up of a basket-like mesh of cellulose fibrils.
- Both types of cell have a nucleus with patches of chromatin and a deeply staining nucleolus.
- The nuclear membrane of each has pores and is continuous with the ER.
- In both, the rough ER forms flattened cisternae with ribosomes attached.
- In both, the smooth ER is not continuous with the nuclear membrane.
- In both, the Golgi body is a series of flattened sacs which is formed by vesicles fusing to one side and budding off the other.
- In both, elongated oval mitochondria are surrounded by two membranes with the inner membrane folded.
- Both possess lysosomes which are small circular vesicles 0.2–0.5 µm surrounded by a single membrane.
- Only some plant cells possess chloroplasts, which have a double membrane surrounding the stroma. Grana made up of lamellae/thylokoids act as a stacking system for chlorophyll.
- Many plant cells have a large central vacuole surrounded by a membrane/tonoplast. This is not found in animal cells.
- The cytoplasm of plant cells connects with others via plasmadesmata (cytoplasmic strands).

b Exchange the eyepiece of the microscope for a graticule with a scale. Place the micrometer slide on the stage and focus on it. Measure how many eyepiece units (EPU) fit into one division (0.1 mm = 100 µm) on the micrometer.

e.g. $8\,EPU = 100\,µm$
$1\,EPU = 100/8 = 12.5\,µm$

Then place the leaf section on the stage and focus with the same objective on the guard cell. Using the eyepiece graticule, measure its width in eyepiece units and convert this into micrometres.

e.g. guard cell $= 2\,EPU = 2 \times 12.5\,µm = 25\,µm$

c i Higher magnification (times enlarged).
Higher resolution (clarity of image).
ii Four of:
- The material must be placed in a vacuum and then treated with a fixative (for example, osmium oxide) to prevent the structure changing.
- It is then dehydrated (removal of water) using ethanol.
- Then cleared and embedded in resin or similar material.
- It can now be sectioned into thin slices using a microtome.
- It may be necessary to stain the sections using an electron dense solution.
- The sections are supported on a copper grid.

9 *An important concept in biology is that many structures are organised to raise their surface area to volume ratio. This often improves their efficiency.*

a Two of:
- Microvilli will **increase the surface area** available.
- For absorption for active uptake. *or*
- To increase the rate of diffusion of metabolites in liver cells or cells of the ileum.

b Cristae increase the surface area of the inner membrane of the mitochondrion; giving more sites for the attachment of enzymes of the respiratory chain and the ATP granules.

10 *(Essay hints: To obtain high marks in essays you need to organise your facts and arguments into a logical order. This can be done as rough notes and then crossed out later. Time must be spent thinking about the precise wording used in the essay title and then ensuring that the points made are relevant and explained in a scientific manner. Definitions and examples are often a source of extra marks. Care should be taken with grammar and spelling.)*

a *This essay requires an overview of eukaryotic cell structure (cells with membrane-bound organelles). Emphasis needs to be placed on the importance of the structure and functions of the fluid mosaic membrane which surrounds/makes up many of the cell organelles.*

A simple diagram of the membrane structure with its phospholipid bilayer and proteins with hydrophilic channels, plus branching glycoproteins related to their functions (partial permeability, facilitated diffusion, recognition and adhesion) could be included. You need to summarise the structure of mitochondria giving details of how they are adapted and which processes occur in the cristae and matrix. This treatment needs to continue for the nucleus, ER, Golgi, ribosomes, microtubules and chloroplasts (plants only).

b Five of:
- The advantages of a eukaryotic type of organisation are that the organelles are surrounded or made up of membranes which allow separate processes to continue within.
- This enables metabolic processes to be organised in a more efficient way.
- Enzymes embedded in the membranes are often arranged in the same sequence as the metabolic pathway.
- This increases the rate of reactions, as the products of reaction one will be in close proximity to enzyme two, etc.
- Membranes control the rate of entry and exit of reactants, therefore controlling the rate of reactions.
- Harmful reactants and enzymes can be isolated within the cell.
- Eukaryotic cells are often larger than prokaryotic cells but the many internal membranes increase the area to volume ratio.

2 Biological molecules

Classes of food substances

Carbohydrates

Carbohydrates are compounds containing the elements carbon, hydrogen and oxygen only. The hydrogen and oxygen are in the same proportions as in water (2:1).

The simplest carbohydrates (those with much smaller molecules) are **sugars**. The simplest sugars consist of a short chain or a ring of no more than six carbon atoms. They are called **monosaccharides** and have either an **aldehyde** or a **ketone** group. Aldehydes and ketones are both reducing groups – they reduce Benedict's solution when heated and a yellowish-red precipitate is formed.

Glyceraldehyde (the simplest sugar) is a **triose**. It has *three* carbon atoms in its molecule (Figure 2.1).

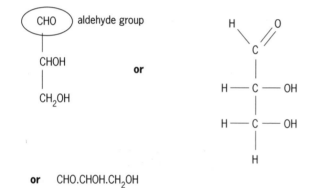

or   CHO.CHOH.CH₂OH

Figure 2.1 A molecule of glyceraldehyde

Glyceraldehyde has an aldehyde group (CHO) at one end of the chain. There is a double bond to the oxygen atom. All carbohydrates have twice as many hydrogen as oxygen atoms (this will mean a number of 'H' and 'OH' groups in approximately equal numbers). Some molecules have the same formula. These are known as isomers. Fructose is a **ketose** with the same formula ($C_6H_{12}O_6$) and a similar structure to that of glucose but instead of an aldehyde group it has a ketone group (C=O). Unlike an aldehyde, no hydrogen atom is attached to the carbon atom.

The structure of glucose is similar to glyceraldehyde but it has more carbon atoms and, hence, more H and OH groups. When glucose is not in solution the oxygen atom attached to C_5 very easily links to C_1. This gives a single ring structure (hence the name **mono**saccharide) and the aldehyde group, although present, is no longer obvious. This structure is called a pyranose ring because of its shape.

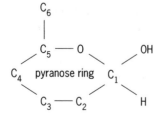

Figure 2.2 A molecule of glucose (H and OH groups attached to the carbon atoms only shown on C_1)

Glucose exists in two forms, alpha and beta (Figure 2.3).

Figure 2.3 Alpha (α) and beta (β) forms of glucose

Monosaccharides are the basic units (**monomers**) of which other carbohydrates are composed. These monomers can be linked by **condensation** reactions (the removal of a molecule of water – see Figure 2.4) to produce biological polymers including sucrose, glycogen and cellulose. The link is called a **glycosidic link**.

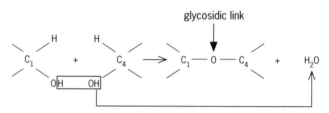

Figure 2.4 Two molecules of α glucose are condensed to give one molecule of maltose

Disaccharides

- **Maltose** – formed by the condensation of two molecules of alpha glucose. Maltose is still a reducing sugar as only one of the carbonyl (aldehyde or ketone) groups is involved in the **1–4 glycosidic link**.
- **Sucrose** – formed by the condensation of one molecule of alpha glucose with one molecule of beta fructose. Sucrose is *not* a reducing sugar – both carbonyl groups are used in the reaction. Sucrose is used as a sweetener and is the sugar obtained from sugar cane and sugar beet.
- **Lactose** – formed by the condensation of one molecule of galactose (a structural isomer of glucose) with one molecule of glucose. It occurs in mammals' milk.

The **hydrolysis** of disaccharides involves the *addition* of a molecule of water and the formation of two molecules of the appropriate monosaccharide(s). It is the opposite of condensation.

Polysaccharides

- **Starch** – a polymer of alpha glucose. Several hundred to a few thousand molecules of alpha glucose give two forms of starch:
  - **Amylose** – formed of glucose joined by unbranched 1–4 glycosidic links wound into a helix.
  - **Amylopectin** – formed of 1–4 glycosidic links plus 1–6 glycosidic links approximately every 25 rings. This gives a branched structure.

  Starch is the main food storage compound in plants. Because it is insoluble, it does not affect the water potential of cells (see **4** The transport of substances across cells and through membranes).
- **Glycogen** – found in muscle and liver cells is similar to amylopectin.
- **Cellulose** – found only in plants. A polymer of several hundred to a few thousand beta-glucose rings. The 1–4 glycosidic links produce very long unbranched fibres. The cellulose cell walls of plants are made of many fine fibres of cellulose in parallel bundles plus some in a criss-cross pattern. This gives a strong structure which may be further strengthened by lignin, for example, xylem tissue.

Fats (also called lipids and triglycerides)

Fats are formed from two different types of molecule:

- A glycerol molecule.
- A long chain of carbon atoms to which hydrogen atoms are attached and with a COOH (acidic) group at one end. These are known as fatty acids and differ from each other both in the length of the hydrocarbon chains and the number of single or double bonds there are between the carbon atoms. An example is stearic acid, $C_{17}H_{35}COOH$.

Three fatty acids react with the three OH groups of one glycerol molecule to give one molecule of **triglyceride** (fat). The reaction is reversible and involves condensation or hydrolysis. Because three fatty acids are needed to form one molecule of triglyceride, three molecules of water will be involved.

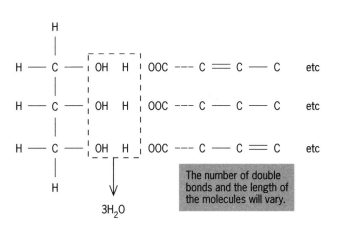

$3H_2O$

Figure 2.5 Formation of a fat molecule by condensation

The carbon chains may be unsaturated or partially saturated. An unsaturated fat is one in which some of the bonds between the carbon atoms are double bonds. Single bonds allow more hydrogen atoms to be attached to the carbon atoms.

Fats contain the elements carbon, hydrogen and oxygen but only a small amount of oxygen. They are important storage molecules as they contain about twice as much energy per gram as carbohydrate molecules. In birds and mammals fats may be important aids in heat insulation (see also **15** The respiration of fats, page 79).

Proteins

Proteins are polymers of hundreds or thousands of amino acid molecules of which there are 20 different ones found in proteins. They are compounds of carbon, hydrogen, oxygen and nitrogen (three also contain sulphur). Proteins and amino acids are amphoteric (they have both acidic and basic properties) and can act as **buffers**. The amino group is basic and the carboxyl group is acidic. In strongly acid conditions ionisation of the carboxyl group is suppressed. At the same time the amino acid groups attract hydrogen ions and become positively charged. In very alkaline solutions the reverse applies.

The simplest amino acid is called glycine and its formula can be written as:

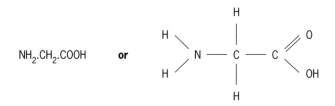

$NH_2.CH_2.COOH$ **or**

Very large protein molecules are formed because the basic end (NH_2) of one molecule reacts with the acidic end (COOH) of another. For example, glycine + glycine = di-glycine (Figure 2.6).

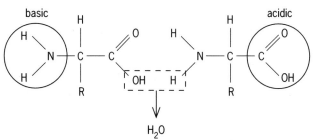

H_2O

Figure 2.6 Formation of a molecule of di-glycine

R = the rest of the molecule. In glycine it is hydrogen but it may be complex – acidic, basic, neutral or polar (soluble). The condensation of amino acid molecules produces chains of amino acid groups called polypeptides which form protein molecules. The CONH link formed between two amino acid molecules is called the **peptide bond** (Figure 2.7).

protein structure

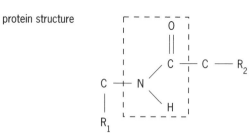

Figure 2.7 The peptide bond

Fibrous proteins consist of long parallel chains of polypeptides (for example, as found in muscles and tendons); **globular** proteins are folded into spherical shapes which are soluble (for example, enzymes); **conjugated** proteins have non-protein groups joined to them (for example, haemoglobin).

There are many different proteins, all containing different combinations of amino acids. Polypeptides have three types of linkage which are responsible for giving the shape of the molecule: ionic bonds (weak), hydrogen bonds (very weak) and disulphide bonds (strong) (see Figure 2.8).

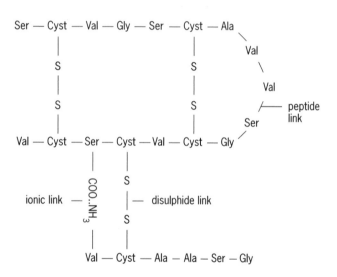

Figure 2.8 Polypeptide and disulphide links

- **Primary structure** – the shape and properties of the protein are determined by the sequence of amino acids in the chain.
- **Secondary structure** – due to hydrogen bonding which produces a spiral known as the alpha helix.

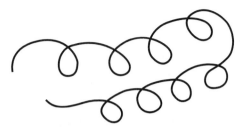

Figure 2.9 Secondary structure of proteins

- **Tertiary structure** – the helix is bent and twisted due to the effect of all three types of bonding.

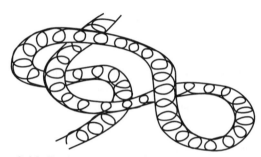

Figure 2.10 Tertiary structure of proteins

- **Quaternary structure** – a large complex molecule with a combination of several different polypeptides together with non-protein (prosthetic) groups. These prosthetic groups are often metal ions which are essential for the functioning of the protein. An example is iron in haemoglobin (see also **33** Oxygen and carbon dioxide transport in mammals).

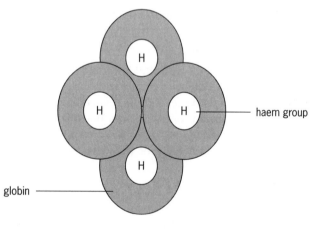

Figure 2.11 Diagram of human haemoglobin

1 a How many carbon atoms do the following carbohydrates contain?
  i glyceraldehyde
  ii glucose
  iii sucrose (3)

b i Albumin is a water-soluble protein. Compare albumin with starch by completing the table. (3)

	Albumin	**Starch**
monomer (a single building block)		
details of identifying test		
colour change in positive result		

  ii Give a general formula for a disaccharide. (1)
c i With the aid of a diagram similar to the one in Q2, show the chemical breakdown of a molecule of sucrose to produce one molecule of glucose and a molecule of fructose. (2)
  ii Name the chemical process by which a disaccharide is broken down into monosaccharides. (1)
  iii Give a difference between the structure of maltose and lactose. (1)
d Explain the term 'glycosidic bond'. (1)

2 The diagram below shows a chemical reaction.

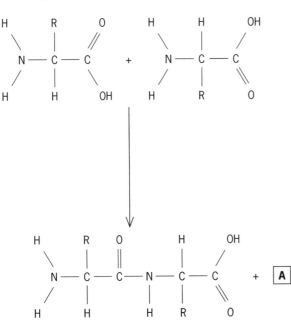

For this reaction:
i Name the type of product formed. (1)
ii Name molecule **A**. (1)
iii What type of chemical reaction is this? (1)
iv What type of enzyme would catalyse this reaction? (1)

3 Sucrose is a disaccharide. The diagram shows the structure of a molecule of sucrose.

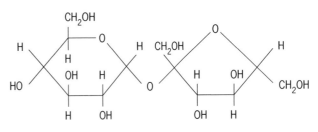

i Use the diagram to explain why sucrose is classified as a carbohydrate. (1)
ii Explain why sucrose will produce a positive result with Benedict's test only after it has been boiled with a dilute acid. (2)

AQA (AEB), Paper 1, January 1999

4 The diagram shows the tertiary structure of a molecule of the enzyme RNAase.

position **A**

O = C

HO

● sulphur-containing amino acid

a What is the name of the chemical group found in position **A**? (1)
b i Explain what is meant by the *tertiary structure* of a protein. (1)
  ii The chemical mercaptoethanol breaks disulphide bonds (bridges). Describe and explain what would happen to the enzyme activity of RNAase if it were treated with mercaptoethanol. (3)

AEB, Paper 1, January 1997

5 The diagram below shows how some biological molecules may be separated from each other.

   a Name element **X**. (1)

   b Which of the molecules **A** to **D** could be:

     i glucose (1)

     ii glycogen (1)

6 a Peanuts contain stored lipids. Describe a test that you could use to show the presence of lipids in peanuts. (1)

   b Compare the monomers of lipids with those of polysaccharides and proteins. (2)

7 Write an account of triglycerides giving details of their structure, functions and location in both mammals and plants. (10)

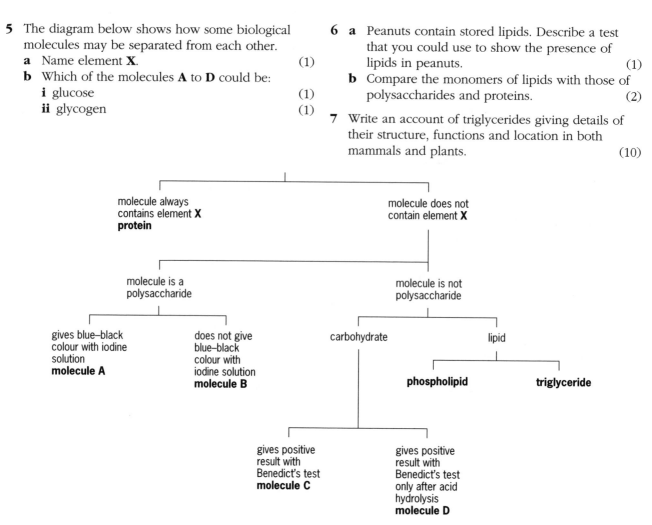

AQA, Specification A Specimen Paper, June 1999 for 2001 exam

1 a i 3 **ii** 6 **iii** 12
  b i

	Albumin	**Starch**
monomer	amino acid	monosaccharide (glucose)
details of identifying test	add dilute copper sulphate solution and dilute sodium hydroxide solution	add dilute iodine in potassium iodide solution
colour change in positive result	the solution turns pale purple	the solution turns blue–black

  ii $C_x(H_2O)_y$
  c i

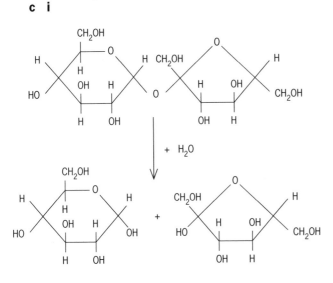

  ii Hydrolysis.
  iii Maltose consists of two glucose units joined together whereas lactose is made up of one glucose and one galactose subunit.
  d A glycosidic bond is a linkage via an oxygen molecule formed by a condensation reaction between monosaccharide units.

2 i dipeptide
  ii water
  iii condensation
  iv hydrolases/peptidases/proteases

3 a i It contains carbon, and the ratio of hydrogen:oxygen is a 2:1 ratio/the same proportions as are found in water.
    ii Two of:
    • It needs to be hydrolysed/the glycosidic bond must be broken.
    • The product is a reducing sugar/glucose/fructose/monosaccharide.
    • This frees the aldehyde/carbonyl/ketone group.

4 a amino group/amine/NH_2
  b i The tertiary structure is the irregular folding of the polypeptide chains to form a globular shape held together by hydrogen bonds, disulphide bridges and ionic bonds.

ii Breaking of the disulphide bonds would alter the globular shape of the enzyme and its active site. The substrate would no longer fit in the active site. This would result in loss of catalytic activity of the RNAase/enzyme will not function/functions slower.

5 a nitrogen
  b i molecule **C**
    ii molecule **B**

6 a The test should be performed using dry, grease free apparatus. Ethanol should be added to the ground up peanut and the tube shaken to dissolve the lipids. 5 cm of water is then added and the tube is shaken. A milky emulsion indicates the presence of lipid.
  b Lipids have two types of monomers, fatty acids and glycerol, whereas both proteins and polysaccharides have only one type of monomer (amino acids and monosaccharides respectively).

7 *In answering such a question it is often useful to underline the key words and then to constantly check back to ensure that the points you make are relevant. In this case you would concentrate on triglycerides, structure, function and location.*

Structure:
• Triglycerides are formed by a condensation reaction removing three molecules of water.
• Between one molecule of glycerol and three fatty acid molecules.
• Diagram showing structure (refer to Figure 2.5 in the text).

Functions – Five of:
• Triglycerides are useful stores of energy.
• They contain on average twice as much energy per gram as carbohydrates/are lighter to carry around than the energy equivalent of starch.
• This makes them suitable for motile animal and plant structures involved in dispersal, such as seeds.
• They are insoluble in water which means they do not take part in the chemical reactions of the cell until they have been hydrolysed into more metabolically active units. This makes them very suitable storage compounds.
• Triglycerides have a high hydrogen content which results in the formation of useful amounts of metabolic water when oxidation occurs during respiration.
• They have insulating properties.
• They can form a protective layer.

Location – Two of:
• Mammals store lipid in adipose tissue under the skin.
• Seeds which require large energy reserves, for example, castor oil, have triglyceride reserves (in the endosperm).
• Animals (such as camels) living in arid conditions store lipid in their hump (for metabolic water).
• There is a thick subcutaneous layer (under the skin) in whales (insulation).
• There is a layer of lipid around mammalian kidneys (protective).

3 Plasma membranes

A phospholipid is a glyceride in which one of the fatty acids has been replaced by a phosphate group which is joined to the glycerol part of the molecule.

The phosphate group combines readily with water because it is polar while the fatty acid chains are non-polar.

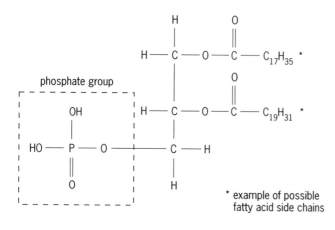

Figure 3.1 A molecule of a phospholipid

Phospholipids are important in cellular and subcellular membranes (for example, the membranes of mitochondria and the endoplasmic reticulum).

The structure of a phospholipid is shown more simply in Figure 3.2.

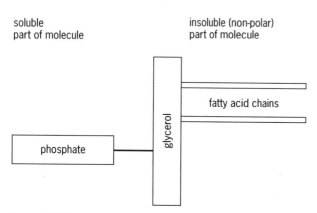

Figure 3.2 Diagram of the structure of a phospholipid

There is a great difference in solubility between the two parts of phospholipid molecules. Because of this, at an air/water or oil/water interface the molecules form layers and thus form a membrane as shown in Figure 3.3.

The **hydrophilic** 'phosphate heads' of these molecules attract water whereas the hydrocarbon 'tails' of the fatty acids are **hydrophobic** and repel water.

If lecithin (a typical phospholipid) is dissolved in ether and a very small amount is carefully poured on water, the ether evaporates and leaves a single layer of the phospholipid on the surface of the water. The 'head' of each phospholipid molecule dissolves in the water whilst the 'tails' stand out of the water (Figure 3.3).

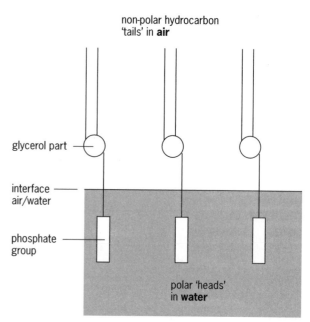

Figure 3.3 A phospholipid monolayer

If a second layer of phospholipid is then very carefully poured on top of the first, a double layer is formed and the 'tails' of the second layer are attracted to the 'tails' of the first layer. Such a bilayer occurs in the structure of cell membranes.

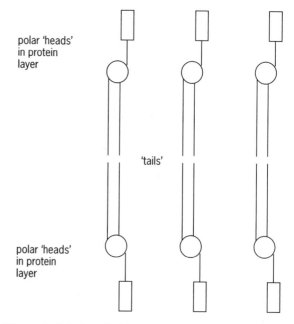

Figure 3.4 A phospholipid bilayer

Surfactant (a complex compound of a phospholipid and a protein) present on the surface of the cell membranes of the alveoli in the lungs reduces surface tension. It makes the meniscus more easily stretched and prevents the alveoli from collapsing. Surfactant is produced by fetuses at 24 to 30 weeks' gestation and may therefore be absent in premature babies who then have to be put on a ventilator until the body produces surfactant – usually after 3 or 4 days.

The cell membrane

The cell membrane acts as a boundary between the cell and its environment. In the laboratory, experiments are often performed using Visking tubing to demonstrate the nature of the cell membrane but, unlike Visking tubing, the cell membrane is a living structure and a functioning organelle. It is not simply semi-permeable like Visking tubing, allowing smaller molecules to pass through but excluding larger ones. Whilst some substances may pass freely in and out of the cell, others may be excluded on one occasion but allowed to pass freely at another time. The cell membrane is best described as being **selectively permeable**.

The cell membrane is composed almost entirely of phospholipids and proteins. Singer and Nicholson suggested that the cell membrane had a fluid structure because the phospholipid part is capable of movement and the protein parts float within it. This idea of the structure of the cell membrane is called the **fluid mosaic model** (Figure 3.5).

The proteins have two parts to their molecules:

- a hydrophobic part which is buried in the 'tails' of the lipid layer
- a hydrophilic part which is involved in a variety of activities.

There are thousands of different proteins associated with cell membranes. Some of them move freely and others are fixed in one place. Some extend right across the lipid layer (integral proteins) whilst others only penetrate part of the way or remain outside of the lipid layer (peripheral proteins).

Some proteins are **carrier proteins**. Their function is to bind with certain ions and molecules and enable them to pass through the cell membrane by **facilitated diffusion** or by **active transport** (see **4** The transport of substances across cells and through membranes). During active transport, energy from ATP is used to actively move materials in or out of cells. These are **active carrier systems**. **Channel proteins** form pores which allow substances such as oxygen and carbon dioxide to pass through the cell membrane by **simple diffusion**.

Other proteins have different functions:

- **Glycoproteins** (proteins with a carbohydrate component) are part of the immune system and are important in cell recognition.
- **Specific receptors** make cells sensitive to a particular chemical (for example, a hormone).
- **Enzymes** – especially on the internal cell membranes.

Cholesterol is a steroid that occurs in the membranes of animal cells (as well as in the blood) but not plant cells.

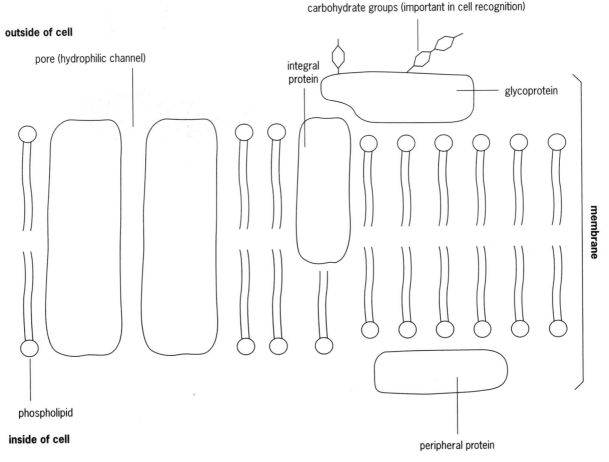

Figure 3.5 Fluid mosaic model of the cell membrane

1 a The molecules listed below may be found in eukaryotic cells.

glycogen glycoprotein histones cytochromes phospholipids RNA DNA

Choose two of the molecules listed above that would normally be found in:
i the cell membrane
ii a chromosome (4)

b Explain the biological significance of the following:
i The nuclear envelope has many pores on its surface.
ii The membrane of some cells is folded into microvilli. (2)

2 Early research on the structure of the cell membrane showed that lipid-soluble compounds passed rapidly into cells. The membrane was found to be selectively permeable to mineral ions, sugars and amino acids. Further work demonstrated that all membranes have the same basic structure but can differ greatly in the types of lipid and protein they contain. Many of the specialised proteins present provide a means of communication between cells and molecules in their environment.

a Apart from lipid solubility, suggest two factors which could affect the rate of penetration of a molecule through the membrane. (2)

b Describe how the structure of the cell membrane is related to:
i Its selective permeability. (5)
ii Its communication with molecules in the cell's environment. (2)

c Describe how prokaryotes and eukaryotes differ in terms of the membrane-bound structures they contain. (3)

NEAB, BY01, June 1997

3 The table below refers to components of the cell surface membrane (plasma membrane) and their roles in transporting substances across the membrane.

Complete the table by inserting the appropriate word or words in the empty boxes.

Component	Subunits	Chemical bond between subunits	Role in transport
phospholipid	fatty acids, glycerol and phosphate		
carbohydrate side chain			receptor
protein		peptide	

London, Paper B1, January 1997

4 The diagram represents a phospholipid molecule.

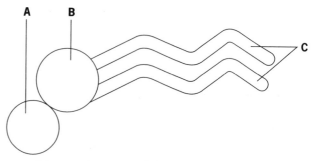

a i Name the parts of the molecule **A**, **B** and **C**. (1)
ii Explain how the phospholipid molecules form a double layer in a cell. (2)

b Cell membranes also contain protein molecules. Give two functions of these protein molecules. (2)

NEAB, BY01, March 1998

5 a Using the following symbols draw a simple diagram to show how phospholipids and proteins make up the structure of the cell surface membrane. (3)

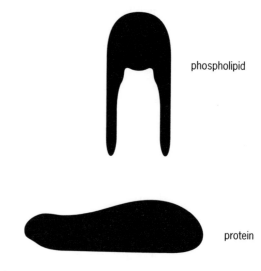

phospholipid

protein

b Why is the cell membrane described as fluid? (1)

6 Give an account of the structure and function of cell membranes. (10)

1 a i The cell membrane would contain phospholipids and glycoproteins.
ii A chromosome would contain histones and DNA (and RNA whilst being transcribed).

b i The nuclear envelope has pores to allow large molecules such as mRNA to pass from the nucleus into the cytoplasm.
ii Microvilli increase the surface area of the cell membrane to facilitate absorption.

2 a Two of:
- size of molecule
- its polarity/water solubility/electrical charge
- temperature/pH
- concentration
- recognition by protein receptors/carriers
- ATP/oxygen/inhibitors linked to active transport
- thickness of membrane
- surface of membrane
- surface area of membrane.

b i Five of:
- The bimolecular phospholipid layer.
- Acts as a barrier to polar (water soluble) molecules/glucose/amino acids.
- Proteins form channels/pores through the membrane.
- Size/shape/hydrophilic groups of pore determines selectivity.
- Facilitated diffusion can occur through these pores.
- Proteins act as carrier molecules.
- Proteins may be involved in active transport.
- Enzymes linked with active transport.
- Cholesterol affects permeability/fluidity.
- Short-chained/unsaturated fatty acids increase permeability.
ii Glycoproteins/glycolipids act as recognition sites. Proteins in membranes act as receptor sites.

c Three pairs from:

Prokaryotes	Eukaryotes
no nucleus	nucleus
no nuclear membrane	nuclear membrane
no mitochondria	mitochondria
mesosome	no mesosome
no endoplasmic reticulum	endoplasmic reticulum
no Golgi body	Golgi body
no lysosome	lysosome
no organised chloroplast	chloroplasts in photosynthetic cells

3

Subunits	Chemical bond between subunits	Role in transport
fatty acids, glycerol and phosphate	ester	reference to vesicle formation/cytosis/chylomicrons/reference to diffusion receptor
monosaccharide/name of a monosaccharide/reference to pentose or hexose	glycosidic	
amino acids	peptide	reference to channels/pores allowing passage/facilitated diffusion/receptor/enzyme/carrier/active transport

4 a i A – phosphate/phosphoric acid; **B** – glycerol; **C** – fatty acid
ii Two of:
- Phospholipids have polarity/heads out/tails in.
- The heads are hydrophilic and arrange themselves towards water.
- The tails are hydrophobic and arrange themselves away from water.

b Two of:
- carriers/for active transport
- receptors
- antigens
- enzymes
- form channels.

5 a

b Both phospholipids and proteins can move by diffusion within the membrane.

6 *Start with a description of the fluid mosaic model and its component molecules.*

Ten of:
- The membrane controls the entry and exit of substances into the cell/it is partially permeable.
- The polar nature of the phospholipids with their hydrophobic tails and hydrophilic heads forms the basis of the membrane, allowing lipid-soluble molecules to diffuse into and out of the cell.
- The proteins may be structural.
- Proteins may function as enzymes, sometimes being arranged in linear sequence on one side of the membrane to catalyse a particular metabolic pathway.
- Some proteins act as carrier molecules transporting specific substances across the membrane.
- Ion channels in proteins allow facilitated diffusion of certain hydrophilic substances.
- Some of the proteins may act as receptors for hormones.
- Proteins may act as antigens in the immune response.
- Active transport of certain molecules or ions against a concentration gradient requires ATP and involves sodium pumps.
- The two sides of the membrane may differ in their structure and properties.
- The outside of the cell membrane often has branching glycoproteins which may aid cell adhesion during tissue formation.
- Glycolipids may also be present. Both of these types of molecules contain polysaccharides.
- Cell organelles are surrounded by this type of membrane in eukaryotic cells, permitting the separation of different areas of the cells and allowing their own particular metabolic processes to proceed unhindered.

4 The transport of substances across cells and through membranes

Aqueous solutions and water

A water molecule is slightly **polarised** – it has negative and positive ends which are slightly charged and which react differently with ions and charged molecules. Figure 4.1 shows how the electrical charges are separated. Such a separation is called a **dipole**.

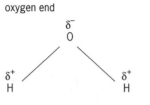

Figure 4.1 Charges on a molecule of water

This polarity of water gives it the following important properties which have an effect on living organisms.

- Electrovalent compounds readily dissolve in water. The charged ions of the solute (for example, K^+ and Cl^-) are attracted to the pole of the water molecule which has the opposite charge.
- Because the charges in a water molecule are separated there is a tendency for **hydrogen bonds** to be formed. There is an attractive force between a hydrogen atom of one molecule and part of another molecule such as the OH group of sugars, the $C=O$ group of organic acids and the NH group of proteins. Molecules go into solution when they become surrounded by water molecules.
- Organic molecules with a high mass (for example, polysaccharides and proteins) attract water molecules to form a colloid but not a true solution. Cytoplasm and blood are examples of colloidal suspensions.

More properties of water which are important for living organisms are as follows.

- Ice is less dense than water and therefore floats. The ice formed then insulates the water below from further heat loss and prevents it from freezing solid.
- Water has a high **latent heat of vaporisation**. This is the energy that is required to evaporate water and it explains why water has a cooling effect and prevents terrestrial organisms from overheating.
- Water heats up and cools down more slowly than other compounds with a comparable relative molar mass. This is because water has a high **specific heat capacity** (the amount of heat energy required to raise 1 gram of a compound 1 °C). Large volumes of water such as lakes and oceans maintain a steady low temperature which is suitable for aquatic organisms. Terrestrial organisms also consist of a large proportion of water and the high specific heat capacity of water helps to prevent rapid changes in body temperature.

Diffusion

> **Diffusion is the net movement of molecules of a gas or a substance in solution from an area of higher concentration towards an area of lower concentration down a diffusion gradient.**

Every cell in an organism has substances in solution diffusing in and out through its cell membrane.

- Oxygen, in solution, diffuses *from* a nearby blood capillary and the carbon dioxide that the cell produces diffuses *to* a blood capillary.
- The soluble products of carbohydrate and protein food digestion diffuse from the gut, through the lining of the small intestine, and into blood capillaries.
- In green plants water diffuses from the mesophyll cells of the leaves into the intercellular spaces and then diffuses as water vapour through the stomata and into the atmosphere.

Apart from the main transport systems of blood and lymph in animals and the vascular systems of plants, diffusion is the main method by which substances travel within cells, through cell membranes and from cell to cell.

Diffusion is only efficient over very small distances. The following factors determine the rates at which substances diffuse.

- *The concentration gradient.* The steeper the gradient (the greater the difference between concentrations) the faster the rate.
- *The distance that the molecules diffuse.* The rate of diffusion is inversely proportional to the square of the distance (**inverse square law**). This means that if the distance is doubled, diffusion takes four times as long.
- *The surface area.* The larger the surface area, the greater the rate of diffusion.
- *The structure of cell membranes.* For example, the larger the number and size of pores in a membrane, the greater the rate of diffusion.
- *The size of the molecules.* Small molecules diffuse faster than larger ones.

According to **Fick's Law**, the rate of diffusion is proportional to:

$$\frac{\text{surface area} \times \text{difference in concentration}}{\text{thickness of membrane}}$$

Facilitated diffusion

In facilitated diffusion the exchange is more rapid. Carrier molecules, which are fat soluble, may assist larger molecules, such as sugars, across membranes. Facilitated diffusion still requires the diffusion gradient to be in the right direction. No energy is involved – the process is **passive**.

Active transport

Substances often have to be moved in or out of cells against the concentration (diffusion) gradient. This is like having to push something uphill. Work has to be done! Carrier molecules may move such substances from one side of the membrane to the other. Examples include:

- the kidney tubules, where chloride and sodium ions are *pumped* from the tubules, against the concentration gradient, into the surrounding tissues
- the root hairs of plants which absorb nutrient ions present in the soil water and accumulate them against the concentration gradient.

Active transport uses the energy from ATP to actively move substances up a concentration gradient.

Osmosis

> **Osmosis is the net flow of water through a membrane, selectively permeable to water but not to solute, down a water potential gradient.**

Osmosis is the special case of the diffusion of water molecules across a selectively permeable membrane. Diffusion is still down a diffusion gradient but because one does not talk about a 'concentrated solution of water', reference is made to the water molecules passing from a lower concentration of *solute* to a higher concentration of *solute*.

The words 'net flow' are used in the above definition to remind you that water is actually diffusing across the selectively permeable membrane in both directions but that more molecules will pass into the more concentrated solution of solute.

Free molecules (gases, solutions and liquids) have a natural kinetic energy which causes them to move at **random**. The more energy they possess (the higher the temperature) the faster they move and the faster will be the rate of diffusion/osmosis.

Because the molecules of both solute and solvent move at random, and only the molecules of water solvent can pass through the pores of the selective membrane, there will be a net movement of water molecules from left to right in Figure 4.2. Remember also that the smaller water molecules diffuse faster than the much larger sucrose molecules.

The total amount of kinetic energy in pure water (or a solution) on either side of a membrane is called the **water potential** (Ψ), which is the potential for water to move out of a solution by osmosis. Pure water at standard temperature and pressure (25 °C at one atmosphere) has a water potential (WP) of zero. Aqueous solutions have a WP of less than zero as they can exert less pressure because the water molecules are impeded in their passage by solute molecules.

When there is a concentration difference between two solutions separated by a selective membrane, a water potential gradient exists. In the simple apparatus shown in Figure 4.3 there is a net movement of water molecules through the Visking tubing and into the glass tube. The pure water in the beaker has a WP of approximately zero and the sucrose solution will have a WP less than zero. Thus, a WP gradient is created.

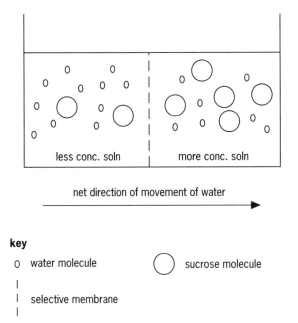

key

o water molecule   ◯ sucrose molecule

|
| selective membrane
|

Figure 4.2 A simple model illustrating osmosis

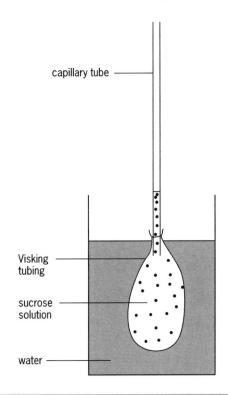

The inward movement of water molecules will continue until the downward pressure of the column of water in the tube equals the osmotic pressure. The steeper the WP gradient, the faster will be the rate of osmosis. As the two solutions on either side of the membrane approach the same concentration (which can never be achieved) the rate of osmosis will slow down.

Figure 4.3 A simple osmometer

Osmosis in plant cells

Plant cells possess the following.

- A resistant cellulose cell wall permeable to aqueous solutions. The cell wall exerts a pressure (rather like a tyre resists the internal pressure of air). This pressure is termed **pressure potential** and has the symbol Ψ_p.
- A cell membrane which is selectively permeable.
- One or more cell vacuoles containing cell sap with dissolved solutes that give it a negative WP. The cell sap has the potential to cause water to move into it across a selectively permeable membrane. This potential is called **solute potential** and has the symbol Ψ_s.

If the cell is full of cell sap so that the cytoplasm is pushed against the cell wall, the cell is said to be **turgid**. If not, the cell is wilting or **flaccid**.

There are three possibilities when a plant cell is placed into an aqueous solution.

- If the cell is turgid, $\Psi_p = \Psi_s$ and $\Psi_{cell} = 0$ and there is no loss or gain of water or size or mass.
- A non-turgid cell will absorb water. When there is no pressure potential then $\Psi_{cell} = \Psi_s$. Water will be gradually absorbed until the water potential equals zero at full turgidity.
- If the concentration of the external solution is greater than that of the cell sap there will be a net loss of water from the cell. When the cell becomes flaccid $\Psi_{cell} = \Psi_s$. As water is lost, the protoplast begins to pull away from the cell wall. This is known as **incipient plasmolysis**. Further loss of water will bring about full plasmolysis.

The relationship between the three is:

$$\Psi_{cell} \quad = \quad \Psi_s \quad + \quad \Psi_p$$

| water potential | solute potential | pressure potential |

Wilting

A loss of turgor in the cells of the leaves, and other soft plant tissues, may lead to wilting. This loss of water may be due to drought or to fungal infections (fungal wilt) which block the xylem vessels. A condition known as **water stress** is due to the rate of water loss by transpiration exceeding the rate of water uptake by the roots. A point may be reached (the permanent wilting point) where the plant cells have collapsed and are unable to recover.

Turgor pressure is particularly important in providing support in non-woody tissues.

Endocytosis

This includes two processes by which some cells enclose a smaller object, such as a food particle or bacterium, within a membrane-bound vesicle.

1. Phagocytosis

Particles, such as food particles, which are too large to be taken in by diffusion or active transport may still be absorbed by cells. A cup-shaped depression is formed and the food material surrounded. The depression is then 'pinched off' to form a vacuole. Lysosomes (sac-like organelles in the cell cytoplasm containing hydrolytic enzymes) fuse with the vacuole and release their digestive enzymes. The digested products are reabsorbed.

This process takes place in a few specialised cells, for example, white blood cells (phagocytes) and some protoctists such as amoeba (see **41** Natural selection and classification).

2. Pinocytosis

This is sometimes called 'cell drinking'. The process is similar to phagocytosis but liquids are taken in instead of solids.

Exocytosis

This is the reverse process to endocytosis in which waste materials are removed from cells. The process needs energy from ATP.

1 Write the letter A, B, C or D to show the one correct answer for each part of the question.

  a The process by which a large molecule of protein leaves a cell is called:
    A – osmosis
    B – diffusion
    C – endocytosis
    D – exocytosis

  b Which of the following statements about factors affecting the rate of diffusion is incorrect?
    A – The rate of diffusion is inversely proportional to the thickness of the membrane.
    B – Large molecules diffuse faster than small molecules.
    C – If the distance is doubled, diffusion will take four times longer.
    D – The greater the surface area, the greater the rate of diffusion.

  c Which of the following is an active process?
    A – The facilitated diffusion of a molecule of glucose into a red blood cell.
    B – The diffusion of oxygen into a red blood cell.
    C – The uptake of sodium ions against a concentration gradient.
    D – The diffusion of carbon dioxide out of an actively respiring cell. (3)

2 Read through the following account of osmosis and osmoregulation. Give an appropriate word or words to complete the account.

For water to enter a living organism by osmosis, the water potential outside the organism must be _____ than the water potential inside. There must also be a _____ membrane, usually the cell surface membrane of a living cell. In freshwater protozoa, a _____ pumps out excess water. This pumping requires energy which is provided by _____. Simple plants such as algae, living in fresh water, do not require a pumping mechanism because the _____ generated by the stretching of their walls counteracts the _____ resulting from the effects of the cell contents. (6)

London, Paper B2, June 1996

3 The water potential of a plant cell (Ψ_{cell}) can be calculated using the following equation.

$$\Psi_{cell} = \Psi_s + \Psi_p$$

where Ψ_s = solute potential
and Ψ_p = pressure potential

  a A plant cell was immersed in pure water until it was fully turgid. It was then placed in a concentrated solution until it was plasmolysed. The table shows some of the values of the potentials in the cell at equilibrium under these different conditions.

Complete the table by writing the missing values in the empty boxes. (4)

Condition of cell	Potential (kPa)		
	Ψ_{cell}	Ψ_s	Ψ_p
fully turgid			+300
plasmolysed		−500	

  b The diagram below shows a fully turgid plant cell.

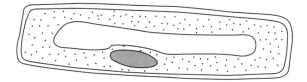

Draw a diagram to show the same cell when it is plasmolysed. Label only the cell surface membrane. (3)

London, Paper B2, January 1997

4 a When some plant cells are immersed in a weak solution of sodium chloride, they are observed to increase in volume.
    Explain the observation in terms of water potential. (4)

  b The Atlantic salmon spends part of its life in the sea and returns to fresh water to spawn. The body of the salmon is covered by a layer of scales which are largely impenetrable to water. The cells lining the mouth and alimentary canal, as well as those making up the gill area, are not covered with scales.
    In an investigation, salmon were caught as they were about to enter the mouth of the river and were transferred to tanks in which the sea water was replaced by flowing fresh water. The fish were observed over a period of 3 days. They stopped drinking and their urine output was four times greater by the end of the observation period.
    i In terms of water control, describe the problem faced by a salmon as it moves from the sea to fresh water. (2)
    ii With reference to the investigation, suggest a reason why the salmon:
    1 – stopped drinking
    2 – produced more urine. (2)

  c Plants and animal cells responded in similar ways to immersion in fresh water. Plant cells, however, are able to survive the treatment whereas animal cells are likely to die unless they are removed within a short space of time. This is due to differences in structure between the two cell types.
    Suggest an explanation for the difference in response of the two cell types. (2)

OCR, Biology Foundation Paper 4801, March 1999

5 a Give two differences between the processes of active transport and osmosis. (2)

b Explain the role played by these two processes in:

i The uptake of materials from the soil by roots. (5)

ii Selective reabsorption in the kidney nephron. (5)

6 Uptake of different mineral ions by cereal plants was measured by growing large numbers of seedlings in nutrient solutions of different mineral ions including calcium, potassium and nitrate.

The concentrations were measured at the beginning and at the end of the experiment for each ion. The results are shown below.

Mineral ion	Ionic concentration (arbitrary units)	
	At start	At end
calcium ion	5.2	5.7
potassium ion	2.9	0.0
nitrate ion	7.2	2.0

During the experiment there is a large change in nitrate concentrations.

Calculate the percentage difference between the nitrate ionic concentration at the start and at the end of the experiment. Show your working. (3)

7 An artificial cell containing a solution of sucrose and glucose was suspended in a solution of four different sugars in a beaker. The diagram shows the concentration in $mol\,dm^{-3}$ of each sugar inside and outside the cell.

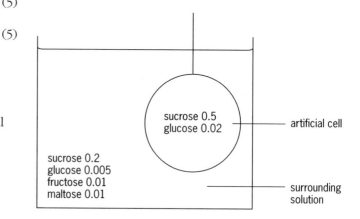

The membrane of this cell is permeable to water and to monosaccharides; it is not permeable to disaccharides.

a At the start of the experiment, which solution would have the lower water potential? Give a reason for your answer. (1)

b How would you expect the volume of the cell to change? Explain your answer. (2)

c As the experiment proceeds, for which solute or solutes will there tend to be net diffusion:

i out of the cell (1)

ii into the cell? (1)

AEB, Specimen Paper, 1997

1 **a** D   **b** B   **c** C

2 higher/greater/less negative;
selectively/partially/differentially permeable;
contractile vacuole;
ATP/respiration/mitochondria/oxidative phosphorylation;
pressure potential/turgor pressure/wall pressure/inward
pressure/Ψ_p; osmotic potential

3 **a**

Condition of cell	Potential (kPa)		
	Ψ_{cell}	Ψ_s	Ψ_p
fully turgid	0	−300	+300
plasmolysed	−500	−500	0

b

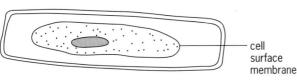

cell
surface
membrane

The cell wall must be shown by a double line and the
nucleus must be in the cytoplasm.
There must be clear separation of the lines for cell wall and
membrane.
The cell membrane must be labelled correctly.

4 **a** Four of:
  • The solute potential Ψ_s inside is lower/more negative
    than outside the cell.
  • Water will move by osmosis into the cell *down* a
    water potential *gradient*.
  • The vacuole/cell contents increase in volume.
  • The cell was not fully turgid to begin with.
  • The contents become diluted/Ψ_s increases.

  b **i** Two of:
  • The salmon is moving from a more concentrated salt
    solution to a less concentrated/from low Ψ to high
    Ψ/the water potential of the fish is now lower than
    the surrounding water.
  • There will be a net movement of water into the cells
    of the fish/the fish now needs to get rid of water.
  • Movement of water into the cells occurs in the
    mouth/gills/alimentary canal areas.

  ii 1 One of:
  • Water is already present in excess.
  • It prevents the intake of even more water.
  • Continued drinking could lead to the cells bursting.

  2 One of:
  • It eliminates the excess water taken in.
  • To prevent waterlogging of tissues.

  c Two of:
  • Plant cells have a cell wall/animal cells do not have a
    cell wall.
  • The cell wall prevents unlimited entry of water/allows
    turgidity/stops bursting.
  • In animal cells the cell membrane breaks due to
    excessive water intake.

5 **a** *This question is asking you to compare two processes.*
  Marks will not be awarded for mentioning just one process.
  Two of:
  • Diffusion of water in osmosis is a passive process
    whereas active transport requires energy in the form
    of ATP.
  • Active transport involves movement of substances
    other than water and involves carrier molecules
    whereas these are not used in osmosis.
  • Active transport occurs against the concentration
    gradient whereas it is down the water concentration
    gradient in osmosis.

  b **i** Five of:
  • Water is drawn into the root hair cells by osmosis due
    to their negative water potential;
  • this is caused by the solute concentrations inside the
    cell being higher than those in the soil;
  • and occurs when the cell is not fully turgid/pressure
    potential is less.
  • Water flows down a water potential gradient across
    the root as it is continually drawn up the xylem by the
    transpiration pull.
  • Active transport of minerals such as nitrate is
    necessary to achieve higher concentrations of solute
    than those present in the soil solution.
  • Specific carrier proteins on the root hair cell
    membranes have binding sites where the minerals
    become attached.
  • The mineral is rapidly transported across the
    membrane as the carrier changes its shape/
    conformation.
  • Energy for active transport is supplied by respiration.

  ii Five of:
  • The high solute potential of the blood draws water
    from the proximal tubule by osmosis.
  • Active transport removes glucose from the filtrate.
  • Other solutes such as amino acids/vitamins are
    selectively reabsorbed into the blood;
  • active transport of urea occurs.
  • Active transport of hydrogen ions occurs in the
    nephron.
  • Sodium ions are actively pumped out of the
    ascending limb of the loop of Henle.
  • Water is lost from the filtrate (descending limb) by
    osmosis due to the high concentration of solutes in
    the medulla.
  • Water passes into blood at the distal/second
    convoluted tube/collecting duct by osmosis.

6 Change in nitrate concentration = 7.2 − 2.0 = 5.2

$$\frac{5.2}{7.2} \times 100 = 72.22\%$$

7 **a** The artificial cell would have more solutes/would be
    more negative/have a lower Ψ_s.

  b The volume of the cell will increase; because water will
    tend to move in.

  c **i** glucose
  ii fructose

5 Genes and DNA replication

The discovery of the structure of DNA by Watson and Crick in 1953 was a giant step forward in the discovery of how genes are passed on from generation to generation.

Differences between one species of an organism and another, or members of the same species, is largely due to their proteins. Which proteins are to be made in a cell is determined by the DNA and RNA (both **nucleic acids**).

DNA makes up the chromosomes in the cell nucleus and carries a code which gives instructions for which proteins are to be made. Every somatic (body) cell of an organism carries an identical set of chromosomes.

RNA follows the coded instructions of the DNA and is concerned with protein/enzyme synthesis in the cell cytoplasm.

The structure of DNA and RNA

Both DNA and RNA are **polynucleotides** and the structure of their **mononucleotides** (basic units) is similar. Each has three parts:

- a pentose sugar with five carbon atoms: either ribose (in RNA) or deoxyribose (in DNA)
- a nitrogen-containing base
- a phosphate group.

Each monomer (consisting of these three parts) is a nucleotide. Nucleic acid is a long chain of nucleotides.

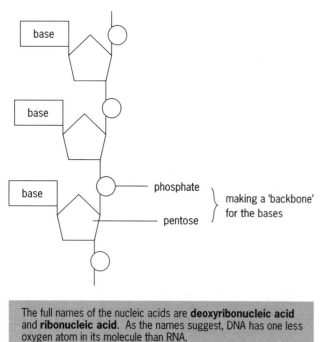

The full names of the nucleic acids are **deoxyribonucleic acid** and **ribonucleic acid**. As the names suggest, DNA has one less oxygen atom in its molecule than RNA.

Figure 5.1 Part of a nucleic acid molecule

There are two types of bases found in DNA:

- purines – adenine and guanine
- pyrimidines – thymine and cytosine.

To help you remember which is which, the three words 'pyrimidine', 'thymine' and 'cytosine' all contain the letter 'y'; the purines do not.

To form DNA, the bases A (adenine), T (thymine), G (guanine) and C (cytosine) join together in long chains which may contain millions of units. RNA is very similar but thymine is replaced by a different pyrimidine called uracil.

The double helix

While James Watson and Francis Crick at Cambridge were working on a three-dimensional model of DNA, Maurice Wilkins and Rosalind Franklin at King's College, London were using X-ray crystallography. The patterns from the X-rays suggested that the molecule had the shape of a double helix which measured 3.4 nm (one nanometre is a millionth of a millimetre) for each complete turn. Watson and Crick already knew, from the work of Erwin Chargaff in 1951, that the proportions of cytosine and guanine, and also adenine and thymine were always the same.

One question was 'What holds the two parts of the double helix together?' Watson discovered that the shapes of the cytosine and guanine bases were such that when paired, hydrogen bonds would hold them together. The same was true of the adenine and thymine bases. This also explained why the number of adenine bases always exactly equalled the number of thymine bases and the number of cytosine bases equalled the number of guanine bases. Adenine always pairs with thymine as does cytosine with guanine. They are known as **complementary pairs**.

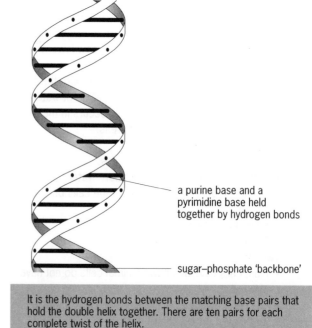

a purine base and a pyrimidine base held together by hydrogen bonds

sugar–phosphate 'backbone'

It is the hydrogen bonds between the matching base pairs that hold the double helix together. There are ten pairs for each complete twist of the helix.

Figure 5.2 The double helix

Hydrogen bonding

An OH group is neutral because the hydrogen part which donated an electron to the oxygen part is positively charged and this balances the now single negative charge of the oxygen part.

If another similar group is nearby, a weak link may be made by the attraction of opposite charges. This is what is called a hydrogen bond.

In nucleic acids, hydrogen bonding occurs where the two chains are able to fit closely together. This happens when an adenine base on one chain corresponds to a thymine base on the other, and similarly with the guanine and cytosine. This arrangement always occurs in DNA molecules.

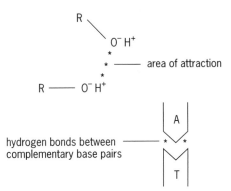

Figure 5.3 Hydrogen bonding

The two chains of a DNA double helix are complementary because of the base pairing. They fit together perfectly and any other arrangement will not work.

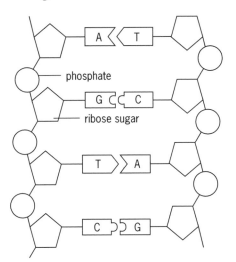

Figure 5.4 Diagram of part of a DNA double helix to show base pairing

At the beginning of this topic we noted that every body cell of a particular animal or plant organism carries an identical set of DNA. This is essential because the DNA codes for the synthesis of enzymes and other proteins, and thus characterises a particular species and individual member of that species. Thus, when a cell divides, the DNA also replicates and forms two identical double helices. This is what is happening when the chromosomes in a cell replicate during cell division.

The hydrogen bonds which hold the two strands of the double helix together form only a weak bond. During DNA replication, the two strands 'unzip' along the line of hydrogen bonds, and separate, starting at one end. The enzyme **helicase** unwinds the helix. Free nucleotides, present in the cell cytoplasm, move in and pair with matching bases forming new hydrogen bonds. The nucleotides thus join together to form a new and identical double helix of DNA.

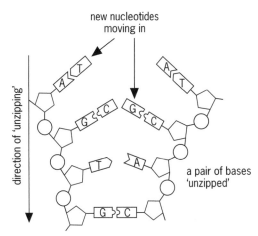

Figure 5.5 Replication of a molecule of DNA

In this way the original molecule of DNA is replicated and two identical DNA molecules are formed. Replication is controlled by an enzyme called **DNA polymerase**.

Each new double helix contains one strand of the original DNA and one strand made up of new material. However, for several years after Watson and Crick had produced their model, biologists were not sure whether the new double helix was formed as explained here, or whether the original double helix remained intact and in some way instructed the formation of a new, identical double helix.

This question was resolved by Meselson and Stahl.

- Some *Escherichia coli* bacteria were grown for several generations in a medium containing 'labelled' nitrogen-15 (^{15}N) which was the only source of nitrogen. Eventually all the bacteria became 'labelled' with ^{15}N.
- Then the bacteria were given only normal nitrogen and the density of their DNA tested as they reproduced.

One of two results was possible:

- *either* half the DNA would have the density expected for ^{15}N and the other half that for ^{14}N. The original double helix would remain intact and consist entirely of ^{14}N. This is termed **conservative replication**
- *or* all the DNA would have the same density which would be the average of the expected results for ^{14}N and ^{15}N. This is **semi-conservative replication**.

As you know, the replication of DNA is semi-conservative.

Note: take great care over spelling thymine and adenine since other biologically important words are similar.

1 Give the letter corresponding to the one correct answer.

   a The base not present in RNA is:
     A – adenine
     B – guanine
     C – thymine
     D – uracil   (1)

   b DNA is not found in:
     A – a gene
     B – a cell membrane
     C – a chromatid
     D – a nucleus   (1)

   c Which of the following base pairs is normally found in DNA:
     A – cytosine and guanine
     B – adenine and guanine
     C – adenine and uracil
     D – cytosine and adenine   (1)

   d The sugar found in RNA is:
     A – ribulose
     B – deoxyribose
     C – ribose
     D – glucose   (1)

2 A DNA molecule is made up of two strands which separate when DNA replicates. Each then acts as a template.

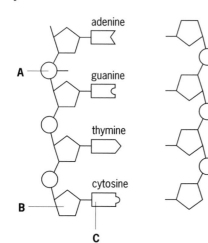

a Complete the drawing of part of the molecule of DNA to show which bases are needed to produce a new DNA molecule. (2)

b Name **A**, **B** and **C**. (3)

c Explain why it is important that DNA is able to replicate. (3)

d Draw a ring on the diagram to enclose one nucleotide. (1)

e Name the type of chemical reaction which occurs when nucleotides join together to form a molecule of DNA. (1)

3 a When a sample of DNA was analysed biochemically it was found that 31% of the nitrogenous bases were cytosine.
   i Calculate the percentage of the base guanine in the sample. Show your working. (3)
   ii Calculate the percentage of the base thymine in the sample. Show your working. (3)

4 The mass of DNA in a liver cell was analysed and found to be 6.60 pg per cell (pg = picogram; $1\,pg = 1 \times 10^{-12}\,g$). What average mass of DNA per cell would you expect in the following cells of the same species:
a a muscle cell
b a mature red blood cell
c an ovum (3)

5 a Compare the structures of RNA and DNA, illustrating your answer with clearly labelled, large diagrams. (8)

b DNA replicates during interphase. Describe this process, illustrating your answer with clearly labelled, large diagrams. (10)

6 Explain how the functions of DNA are related to its structure. (6)

1 a C (thymine)
  b B (cell membrane)
  c A (cytosine and guanine)
  d C (ribose)

2 a

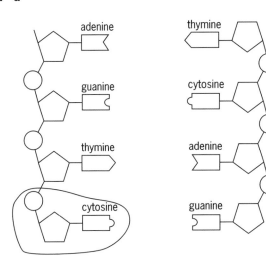

  b **A** – phosphoric acid/phosphate group
    B – deoxyribose/pentose sugar
    C – nitrogenous base
  c • The base sequence of DNA codes for the type and order of amino acids joined together at the ribosome to form a specific polypeptide/protein.
    • The form and function of the proteins results in the characteristics and specific functions of the organism.
    • It is therefore essential that exact copying of the genetic information (DNA) should occur so that each new cell formed by mitosis will have its own full complement of genes identical to the parent cell.
  d See answer to Q2a.
  e condensation (dehydration)/polymerisation

3 a i Cytosine pairs with guanine, therefore there are equal amounts of each.

$$\%C = \%G = 31\%$$

  ii Total bases must be 100%. Adenine pairs with thymine, therefore there are equal amounts of each.

$$100\% - 62\% \,(C + G) = 38\% \,(A + T)$$

  Therefore thymine $= 38\% \div 2 = 19\%$

4 *Note: all the diploid body cells of an organism contain the same number of chromosomes and therefore the same amount of DNA. Mature red blood cells are an exception since there is no nucleus present. Sex cells/gametes are haploid containing only one chromosome from each pair and therefore half as much DNA.*

  a 6.60 pg per cell
  b 0 pg per cell
  c 3.3 pg per cell

5 a *The key word here is compare. You will lose marks if you merely describe RNA and then describe DNA.*

  Key points to bring out are:
  Both consist of base/pentose sugar/phosphoric acid and nucleotides.
  Three of the nitrogenous bases are the same: adenine, guanine and cytosine.

RNA
  • The sugar is ribose.
  • Fourth base is uracil.
  • A single chain that can be folded back on itself as in tRNA.
  • In tRNA there is some pairing of complementary bases (A–U, C–G) but some bases remain unpaired forming the anticodon.
  • Clear diagram to be included (see Figure 5.1 – pentose to be labelled as ribose).

DNA
  • The sugar is deoxyribose.
  • Fourth base is thymine.
  • Two chains forming a double helix.

  • In DNA all the complementary bases pair up (A–T, C–G).

  • Clear diagram to be included (see Figure 5.4).

  b The enzyme helicase catalyses the unwinding of the DNA helix exposing the bases of the nucleotides.
    Complementary nucleotides attach to the DNA template.
    This is catalysed by DNA polymerase (in the 5'–3' direction).
    Adenine pairs with thymine, cytosine with guanine.
    The bases pair by means of hydrogen bonds.
    Replication occurs in opposite directions on the two strands/antiparallel.
    DNA ligase then catalyses the joining of the fragments (from the 3'–5' strand).
    The two DNA helices are then complete, each one being an exact copy of the base sequence of the original.
    Each DNA molecule consists of one original chain of nucleotides and a new complementary chain (semi-conservative replication).
    Replication occurs during interphase.
    Large clear diagrams should be included (see Figures 5.2, 5.4 and 5.5).

6 DNA has two main functions.

  • It is a **code** for all the information required by the cell. It is a stable molecule, which is very important if it is to maintain the information safely without change.
    Each of the triplets of bases acts as a code for the 20 different amino acids.
    Each cistron/gene is a DNA code for a particular sequence of amino acids which form a particular polypeptide or protein.
    The double helix is able to unzip, allowing transcription of genes.
    There are codes for 'start' and 'stop'.
  • The DNA must be capable of accurate **replication**. Its double helix structure allows the two strands to act as templates when unzipping of the H bonds occurs.
    Complementary bases of nucleotides pair with each template, forming two identical copies of the original DNA.

The genetic code and protein synthesis

Every organism consists of millions of individual cells, each of which is unique for a particular animal or plant, and has its own genetic instructions. Organisms have many different types of cells with different structures and enzymes to perform a variety of tasks. DNA contains the information needed to synthesise the enzymes, in the form of a code.

Enzymes are proteins, and enzymes control the cell's living processes. DNA uses the sequence of base pairs in its molecule to determine which amino acids are combined in what order to form specific proteins.

The genetic code

There are 20 different amino acids found in proteins, although over 80 are known to occur. There are, therefore, at least 20 possible amino acids that need a code, and there are four bases which need to be arranged along the helix to provide that code. A **triplet code** gives $4 \times 4 \times 4 = 64$ possible combinations which is more than enough. Thus a sequence of three bases along a length of DNA codes for a specific amino acid or, in the case of what are called 'nonsense triplets', a beginning code or an end code for a sequence of amino acids that will form a protein.

A **triplet** of bases that codes for a specific amino acid is called a **codon**.

A dictionary of codes has been compiled. Because the DNA molecule is so large most of the work has been done with **messenger RNA** (dealt with later in this topic) which has much smaller molecules and is simpler to work out. In all types of RNA, uracil replaces thymine as one of the four bases.

For example, in the table below, the first codon in the first column on the left is AUU. This stands for three consecutive bases (adenine, uracil and uracil, in that order) which together code for the amino acid called isoleucine. Remember that uracil replaces thymine in RNA.

However, the code is said to be **degenerate**. For much of the time only the first two bases seem to matter in determining which amino acid is produced. You will see, for example, that all codons beginning with the two bases G and U code for the amino acid valine, whatever the third base may be. One advantage of this to the organism may be that a mutation causing a change in the third base of a codon might still result in the same amino acid being produced. For example, any triplet starting with the bases C and U, in that order, will produce the amino acid leucine.

Protein synthesis

There are three main stages in the synthesis of proteins.

Transcription

- A specific region of the DNA molecule which codes for the required protein *unwinds* as the hydrogen bonds between the base pairs are broken (Figure 6.1).
- The bases along the strands are thus exposed. There is a difference between the two strands which are called **5 prime** (5′) and **3 prime** (3′). The numbers refer to the number of the carbon atom in the pentose sugar to which the phosphate group is attached.
- Complementary nucleotides pair with the exposed bases of the 5 prime strand. The enzyme **RNA polymerase** binds to the unwound portion of the DNA; the complementary nucleotides join by condensation reactions and the complete mRNA molecule becomes detached.

   The transcribed messenger RNA molecule is thus a code for that portion of DNA and each group of three DNA nucleotide bases codes for a specific amino acid.
- The mRNA molecule passes easily through a pore in the nuclear membrane into the cell cytoplasm. The DNA double helix structure is then restored.

1st base	2nd base								3rd base
A	AUU	Ile	ACU	Thr	AAU	Asn	AGU	Ser	U
	AUC	Ile	ACC	Thr	AAC	Asn	AGC	Ser	C
	AUA	Ile	ACA	Thr	AAA	Lys	AGA	Arg	A
	AUG	Met*	ACG	Thr	AAG	Lys	AGG	Arg	G
G	GUU	Val	GCU	Ala	GAU	Asp	GGU	Gly	U
	GUC	Val	GCC	Ala	GAC	Asp	GGC	Gly	C
	GUA	Val	GCA	Ala	GAA	Glu	GGA	Gly	A
	GUG	Val	GCG	Ala	GAG	Glu	GGG	Gly	G
U	UUU	Phe	UCU	Ser	UAU	Tyr	UGU	Cys	U
	UUC	Phe	UCC	Ser	UAC	Tyr	UGC	Cys	C
	UUA	Leu	UCA	Ser	UAA	c.t.	UGA	c.t.	A
	UUG	Leu	UCG	Ser	UAG	c.t.	UGG	Try	G
C	CUU	Leu	CCU	Pro	CAU	His	CGU	Arg	U
	CUC	Leu	CCC	Pro	CAC	His	CGC	Arg	C
	CUA	Leu	CCA	Pro	CAA	Gln	CGA	Arg	A
	CUG	Leu	CCG	Pro	CAG	Gln	CGG	Arg	G

*Sometimes this codon is used to indicate the beginning of synthesis of a polypeptide.
c.t., codons which code for the termination of polypeptide synthesis.

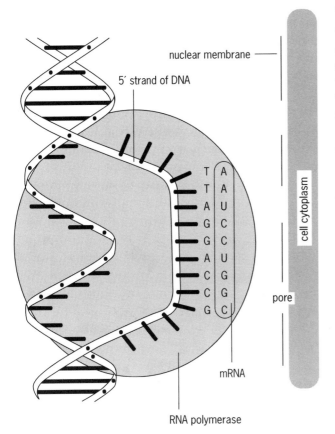

Figure 6.1 Transcription of DNA

Amino acid activation

Amino acids in the cytoplasm combine with **transfer RNA** molecules, using energy from ATP, to form amino acid–tRNA complexes. The tRNA molecules are specific for each amino acid and so there are at least 20 different ones. The point of attachment is always a free end of the tRNA ending in CCA.

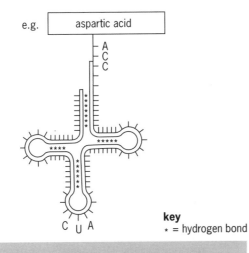

key
* = hydrogen bond

All tRNA molecules have a shape related to their function. It consists of a single strand of RNA folded as shown. Where the strand doubles back on itself the bases are complementary. They all have three bases, called an **anticodon**, which are complementary to the mRNA code for a specific amino acid.

Figure 6.2 A molecule of transfer RNA

Translation

The mRNA binds to two or more ribosomes in the cytoplasm to form a **polysome**. Then a tRNA–amino acid complex which has an anticodon complementary to the first codon on the mRNA attaches to it. Likewise the second codon attracts its complementary anticodon. A ribosome acts like a workbench, holding all the parts together until the two amino acids have formed a peptide link. Then the ribosomes move on until a third amino acid has joined to the second. This continues until the polypeptide has been synthesised. As the ribosomes move on, the earlier tRNA molecules are freed to combine with more amino acid molecules in the cytoplasm. Further ribosomes usually move along the mRNA immediately behind the first and produce molecules of the same polypeptide.

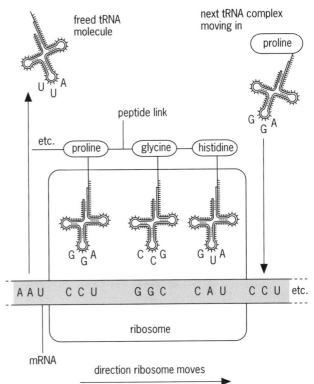

Figure 6.3 Translation of the mRNA molecule

THE GENETIC CODE AND PROTEIN SYNTHESIS

1 Answer the following questions about protein synthesis.
   a What type of bonds link the amino acids together to form polypeptides?
   b Which organelle in the cell is the site of protein synthesis?
   c Name the triplet of bases on mRNA that code for a specific amino acid.
   d What name is given to the three unpaired bases on the tRNA which are complementary to a base triplet on mRNA?
   e Which molecule supplies the energy to attach amino acids to their specific tRNA molecule? (5)

2 The diagram below shows the sequence of bases in a short length of mRNA.

   A U G G C C U C G A U A A C G G C C A C C A U G

   a i What is the maximum number of amino acids in the polypeptide for which this piece of mRNA could code? (1)
      ii How many different types of tRNA molecule would be used to produce a polypeptide from this piece of mRNA? (1)
      iii Give the DNA sequence which would be complementary to the first five bases in this piece of mRNA. (1)
   b Name the process by which mRNA is formed in the nucleus. (1)
   c Give two ways in which the structure of a molecule of tRNA differs from the structure of a molecule of mRNA. (2)
   AQA (NEAB), BY02, March 1999

3 The diagram shows the base sequence on a length of mRNA. The key indicates which amino acids they represent.

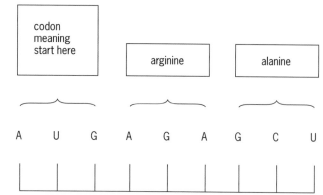

   a The amino acid arginine is shown here as AGA but can also be represented by the codon AGG.
      i Explain why the substitution of A by G in this case does not affect the amino acid represented. (2)
      ii The codons in the mRNA sequence do not overlap. Explain the advantage of this. (2)

   b What would be the anticodons which correspond to the codons shown in the diagram for:
      i arginine (1)
      ii alanine? (1)
   c Describe the part played by tRNA molecules in protein synthesis. (3)

4 The diagram shows part of the sequence of events which take place when a protein is synthesised.

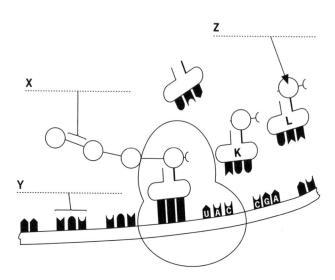

   (Adapted from: Prentis Orbis S., Biotechnology, a new industrial revolution. George Braziller, 1985.)

   a i Name the step in protein synthesis which is shown in the diagram. (1)
      ii In which cell organelle do these events take place? (1)
   b On the dotted lines in the diagram label the three features indicated. (3)
   c If the features shown in the diagram were chemically analysed, how many types of nucleic acid would you expect to find? (1)
   d State the base sequence on molecules **K** and **L**. (2)
   Welsh, Paper B2, June 1997

5 a Name the type of bond that holds together the two strands of nucleotides in a DNA molecule. (1)

   Genetic drugs are short sequences of nucleotides. They act by binding to selected sites on DNA or mRNA molecules and preventing the synthesis of disease-related proteins. There are two types:

   • triplex drugs are made from DNA nucleotides and bind to the DNA forming a three stranded helix
   • antisense drugs are made from mRNA nucleotides and bind to mRNA.

   b Name the process in protein synthesis which will be inhibited by:
      i triplex drugs (1)
      ii antisense drugs (1)

c The table shows the sequence of bases on the part of a molecule of mRNA.

Base sequence on coding strand of DNA	Base sequence on mRNA	Base sequence on antisense drug
	A	
	C	
	G	
	U	
	U	
	A	
	G	
	C	
	U	

Complete the table to show:

i The base sequence on the corresponding part of the coding strand of a molecule of DNA. (1)
ii The base sequence on the antisense drug that binds to this mRNA. (1)

AEB, Paper 2, Summer 1998

6 a A gene coding for a polypeptide was shown to contain 666 bases. Estimate the number of amino acids this polypeptide contained. Explain your answer. (2)
b Complete the table to show where nucleic acids are normally found in a eukaryotic cell.

Nucleic acid	Location
DNA	
mRNA	
tRNA	
rRNA	

(4)

7 Explain:
a How a molecule of DNA codes information for the synthesis of a protein. (3)
b How this information is copied to form messenger RNA. (5)
c How a polypeptide is synthesised from the information in the RNA code. (8)

(+2 for relevance and clarity)

1 **a** peptide bonds
   b ribosome
   c codon
   d anticodon
   e ATP

2 **a** **i** 8 **ii** 6 **iii** T A C C G
   b transcription
   c Two of:
- tRNA has a 'clover leaf' shape.
- tRNA has a standard length, mRNA is variable.
- tRNA has an amino acid binding site.
- tRNA has anticodon available/mRNA has codons.
- tRNA has hydrogen bonds between the base pairs.

3 **a** **i** The code is degenerate. The first two bases are the most important in determining which amino acid is produced/the third base is less critical.
     ii Two of:
- Substitution of a base/mutation will only change one amino acid not two.
- Overlapping could cause spatial difficulties during translation.
- There is less restriction on the order of amino acids since the last base of one codon has no effect on the next codon.

   b **i** UCU **ii** CGA
   c The tRNA molecule attaches to its own specific amino acid, transporting it to the ribosome.
The tRNA has an anticodon/three unpaired bases which pair with the complementary mRNA codon.
The amino acids link by peptide bonds at the ribosome.

4 **a** **i** translation **ii** rough endoplasmic reticulum/ribosome
   b **X** = peptide bond; **Y** = codon; **Z** = amino acid
   c Three (rRNA, tRNA and mRNA)
   d **K** = AUG; **L** = GCU
These are complementary to the mRNA codons shown in Figure 6.3, page 29.

5 **a** hydrogen bond
   b **i** transcription **ii** translation
   c

Base sequence on coding strand of DNA	Base sequence on mRNA	Base sequence on antisense drug
T	A	U
G	C	G
C	G	C
A	U	A
A	U	A
T	A	U
C	G	C
G	C	G
A	U	A

6 **a** Three bases code for one amino acid:

666/3 = 222 bases

Note: A good student would however recognise that there are stop and start codes so that the total number of amino acids would be less than this.
   b

Nucleic acid	Location
DNA	chromosomes of the nucleus
mRNA	formed in the nucleus, passes out through nuclear pores and attaches to ribosomes in the cytoplasm
tRNA	cytoplasm
rRNA	forms in the nucleolus and makes up the structure of ribosomes

7 **a** Three of:
- A length of DNA corresponding to one gene codes for a polypeptide or protein.
- The nitrogenous bases of the DNA are arranged in a particular sequence.
- There are four different bases – adenine, guanine, cytosine and thymine.
- A triplet of bases codes for each amino acid and there are also triplet codes for 'start here' and 'stop'.
- The order of the base triplets on the DNA will determine the order of the amino acids making up the polypeptide or protein.

   b This information contained in the DNA now has to be transcribed into mRNA.
Firstly the histones which are associated with the DNA are removed so that the relevant cistron of DNA can unwind.
The hydrogen bonds holding the two strands of DNA together now break so that the sequence of bases on one of the strands can act as a template.
Complementary nucleotides fit into place along the template so that G pairs with C and U pairs with T.
In this way an mRNA molecule is formed, catalysed by RNA polymerase.
   c The messenger RNA travels from the DNA in the nucleus, through the nuclear pores to the ribosome. It becomes attached to the ribosome.
Translation occurs at the ribosome as the codons (triplets of bases) bind to the complementary anticodons on the tRNAs.
Each tRNA is carrying its own particular amino acid and these are joined together by means of peptide bonds in the order specified by the codons on the mRNA.
The tRNA is released and can collect another amino acid.
The ribosome moves along the mRNA strand exposing two codons at a time.
Translation continues along the mRNA from the 5' end to the 3' end until the stop codon is reached.
Several ribosomes may work their way along one strand of mRNA thus producing several copies of the polypeptide molecule.

7 Gene expression and control

> **A gene is that part of a DNA molecule which codes for one polypeptide.**

Some genes are required constantly but others are only needed at certain times. They need, therefore, to be switched on and off. This can be achieved by two methods:

- **Enzyme induction** – genes are switched on when the polypeptide/enzyme they are coded for is required.
  EXAMPLE: *Escherichia coli* (bacterium) usually respires glucose but if only lactose is made available, it will start to produce an enzyme which enables it to respire lactose.
- **Enzyme repression** – although normally switched on, some genes can be switched off in certain circumstances.
  EXAMPLE: *E. coli* makes tryptophan (an amino acid) using an enzyme called tryptophan synthetase. If tryptophan is supplied in the growing medium, then the gene is switched off and production of tryptophan ceases.

Genetic engineering

This is a phrase on most people's lips at present. Already techniques of gene manipulation are sufficiently advanced for society to have to make big decisions. Genetic engineering promises benefits for mankind but poses considerable moral and social questions as well.

There are three areas in which genetic engineering has a particular value.

Recombinant DNA techniques

Genes can be inserted into bacteria (for example, *E. coli*), making them produce quantities of a required substance such as insulin. Insulin produced for diabetics in this way is much safer for the patient than extracting it from animal pancreas. This **recombinant DNA technique** involves:

- Messenger RNA (mRNA) is first extracted from human pancreas.
- The extracted mRNA is treated with the enzyme **reverse transcriptase** to make copy DNA. For example, the bases UACCG in mRNA would be transcribed to ATGGC in the copy DNA.
- The genes which are required are removed from the DNA by the enzyme **restriction endonuclease**. A piece of 'donor DNA' is produced. The enzyme acts like a pair of 'biological scissors'.
- A plasmid (a piece of circular double-stranded DNA which is generally separate from a bacterial chromosome but which nevertheless carries genes) is opened up at a specific point, again by the enzyme **restriction endonuclease**.

- The restriction endonuclease opens up a plasmid in such a way as to leave 'sticky' end pieces to which the donor DNA, when inserted, attaches with the aid of the enzyme DNA ligase. This process is known as **gene splicing**.
- The plasmids (which now carry the piece of donor DNA) are used as vectors (carriers) by exposing them to the bacteria. The plasmids are highly infective but sometimes the bacteria are treated with calcium salts which increases the permeability of the bacterial membrane and hence the incorporation of the plasmids into the bacterial cells.
- The plasmids replicate within the bacterium and several hundred may be formed.
- Large numbers of bacteria can quickly be cultured and as many as 1×10^9 can be produced in $1\,cm^3$ of culture medium in 24 hours.
- When a bacterial cell divides, the plasmids also replicate and are passed on to the daughter cells.
- The culture can be treated to remove all the unwanted bacterial cells which do *not* contain the donor DNA by attaching a gene to the donor DNA which gives resistance to an antibiotic (for example, tetracycline). By then treating the culture with the antibiotic all the unwanted bacteria will be killed and the productivity of the culture will be greatly increased.

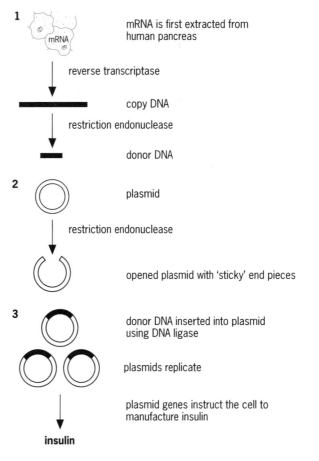

Figure 7.1 Diagram illustrating the recombinant technique for the production of insulin

33

By using similar recombinant DNA techniques, the enzyme chymosin (rennin) which makes milk clot, has been produced by genetically modified yeast and used instead of cow's rennin in cheese making. Traditionally, the enzyme was extracted from the stomachs of calves after slaughter but the calf gene for chymosin has been copied and inserted into yeast cells which produce pure chymosin. The yeast-derived enzyme has several advantages.

- It is more acceptable to vegetarians.
- It produces cheese of a more uniform quality.
- The problems caused by a worldwide shortage of rennet (preparation of rennin from the stomachs of calves) has been solved.

Human growth hormone and hepatitis B vaccine are examples of medically useful substances that have been similarly produced.

Improvement of characteristics

Genes which have been engineered can be used in animals and plants to improve their characteristics; for example, protein quality, rate of growth, less fat in meat and inbuilt pesticides in plants.

Tomatoes can be genetically modified to improve flavour and delay ripening. They are picked when they are still green and so remain firm during storage and transportation. They are then treated with ethylene to turn them red. The enzyme pectinase, which is found naturally in the fruit, causes the tomatoes to soften. Researchers removed the gene that codes for pectinase, copied it, and then replaced it backwards. This technique 'switches off' the pectinase and the rate of softening is slowed; the modified tomatoes remain longer on the plant to develop their full flavour, but stay firm enough for transportation to market.

Genetic diseases

Genetic diseases such as cystic fibrosis, haemophilia and sickle-cell anaemia, among many others, may one day be cured by implanting engineered genes into the relevant somatic (body) cells to allow normal functioning. Plants can be altered to enable them to fix atmospheric nitrogen.

Note that where engineered genes are used to treat someone with a genetically inherited disease, only that person will be cured. The sex cells will not have been affected and so the disease may be transmitted to the next or subsequent generations. If the disease is to be eliminated, then it is the germ cells that would have to be treated. This would be a major step involving both moral and ethical considerations.

There is a growing concern that some of these achievements might result in undesirable genes entering wild populations (see also **38** The continuity of life: Variation).

Genetic screening

Genetic diseases can be traced through family trees. A classic example is the case of haemophilia among descendants of Queen Victoria. Provided that there is sufficient information about the family tree, genetic diseases such as haemophilia, cystic fibrosis and some forms of muscular dystrophy can be predicted with a fair degree of accuracy. Where there is such a history of disease in a family, genetic counsellors can advise couples wishing to marry, or have children, of the risks of the disease being inherited. Inheritance is covered in **39** (The continuity of life: Monohybrid and dihybrid inheritance) but the following is an example of how the inheritance of cystic fibrosis can be predicted.

The gene for cystic fibrosis is recessive and autosomal (not on the X or Y chromosome). Cystic fibrosis affects about 1 child in 2000. Mucus-secreting glands, especially in the lungs, produce an abnormally thick mucus causing intestinal blockage and pneumonia which can be fatal. There can be a prenatal test at 9–11 weeks and if the test proves positive, the parents have the choice of whether or not to terminate the pregnancy.

M is the gene for normal mucus production.
m is a recessive gene for the production of the abnormal mucus associated with cystic fibrosis.

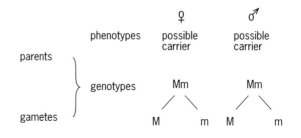

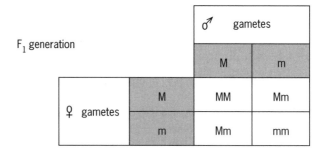

Figure 7.2 Inheritance of cystic fibrosis

From the Punnett diagram in Figure 7.2 we can see that there are the following possible ratios:

25% normal children with the genotype MM
50% normal children but carriers of the gene (Mm)
25% children with cystic fibrosis (mm).

Genetic fingerprinting

This technique is a diagnostic tool which identifies people from the 'information' they provide. Almost any cell of the body can be used to provide the 'information' – skin, a spot of blood or a few sperm for example. The process involves the following.

- Separating the DNA from the sample material.
- 'Cutting' the DNA into sections using the enzyme **restriction endonuclease** which cuts the DNA at specific base sequences and also cuts both of the DNA strands at the same time.
- Separating the fragments using electrophoresis (a method of separating particles which have different electrical charges).
- After separation, transferring the pieces of DNA to a nylon membrane.
- Joining radioactive probes to some specific portions of the DNA and washing off the rest of the DNA.
- Placing the pieces of DNA, still attached to the membrane, next to a sheet of X-ray film.
- Exposing the film to radioactive probes on the DNA. When the film is developed a pattern of dark and light bands appears.

These patterns are unique for each individual except identical twins. They can be used to determine, for example, if a suspected rapist is guilty; the relationship between an immigrant and a resident of the country; in paternity suits to determine who is the true father of a child.

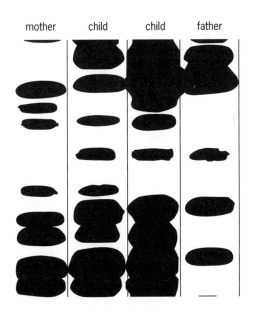

White blood cells are taken from the mother and the possible father. From the pattern of bands of the child, those bands which correspond with the mother's are subtracted. A true father possesses all the remaining bands in the child's genetic fingerprint.

Figure 7.3 Genetic fingerprints in a paternity suit

Figure 7.3 shows that the man is the father of both children.

Variation

Variation may be due to 'nurture' or to 'nature'.

'Nurture' is the effect of the environment upon an organism. Such effects include the supply of food for animals, or light intensity in the case of green plants.

Variation due to 'nature' is genetic variation. Within a species, genetic variation may be **continuous** or **discontinuous**.

Height of human adults is an example of continuous variation, because there is a graduation of heights within a population from one extreme to the other. If the varying heights of a number of adults were plotted, the peak would represent the average height of the population. This characteristic is controlled by the *combined effect of a number of genes* and is thus said to be a **polygenic character.**

The four blood groups, A, B, AB and O in humans, is an example of discontinuous variation. There are three alleles controlling the human blood groups and even with various combinations, one's blood group can be only one of the four types. There are no intermediaries.

How does variation occur?

Genetic variation may be the result of either the reshuffling of genes or the result of mutations.

The reshuffling of genes will be dealt with in **37** The continuity of life: Meiosis. Most gene mutations occur during cell division in somatic cells (body cells but not sex cells) and therefore are not passed on from one generation to the next. Mutations which occur in the formation of gametes (during meiosis) can be inherited and will produce variations between individuals.

> **A mutation is any change in the structure or the amount of DNA of an organism.**

A gene mutation is the result of a change in the structure of DNA at a single point (**locus**). Such a change produces an incorrect series of amino acids in a polypeptide during protein synthesis. If the protein is an enzyme it could have a different molecular shape which would alter the active site and prevent it from catalysing its reaction. Such a change may have a dramatic effect on an organism, for example, the absence of skin pigments would produce an albino.

There are a number of forms of gene mutation:

- **duplication** – a length of DNA becomes repeated
- **deletion** – a piece of DNA is removed
- **addition** – one or more extra nucleotides are added
- **substitution** – one nucleotide is replaced by another with a different organic base
- **inversion** – a piece of DNA becomes separated and then rejoins in the original position but the other way round
- **translocation** – non-homologous chromosomes break and exchange pieces of DNA.

1 The histogram shows the heights of wheat plants in an experimental plot.

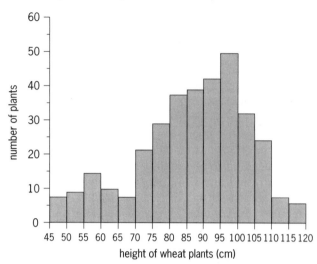

a What evidence from the data suggests that there were two strains of wheat growing in the experimental plot? (1)

b **i** Which type of variation is shown by the height of each of the strains of wheat plants? Give the reasons for your answer. (2)

ii Explain why the height of the wheat plants varies between 45 cm and 120 cm. (1)

NEAB, BY02, June 1997

2 Read through the following passage on gene technology (genetic engineering), then write the most appropriate word or words to complete the passage.

The isolation of specific genes during a genetic engineering process involves forming eukaryotic DNA fragments. These fragments are formed using _____ enzymes which make staggered cuts in the DNA within specific base sequences. This leaves single-stranded 'sticky ends' at each end. The same enzyme is used to open up a circular loop of bacterial DNA which acts as a _____ for the eukaryotic DNA. The complementary sticky ends of the bacterial DNA are joined to the DNA fragment using another enzyme called _____. DNA fragments can also be made from _____ template. Reverse transcriptase is used to produce a single strand of DNA and the enzyme _____ catalyses the formation of a double helix. Finally new DNA is introduced into host _____ cells. These can then be cloned on an industrial scale and large amounts of protein harvested. An example of a protein currently manufactured using this technique is _____.

London, Paper B1, Specimen Paper, 1996

3 The diagram below represents a bacterial plasmid.

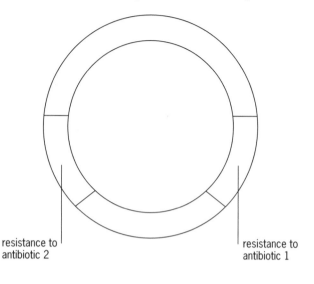

resistance to antibiotic 2

resistance to antibiotic 1

a What is a bacterial plasmid? (2)

b Explain how such a plasmid is used in genetic engineering. (3)

c Explain how enzymes are used to insert a foreign gene into a plasmid. (3)

d Explain why plasmids with genes for antibiotic resistance are often used. (3)

4 Genetic fingerprinting is becoming commonly used.

a Explain the theoretical basis of the process. (4)

b Give two applications of genetic fingerprinting. (2)

c Outline the steps involved in genetic fingerprinting. (9)

5 Cancer may be treated by chemotherapy. This involves using drugs which kill cancer cells but have no effect on normal healthy cells. Unfortunately, cancer cells develop from normal cells so the two types of cells are similar to each other. Trials have begun which involve adding a new gene to the normal cells in the body. This gene makes a protein which protects these healthy cells against the drug being used. The cancer cells do not produce this protein, so they are killed.

a Describe the features of a gene which enable it to code for a particular protein. (4)

b Explain how enzymes and vectors may be used to isolate genes and insert them into another organism. (6)

c Describe how the new protein is made once the gene has been inserted into the cell. (7)

(+3 for quality of language)

AQA (AEB), Module 2, January 1999

1 a The graph shows two normal distributions, that is, two peaks.

b i The type of variation is continuous for each strain of wheat plants.
Each contains a range of heights (there is a normal distribution within each strain).
ii Height will be affected by both genetic and environmental factors/it is controlled by polygenes.

2 restriction/endonuclease; vector or carrier; ligase; mRNA; polymerase; yeast or bacterial or prokaryotic; insulin or growth hormone or interferon.

3 a A bacterial plasmid is a small circle of double-stranded DNA in addition to the main chromosome of the bacterium. It is able to replicate independently.

b Foreign DNA (a gene) is inserted into the plasmid.
The plasmid is then used as a vector and is introduced into the host cell.
The DNA is then cloned producing many copies of the foreign DNA.

c Restriction endonucleases are used to cut DNA at specific recognition sites leaving sticky ends.
Complementary sticky ends on the DNA segment will base pair with sticky ends of a plasmid cut by the same endonuclease.
They are sealed together using DNA ligase.

d After the plasmids have been inserted into bacteria, the bacteria can then be plated onto agar containing antibiotics. Bacteria which have not taken up the plasmid will not have antibiotic resistance and will not grow on antibiotic 1. The gene is spliced into the antibiotic 2 resistant gene of the plasmid so that bacteria containing the plasmid plus inserted gene will be sensitive to agar containing antibiotic 2.

4 *It is important to distinguish between the points needed to answer parts **a** and **c**. Part a requires the explanation of the underlying processes and the principles they are based on, whereas **c** requires the details of the method.*

a Repetitive DNA pieces called minisatellites are unique in their distribution in each individual. Fragments of DNA are labelled by radioactive probes. The fragments of DNA are then separated by gel electrophoresis. The distance travelled is proportional to the electrical charges on the fragments. X-ray film is used to determine the position of the DNA fragments.

b Two of:
- To compare samples of DNA from a suspect with samples of blood, semen etc found in criminal investigations.
- In paternity investigations.
- To investigate evolutionary links.

c Human DNA is cut into segments using specific restriction endonucleases. The fragmented DNA is then loaded into a well in the agar gel and a potential difference is applied. The fragments separate by gel electrophoresis. A thin nitrocellulose filter is used to take up the separated fragments. The filter is then washed with a radioactive P viral probe. This has been prepared using the same restriction nucleases to form sticky ends. Complementary sticky ends adhere. The filter must be well rinsed and is then placed on X-ray film. Radiation affects the film giving a banding pattern which is unique for each individual.

5 *Do not be put off by such questions. Although you may not have heard about the gene therapy for cancer before, the answers required are straightforward, merely requiring a knowledge of gene manipulation and protein synthesis.*
Answers should be written in continuous prose. Credit will be given for biological accuracy, the organisation and presentation of the information, and the way the answer is expressed.

a Four of:
- A gene is a length of DNA.
- A gene is a sequence of bases/chain of nucleotides.
- Contains a triplet base code on the sense/coding strand.
- A triplet of bases codes for each amino acid.
- The code is non-overlapping/degenerate/has stop and start codes.
- A sequence of the triplets of bases codes for a protein.

b Six of:
- Restriction enzymes cut the gene DNA at specific base sequences.
- The same restriction enzyme also cuts the DNA (of the vector).
- The gene is then inserted into the plasmid/virus/agrobacterium.
- The enzyme DNA ligase is used to join the two pieces of DNA together/form recombinant DNA.
- The complementary sticky ends pair up.
- The vector is needed to insert the DNA into the host/second organism.
- Another method is to use reverse transcriptase to make copy DNA from mRNA.

c Seven of:
- The DNA unwinds/unzips.
- This involves the breaking of hydrogen bonds.
- Free mRNA nucleotides assemble.
- Complementary base pairing occurs with the strand of DNA.
- Polymerase catalyses the addition of nucleotides to form mRNA.
- mRNA enters the ribosome.
- Specific tRNA molecules become associated with their specific amino acids.
- At the ribosome complementary pairing occurs between the codons of mRNA and anticodons of tRNA.
- The amino acids are joined together by peptide bonds.
- ATP is used to provide energy for attachment of tRNA to amino acid/for peptide bond formation.
- Refer to the gene needing to be switched on.

Quality of language: these marks (one for each) are an important way of improving your grade and so you should take care to organise the facts logically, with correct use of English.

Grammar, punctuation and spelling of an acceptable standard. Appropriate scientific style and correct use of technical terms. Argument clearly and logically presented.

8 How cells divide – mitosis

When somatic (body) cells divide during mitosis, each of the two new cells formed receives an exact replica of the original **diploid** number of chromosomes (shown as 2n). In humans there are 23 *pairs* of chromosomes in each somatic cell, one of each pair originating from the father and the other from the mother. This type of cell division occurs during growth and tissue repair but *not* during the formation of gametes (sex cells).

Microscope slides of stained preparations showing mitosis are prepared from actively growing and dividing tissues. Such tissues are difficult to isolate in animals, but in plants, whose growth is apical (at the tip of roots and shoots), slides can easily be prepared from the gently squashed tips of roots. It is convenient for our studies that mitosis in plants and animals is virtually identical.

An important point about mitosis is that the hereditary material (genes) present in the original cell replicates and is then divided equally between two daughter cells so that each of the new cells is identical to the parent cell.

You need to be able to explain what is taking place at each stage of mitosis and recognise these stages from diagrams and micro-photographs. Do not try to learn a series of diagrams by heart, but rather make sure that you understand the significance of each phase.

Interphase

This is the stage between visible cell divisions. Nothing seems to be happening in the nucleus which appears as a roughly spherical granular body containing one or more dense nucleoli. However, despite this apparent inactivity there is much happening:

- The DNA replicates (**5** Genes and DNA replication).
- New ribosomes, for the synthesis of nuclear proteins and spindles are made.
- The **mitochondria**, which contain DNA, divide.
- The **centrioles** divide.

All of the above are necessary prerequisites for any cell division (mitosis or meiosis).

During both mitosis and meiosis (**37** The continuity of life: Meiosis) the nucleus divides first and then the cytoplasm.

Prophase

This is a preparation stage for the division. Several things need to happen and prophase is usually divided into early and late stages.

Early prophase

Early prophase begins when the chromosomes become visible and can be stained with biological stains.

Late prophase

- The threads become shorter and thicker, disentangle, and can now be seen as individual chromosomes.
- The **nucleolus** disappears.
- The **centrioles**, which divided during interphase, move to the opposite poles of the cell and lay down a **spindle** of **microtubules** as they migrate. A mass of these microtubules also radiate from the centrioles to form **asters**.

In Figure 8.1, only two different chromosomes of the 23 pairs as found in humans have been shown. Before the cell in late prophase can begin to divide into two new cells, each with an exact copy of the original hereditary material, several things need to happen during the next stage, the metaphase.

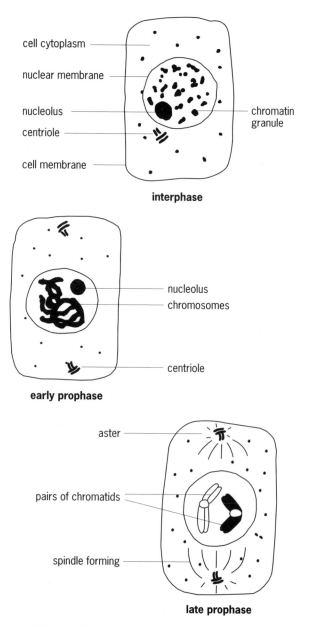

Figure 8.1 Interphase and prophase

Metaphase

- The nuclear membrane disappears and the spindles connect with the asters at the opposite poles rather like 'lines of longitude' on a globe.
- The chromosomes align themselves around the equator and lie parallel to it.

At this stage each chromosome can be seen to consist of two threads called **chromatids** which have been formed by replication of the original chromosomes. The chromatids are joined together at a small portion, called the **centromere**, which does not stain like the rest of the chromosome.

The position of the centromere along each chromosome varies. Each centromere is attached to a spindle.

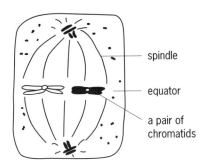

Figure 8.2 Metaphase

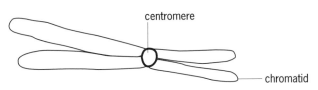

Figure 8.3 A pair of chromatids as seen at metaphase

Figure 8.4 is a drawing of *Drosophila* chromosomes which have been separated.

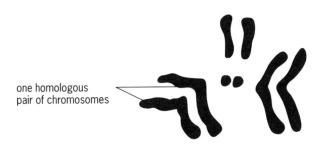

Figure 8.4 Chromosomes of *Drosophila* (fruit fly)

It is possible to arrange chromosomes in **homologous** pairs. That is, there are two of each type. This is known as the **diploid** number and is expressed as $2n$.

There are 46 chromosomes (23 pairs) in human somatic cells but this number is fixed for each species. For example, *Drosophila*, the fruit fly, has eight chromosomes (four pairs) as shown in Figure 8.4.

Anaphase

The chromatids separate and move to the opposite poles with the centromeres leading.

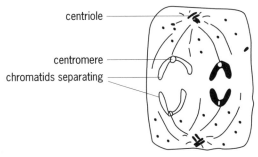

Figure 8.5 Anaphase

Telophase

During telophase the two sets of chromosomes gather at the spindle poles, uncoil and become extended. A new nuclear membrane forms around each of the two groups of chromosomes, the nucleoli re-form and the nucleus takes on the appearance that it had previously at interphase. The asters and spindles also disappear and the chromosomes can no longer be identified. The cytoplasm then cleaves during **cytokinesis** to form two new cells.

Figure 8.6 Early telophase

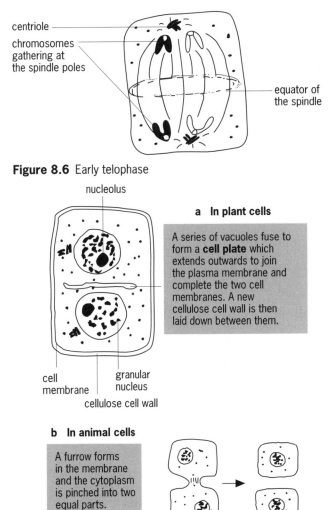

Figure 8.7 Cytokinesis

HOW CELLS DIVIDE – MITOSIS

1 Read through the following passage about mitosis and then write a list of the most suitable words to fill the gaps.

During mitosis a _____ divides to form daughter cells with _____ composition. Replication of DNA occurs during _____. In the first stage which is called _____ the chromosomes condense and can be seen to consist of two _____ joined together by the _____. Then the nuclear membrane disappears and a _____ forms in the cell. The chromosomes line up on the _____ of the cell at the stage called _____. Daughter chromosomes separate, travelling to opposite poles. Nuclear division is followed by _____. (10)

2 The photographs show four different stages in mitosis.

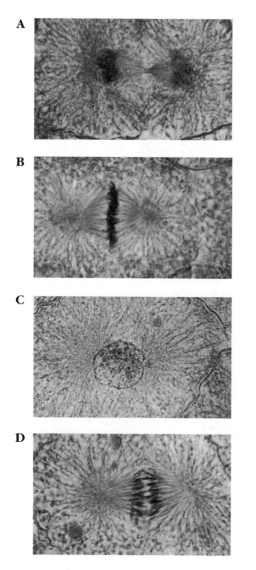

A

B

C

D

a Give the correct sequence of the four photographs:

i ii iii iv (1)

b Explain how the mitotic division of a parent cell results in the daughter cells containing identical copies of the parental genetic information. (3)

3 The flow chart shows one way in which a chromosome preparation may be obtained.

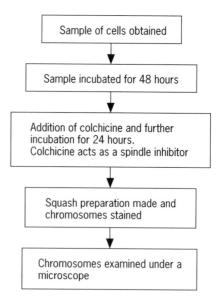

Sample of cells obtained

↓

Sample incubated for 48 hours

↓

Addition of colchicine and further incubation for 24 hours.
Colchicine acts as a spindle inhibitor

↓

Squash preparation made and chromosomes stained

↓

Chromosomes examined under a microscope

a Explain why it is necessary to incubate the cells for 48 hours before adding colchicine. (2)

b **i** What is the function of the spindle in mitosis? (1)

ii Suggest why a spindle inhibitor like colchicine was added in making this preparation. (1)

iii The total amount of DNA in the nucleus of these cells immediately before cell division was 6.8pg. How much DNA would there be in this cell after treatment with colchicine? Explain your answer. (2)

AEB, Paper 2, Summer 1997

4 The diagram shows the life cycle of a water flea.

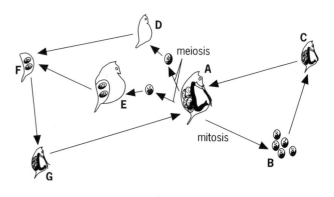

In favourable conditions all the animals in a population are females (**A**). These females produce eggs (**B**), by mitosis, which develop into young females (**C**) without being fertilised. In unfavourable conditions, eggs produced by meiosis develop directly without fertilisation into either males (**D**) or females (**F**). The eggs produced by females (**E**) are fertilised by sperm from the males, then released in a protective case (**F**) which enables them to survive unfavourable conditions. When favourable conditions return these eggs develop into young females (**G**).

a **i** Explain why the eggs in female **F** must be produced by mitosis. (1)

ii Complete the table to show the number of chromosomes at each stage in the life cycle.

Stage in life history	Chromosome number
A	2n
C	
D	
E	
G	

(2)

b Explain why the females **G** are genetically different from each other, but females **C** are genetically the same. (3)

c Explain in terms of natural selection why it is advantageous to an organism to have a sexual stage as well as an asexual stage in its life history. (3)

NEAB, BY02, June 1997

5 You are required to design an experiment to test the hypothesis that caffeine is a chemical that inhibits spindle formation during mitosis.

You are provided with onion bulbs, a standard solution of caffeine and any other materials and apparatus you may require. An onion bulb may be encouraged to grow its roots by supporting it with its base just dipping into water as shown in the diagram.

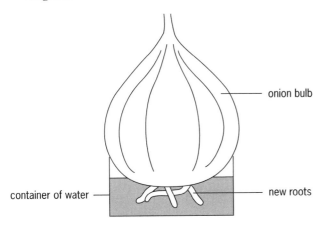

onion bulb

container of water — — new roots

a Explain why, if spindle formation is inhibited, cells will remain in metaphase. (2)

b Describe:
i How the experimental bulb and the control bulb would be treated. (2)
ii How you would prepare a slide to show cells dividing mitotically. (3)
iii Exactly what you would need to record. (2)

c Explain why an appropriate statistical test should be carried out. (1)

AEB, Paper 1, Summer 1997

6 **a** Name the stage in mitosis during which DNA replication takes place. (1)

b The diagram below shows the changes in DNA content of a cell during one cell cycle.

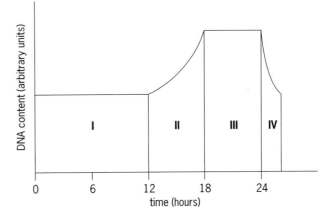

i 1) Which part of the diagram, **I**, **II**, **III** or **IV**, shows when DNA replication is taking place? (1)
2) Using only information in the diagram, explain your answer. (1)
ii During which part of the diagram would you expect the chromosomes to have become visible? (1)
iii 1) Describe **fully** what is happening to the cell during the part of the diagram labelled **IV**. (2)
2) What is the name given to this stage in mitosis? (1)

Welsh, Paper 2, January 1999

7 The graph below shows how the quantity of DNA, measured in arbitrary units, varies with time during the different phases of the cell cycle in an animal cell.

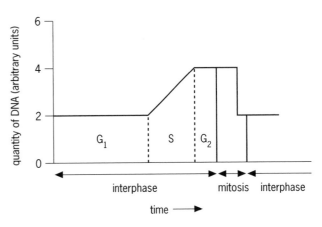

a Interphase is made up of two growth phases, G_1 and G_2, separated by an intermediate phase, S.
i Explain what is happening within the cell during phase S. (2)
ii State one process other than cell growth which occurs during phase G_2. (1)

b Account for the changes in the quantity of DNA in the cell during mitosis. (2)

London, Paper B1, January 1996

1 cell; identical; interphase; prophase; chromatids; centromere; spindle; equator; metaphase; cytokinesis (cytoplasmic division)

2 a The correct order is:
   i photograph C
   ii photograph B
   iii photograph D
   iv photograph A

b Three of:
   - Replication of the DNA occurs (during interphase).
   - Producing two identical chromatids in each chromosome.
   - These chromatids separate at anaphase.
   - One copy going to each daughter cell.

3 a Two of:
   - They are incubated to allow the cells to divide mitotically.
   - To increase the cell number.
   - Chromosomes are only visible during cell division.

b i The function of the spindle is to separate/pull apart the chromosomes/chromatids/separate centromeres.
   ii It will increase the number of cells where chromosomes are visible/stops the cells completing mitosis/cell division/stops chromatids/chromosomes separating.
   iii 6.8 pg. Colchicine would prevent cells completing division/chromosomes not separated.

4 a i Because female **E** is haploid/meiosis produced the egg from which **E** came/no pairing of homologues possible.
   ii

Stage in life history	Chromosome number
A	2*n*
C	2*n*
D	*n*
E	*n*
G	2*n*

b **G** is produced sexually (for example, from gamete fusion/meiosis involved).
   C is produced by mitosis/asexual reproduction.
   Any one cause of variation in meiosis, for example:
   - random assortment of chromosomes in meiosis
   - crossing over
   - random fertilisation.

c Three of:
   - Sexual reproduction produces new variants.
   - Some may be better fitted to survive *changed* conditions.
     Advantages of asexual reproduction, for example:
   - (All female population) allows rapid population growth.
   - Well adapted genotypes reproduced exactly.

5 a Spindle fibres pull chromosomes/chromatids apart; in anaphase/next stage.

b i Experimental bulb should be dipped in caffeine solution and the control bulb in water.
   Everything else/named factor kept constant.
   ii Take 5 mm of root tip; macerate/place in acid; add named stain (for example, acetic orcein); squash to separate cells.
   iii Number of cells in metaphase/prophase and metaphase/all stages.
   Total number of cells dividing.

c To assess the probability of results being due to chance/being significant.

6 a interphase

b *In order to gain marks it is important to describe what has happened to the DNA content rather than saying the graph has increased/decreased.*

   i 1) part **II**
   2) The DNA content at the end of the phase is double that at the beginning/has increased.
   ii part **III**
   iii 1) Two of:
   - The DNA content of the cell is halved.
   - Daughter cells are produced, each containing half the original amount of DNA.
   - The cell divides by constriction/cell plate formation.
   - Chromosomes become long and thin/disappear/become indistinct.
   - Spindle fibres disappear; nuclear envelope reforms; nucleoli reappear.
   2) cytokinesis/telophase

7 a i Two of:
   - the quantity of DNA doubles
   - replication of DNA/chromosomes
   - preparation for mitosis/nuclear division/cell division/asexual reproduction.
   ii One of:
   - mitochondria divide
   - energy stores increase
   - ATP production
   - respiration
   - duplication of the centrioles
   - the spindle begins to form
   - protein synthesis.

b Two of:
   - The DNA content halves/returns to its original level.
   - DNA or chromosomes/chromatids are shared between daughter cells/nuclei;
   - during cell division/cytokinesis.

9 How enzymes work

> **Enzymes are proteins produced by living cells. They act as catalysts and lower the activation energy level at which a reaction takes place.**

The structure of enzymes

All enzymes are globular proteins. They are soluble in water and therefore work in solution in living cells. Some enzymes are conjugated proteins which have a non-protein part called the prosthetic group chemically bound to the enzyme molecule (**2** Biological molecules).

The properties of enzymes

- Enzymes act in very low concentrations.
- Enzymes speed up the chemical reaction rate by lowering the **activation energy level**. The energy needed for a reaction to occur is called activation energy. For living organisms this means that reactions which normally require much higher temperatures are able to proceed in cells at ambient temperatures.
- Enzymes are specific for a substrate – they only act on one or a few closely-related compounds.
- Many enzymes are denatured (destroyed) by temperatures above 40°C, and by strong acids or alkalis.
- Enzyme molecules are usually much larger than the molecules of the substrate on which they work.
- The active group of an enzyme forms an enzyme–substrate complex and is then converted to enzyme–product complex by a new pathway requiring less activation energy. The substrate molecules fit onto the enzyme at the **active site**.
- Some enzymes require the help of a non-protein part, or prosthetic group, before they can work. If only the presence of the non-protein part is needed it is called a **coenzyme**, but if the non-protein part is chemically bound to the enzyme it is called a **conjugated protein**.

Groups of enzymes and their actions

Enzyme group	Reactions catalysed	Example of enzymes
hydrolases	hydrolysis reactions in which molecules of water are added/removed, e.g. hydrolysis of starch to maltose	amylase
oxidoreductases	the transfer of electrons in oxidation/reduction reactions, e.g. the removal of hydrogen in the Krebs cycle	dehydrogenase
ligases	the bonding together of molecules using free energy released during the hydrolysis of ATP	DNA ligase
transferases	the transfer of a chemical group from one molecule to another	phosphorylase

Most reactions catalysed by enzymes are reversible reactions:

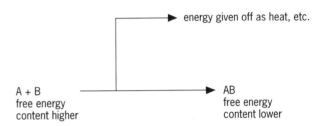

$$\text{glucose 6 phosphate} \underset{75\%}{\overset{\text{hexose phosphate isomerase}}{\rightleftharpoons}} \text{fructose 6 phosphate} \quad 25\%$$

What is meant by activation energy?

A chemical reaction can take place spontaneously only if there is a decrease in **free energy** on the right hand side of the equation. Free energy is energy which is free to do work. For example:

```
                              energy given off as heat, etc.

A + B                                        AB
free energy                                  free energy
content higher                               content lower
```

A reaction may look as though it would be spontaneous but is still very slow, or may not occur. This may be because A and B have energy barriers which must first be overcome. Enzymes work by lowering this activation energy barrier.

Figure 9.1 is a simple illustration of an activation energy barrier. Ball **B** needs to be 'pushed uphill' to a higher level (**activation energy**) before the reaction can take place. Ball **B** will overcome the barrier only when it is released from a greater height.

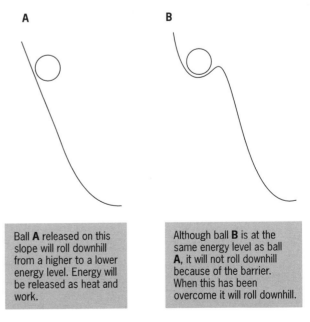

A   **B**

Ball **A** released on this slope will roll downhill from a higher to a lower energy level. Energy will be released as heat and work.

Although ball **B** is at the same energy level as ball **A**, it will not roll downhill because of the barrier. When this has been overcome it will roll downhill.

Figure 9.1 Lowering the activation energy level

The nature of enzyme activity

At temperatures of more than 60°C many enzyme molecules unravel, lose their tertiary and quaternary structures and are denatured. Globular proteins possess an **active site** in their three-dimensional structure. The active site is that part of the surface of an enzyme molecule which has a shape complementary to a specific substrate. This allows the enzyme and the substrate to become temporarily bonded into an enzyme–substrate complex. This is known as the **lock and key mechanism** and it explains why enzymes are specific for a particular substrate and why changes in the shape of the enzyme molecule, due to temperature or pH changes, cause the enzyme to be denatured (Figure 6.2).

In the example below there is a third substance present in the substrate. It is represented by a triangular shape which will not fit into either of the shapes of the two active sites present on the enzyme molecule. Therefore an enzyme–substrate complex cannot be formed. This illustrates what is meant when we say that the active site of an enzyme has a shape which is **complementary**, or otherwise, to that of the substrate molecule. The triangle is *not* a complementary shape.

Two of the substances in the substrate fit onto the active site of the enzyme to form an enzyme–substrate complex. When they are brought close together like this they are more likely to react, with a bond being formed between them. The product is then released and the enzyme's active site is freed for further reactions.

Note that the reaction is reversible and the bond can also be broken when the reaction proceeds from right to left in the equation. As in most enzyme reactions an equilibrium will be reached unless the products of the reaction are removed as soon as they are produced, in which case the reaction would proceed to completion.

How enzyme reactions are controlled in living organisms

Many biological processes involve a series of linked reactions, each of which is reversible and controlled by a specific enzyme. This can be illustrated as:

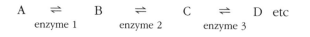

An example of a series of reactions like the one illustrated is **glycolysis**, which is the initial stage in the respiratory pathway (**14** Respiration: Glycolysis).

One of the final products of glycolysis is ATP (adenosine *tri*phosphate). ATP is the means by which the energy locked in food molecules is transferred to the cell for its own needs. Some of the energy released during glycolysis goes into producing ATP from ADP (adenosine *di*phosphate) and inorganic phosphate.

$$ADP + phosphate \rightleftharpoons ATP$$

ATP is not stored in the cell but is made in the cell cytoplasm as it is required (although most ATP is made in the mitochondria during oxidative phosphorylation) and, therefore, the rate of glycolysis needs to be carefully controlled.

High levels of ADP stimulate the production of ATP, and when the level of ATP is sufficiently high further production ceases.

The five steps of glycolysis, from glucose to pyruvic acid, are each controlled by a specific enzyme as illustrated in Figure 9.3.

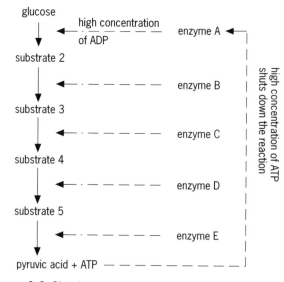

Figure 9.3 Glycolysis

Enzyme A is activated by a high concentration of ADP. The series of reactions is then set in motion; pyruvic acid and some ATP are formed.

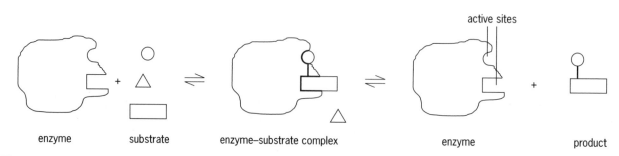

Figure 9.2 The lock and key mechanism

Enzyme A is also inhibited by high concentrations of ATP and so when sufficient ATP has been made, the reaction shuts down until there is a high level of ADP once again.

An enzyme such as enzyme A in our example is called a **regulatory enzyme**. A control system, such as the one above, in which increased concentration of the final product inhibits an enzyme in the chain and halts production, is known as **negative feedback**.

Nomenclature of enzymes

Note that the simple name given to an enzyme is usually the name of the substrate that the enzyme works on, followed by the suffix '-ase'. For example, malt*ase* catalyses the breakdown of malt*ose* into gluc*ose*.

Commercial uses of enzymes

There is a wide range of enzymes used commercially. Examples are given in the table.

Enzyme	Process
chymosin	fermentation of lactose to lactic acid in cheese making
lactase	removal of lactose in preparation of milk drinks
pectinase and cellulase	liquefaction of fruit mash prior to extraction of fruit juice
pectinase	clearing of wines and fruit juices
amylase	preparation of glucose syrup from starch
protease	additive for biological detergents removing hair from hides
trypsin	pre-digestion of baby food

Immobilised enzymes and their use in biotechnology

Enzymes are not always ideal catalysts because they are neither stable nor can they be used at high temperatures or in the presence of many other chemicals such as organic solvents. In a commercial production, profitability will be affected by the stability of the enzyme. Another disadvantage when using enzymes is that it is usually expensive, or impossible, to separate efficiently the dissolved enzymes from the products. However, enzymes have a high substrate specificity and can be used to produce useful compounds of a high purity which are not otherwise easily synthesised.

The use of immobilised enzymes and immobilised cells has been developed to eliminate the disadvantages. When a cell culture is discarded, the required enzyme produced by the cell has to be replaced. To offset this an enzyme is immobilised by trapping or attaching it, or the cells which produce it, onto a stable support such as a mesh of inert material. The enzymes or cells are then incorporated into small beads which may be packed into columns. A solution of the nutrient can then be passed continuously through the column without the need for constantly changing the cells or enzyme.

A naturally occurring immobilised cell system to produce 'quick vinegar' has been in use since 1823 before enzymes or even cells were known. Wooden vats were filled with beechwood shavings on which had developed a film of microbes which oxidised the ethanol (ethyl alcohol) to ethanoic acid (acetic acid). The vats could then be emptied and repeatedly filled with wine to produce the vinegar.

1 Complete the passage on enzyme action by filling in the most appropriate word or words.

Enzymes are very efficient _____, speeding up metabolic reactions. They are globular _____ with an area called the _____ which interacts with the substrate molecules. If the shape of this part of the enzyme molecule is changed the enzyme is _____. This can be caused by extremes of _____ or _____. Enzymes lower the energy _____ to a reaction so that less _____ energy is needed for the substrate molecules to react. (8)

2 The following are types of enzymes:
A – transferases
B – oxidoreductases
C – hydrolases
Complete the following table by using these letters to identify the enzymes which would catalyse the reactions shown. A letter may be used more than once, once or not at all. (4)

Type of reaction	Letter of enzyme catalysing this reaction
H_2O + glycogen $\rightleftharpoons$ glucose	
$2AH_2 + O_2 \rightleftharpoons 2H_2O + 2A$	
$AB + C \rightleftharpoons A + BC$	
$AH_2 + NAD \rightleftharpoons A + NADH + H^+$	

3 Indicate, by using the letter A, B, C or D, the one correct answer in each of the parts **a** to **c**.
  a The value of enzymes to organisms is their ability to:
  A – increase the amount of product from a reaction
  B – bring about reactions which could not otherwise occur
  C – increase the rate at which reactions take place
  D – increase the energy output from reactions
  b An enzyme which splits a molecule into smaller parts using water molecules is called:
  A – hydrogenase
  B – hydrolase
  C – synthesase
  D – oxidase
  c An enzyme functions by:
  A – speeding up reactant molecules to increase the frequency of encounters
  B – bringing the reactants into a specific relationship with one another
  C – adding energy to the system
  D – rapidly removing the products of the reaction (3)
  Welsh, Paper B1, June 1997

4 What do you understand by the following terms when applied to enzymes?
  i organic catalyst (2)
  ii active site (2)
  iii denaturation (2)

5 Complete the following table giving named examples and details of the process catalysed.

Class of enzyme	Named example	Chemical process catalysed
ligase		
oxidoreductase		

(4)

6 Lactose is a disaccharide found in milk. Many adults are unable to digest lactose and suffer intestinal problems if they drink milk. Milk can be treated with the enzyme lactase and this reduces the amount of lactose present. The diagram shows an industrial reactor used to produce lactose-reduced milk.

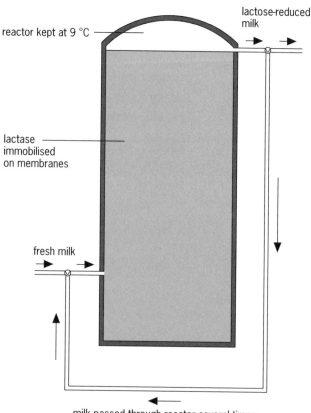

reactor kept at 9 °C

lactose-reduced milk

lactase immobilised on membranes

fresh milk

milk passed through reactor several times

  a i Suggest one advantage of immobilising the lactase used in this reaction. (1)
  ii In terms of your knowledge of the way enzymes work, explain why it is necessary to pass the milk through the reactor several times to reduce the amount of lactose sufficiently. (3)

b The graph shows the change in lactose concentration during the course of the reaction.

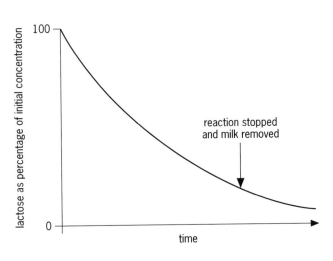

i Explain the change in the rate of the reaction with time. (2)

ii Suggest why the reaction is stopped at the time shown on the graph. (1)

AQA (AEB), Module 1, January 1999

7 a The enzyme pancreatic lipase was incubated with glycerol and a fatty acid, oleic acid. The products of the reaction were analysed by chromatography. The chromatogram is shown below.

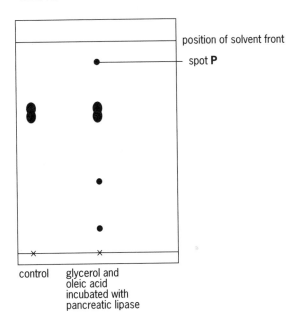

i Describe the control that you would set up to show that this reaction was catalysed by pancreatic lipase. (1)

ii The R_f value of a spot on a chromatogram is the distance moved by the spot divided by the distance moved by the solvent front.

1) Calculate the R_f value of spot **P** on this chromatogram. (1)

2) Explain how spot **P** could be identified from its R_f value. (1)

b Lipase enzymes have important industrial uses. Suggest an explanation for each of the following observations.

i When added to early biological washing powders which also contained protein-digesting enzymes, lipase failed to work in removing fat stains. (2)

ii Different enzymes must be used to remove fatty acids from different positions on triglyceride molecules when making substitutes for cocoa butter. (2)

c In some industrial processes, lipase enzymes are immobilised.

i What are immobilised enzymes? (1)

ii Give two industrial advantages of immobilised enzymes. (2)

AEB, Paper 1 (Section B), January 1997

8 Explain how an enzyme's structure is related to its mode of action. (10)

9 Discuss the part played by enzymes in cell metabolism. (10)

1 catalysts; proteins; active site; denatured; temperature/pH; pH/temperature; barrier; activation

2

Type of reaction	Letter of enzyme catalysing this reaction
$H_2O + glycogen \rightleftharpoons glucose$	C
$2AH_2 + O_2 \rightleftharpoons 2H_2O + 2A$	B
$AB + C \rightleftharpoons A + BC$	A
$AH_2 + NAD \rightleftharpoons A + NADH + H^+$	B

3 a C **b** B **c** B

4 i An organic catalyst is a protein, which speeds up a biochemical reaction, but remains unchanged at the end of the reaction.

ii An active site is a region of the enzyme with its own specific shape where the substrate binds and catalysis occurs.

iii Denaturation results in a distortion of the shape of the active site so that catalysis is prevented (the substrate cannot bind to form an enzyme–substrate complex).

5

Class of enzyme catalysed	Named example	Chemical process
ligase	DNA ligase	synthesis of DNA (macromolecules using energy from ATP to catalyse formation of new chemical bonds)
oxidoreductase	succinic dehydrogenase	oxidation of substrates by transferring hydrogen to coenzymes which become reduced

6 a i One of:
- The enzyme does not contaminate the product.
- It stays in the reactor at the finish.
- The enzyme can be reused.
- It allows the process to occur as a continuous reaction.

ii Three of:
- At low temperatures/9 °C.
- There is relatively little kinetic energy/the molecules are only moving slowly.
- There will be fewer collisions with the enzyme.
- There will be a slower rate of reaction/it will take longer for the lactose to be reduced/some of the lactose will go through unchanged.

or

- The enzyme concentration was limiting.
- The substrate is in excess.
- There is saturation of all the active sites/all the active sites are occupied.
- Some substrate goes through unchanged.

b i There are fewer substrate/lactose molecules; therefore, there is less chance of a collision with the enzyme/forming a substrate–enzyme complex.

ii Economic reasons such as:
low levels of lactose are not harmful/it would take too much time/there is a high cost in removing all the lactose.

7 a i Incubate the glycerol and fatty acid/oleic acid under the same conditions, replacing the enzyme with the same volume of distilled water/or boiled lipase/without enzyme.

ii 1) $R_f = \dfrac{92\,mm}{100\,mm}$

$R_f = 0.92$ (accept 0.90)

2) Enables comparison with standard figures obtained for this substance, for example, from table or with known substance/look up the R_f values in R_f tables for the *particular* solvent.

b i Lipase itself is a protein; protein-digesting enzymes would hydrolyse the lipase.

ii An enzyme has its own specifically shaped active site; the site to fit into the terminal (ester bond) fatty acid substrate might not fit the middle fatty acid substrate.

c i Immobilised enzymes are bound to an inert substance which holds it in a fixed position.

ii Two of:
- The enzyme does not have to be recovered from the solution containing the products/it can be separated more easily.
- It can be used over and over again in a continuous flow process.
- It is more stable.

8 *Explain how the enzyme's structure is related to its mode of action.*

Ten of:
- Enzymes are large globular proteins with an active site of specific shape to form a complex with the substrate.
- This specific three-dimensional shape is achieved when a particular sequence of amino acids is coiled into a helix stabilised by hydrogen linkages, and then the polypeptides are folded into a globular shape stabilised by disulphide, electrovalent and hydrophobic bonds.
- In the lock and key mechanism the shape of the active site and the substrate are complementary.
- In the induced fit hypothesis the active site may be flexible, altering shape so that the substrate is held in the best position for catalysis;
- Less energy is needed for formation or breaking of bond/less activation energy.
- This allows a new chemical pathway to be opened/a substrate–enzyme complex is formed and this is then changed to a substrate–product complex.
- The product is then released leaving the enzyme free to catalyse a new substrate molecule.
- The polar R groups of the amino acids in the enzyme attract water, making the enzyme water soluble.

- The precise shape of the active site of an enzyme can be denatured by extremes of temperature and pH/excessive heat will cause strong vibrations which destroy bonds in the enzyme, thus altering the three-dimensional shape of the active site so that bonding with the substrate cannot occur.
- The shape of the enzyme's active site may be affected by allosteric inhibitors which bind with another part of the enzyme making it inactive; (this fits in with the induced fit hypothesis).
- The active site may be occupied by competitive inhibitors reducing enzyme activity.

9 *Discuss the part played by enzymes in cell metabolism.*

Ten of:
- Metabolism is all the chemical reactions which occur within the organism.
- It consists of anabolic reactions, building up complex compounds from simpler ones, and catabolic reactions in which complex compounds are broken down.
- The rate of chemical reactions is directly proportional to the temperature since this provides the kinetic energy for the molecular movement which will increase the number of collisions between the reactants. The range of temperatures found within living organisms means that reactions would occur far too slowly to provide for the organism's needs.
- Enzymes speed up the rate of reaction, sometimes catalysing a reaction millions of times faster at normal temperatures and pressures.
- An enzyme's presence or absence will determine whether a particular reaction can occur or not/because enzymes are very specific/usually able to catalyse only one particular reaction.

- At the ribosomes mRNA is translated into specific enzymes. They are either continuously transcribed from the nuclear genes, in the case of enzymes catalysing essential reactions such as glycolysis, or only transcribed when the need arises, for example, the presence of lactose will trigger production of lactase in *E. coli* bacteria.
- The rate of reaction will be dependent on the concentrations of the substrate and product molecules present at a particular time. Enzymes catalyse reversible reactions, allowing the amount of product produced to be tailored to the needs of the organism. If, however, conditions change, then the direction of the reaction may be reversed. Each enzyme works best at its own particular pH and this may affect the direction of the reaction, allowing response to differing conditions in different parts of the organism, for example, alimentary canal. Extremes of pH will denature enzymes.
- Control over metabolic pathways can be brought about by end-product inhibition achieved by the mechanism of negative feedback (give example).
- Enzymes for a particular pathway may all be located within a particular organelle, ensuring that the products of one reaction are readily available as substrates for the next enzyme in the pathway. This will speed up the rates of reaction.
- Enzymes may be arranged in sequence along a membrane in linear metabolic pathways. In branched metabolic pathways the enzymes may be free in the cytoplasm, allowing the formation of different end products, depending on the requirements of the cell at a certain time.
- Organelles may differ in their internal conditions so that these are optimal for the enzymes they contain, giving control of the metabolic pathways.

10 Factors affecting enzyme activity

How is enzyme activity measured?

Enzyme activity is affected by change in a number of different factors. Cells contain many different enzymes and so, when studying the activity of an enzyme, it is first extracted and then purified so that the isolated enzyme can be studied under controlled conditions.

> **Enzyme activity is measured as the amount of substrate converted by a known amount of enzyme in a given time.**

Effect of pH on enzyme activity

- Extremes of pH irreversibly denature enzymes (hydrogen bonds are irreparably broken).
- Changes in pH which are less extreme affect enzyme activity temporarily (the change is reversible).
- Every enzyme has a pH at which it is most active. This is the optimum pH for that enzyme.

Enzyme activity is measured in a buffered solution with a pH that is optimum for that enzyme. If you look at Figure 10.1 you will realise that if, for example, an experiment was carried out at pH 3 or pH 11 for any of the three enzymes, then the values for enzyme activity would be at, or near to, zero.

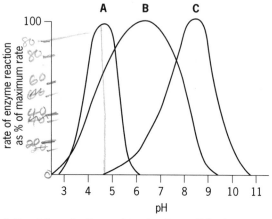

A, B and C are the three points of optimum pH for three different enzymes

Figure 10.1 Affect of pH on enzyme activity

Remember:
- **pH** is a measure of hydrogen ion ($[H]^+$) concentration to the power of 10. An increase of 1 on the pH scale is equal to a ten-fold increase in $[H]^+$ concentration, and an increase of 2 on the pH scale is, therefore, equal to a hundred-fold increase in $[H]^+$ concentration.
- A **buffer** is a substance which when added to a solution will help it maintain a constant pH by bonding to and removing hydrogen ions when their concentration starts to increase, and releasing them when their concentration begins to fall.

Effect of substrate concentration on enzyme activity

- If the concentration of enzyme present remains constant, the rate of reaction is proportional to the concentration of the substrate. In other words: if the concentration of the substrate is doubled then the rate of reaction is doubled; if the concentration is trebled, the rate is trebled and so on.
- This is true until the rate remains constant because all the active sites are occupied. At this point the enzyme is said to be **saturated** with substrate.

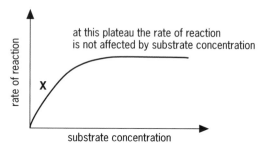

Figure 10.2 Effect of increasing substrate concentration

At point **X** in Figure 10.2, the rate of reaction is proportional to the substrate concentration. The plateau represents the stage of the reaction where any increase in substrate concentration has no further affect on the rate of reaction.

One might have expected the curve in Figure 10.2 to have been the intersection of the two straight lines representing 'rate of reaction' and 'substrate concentration'. However, the curve is not sharp because we are actually dealing with averages of very large numbers of molecules. It is therefore impossible to 'read off' the point at which the rate of reaction is at its maximum. Because of this, we measure that concentration of substrate at which **the reaction rate is at half of its maximum**. This involves much less error than trying to measure the gradual slope as it reaches its plateau.

This point at which the reaction is at half of its maximum is called the Michaelis constant (K_m). It is a constant because it is characteristic for each combination of enzyme and substrate (just as the melting point of a solid is constant for any pure sample).

Note that more than one enzyme may have the same Michaelis constant.

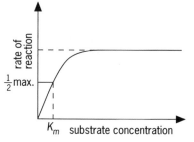

Figure 10.3 The Michaelis constant

Effect of enzyme concentration

The amount of substrate converted is proportional to the amount of enzyme present. Double the amount of enzyme; double the rate of reaction.

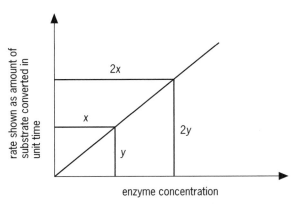

Figure 10.4 Effect of enzyme concentration on the rate of reaction where there is no shortage of substrate

Effect of temperature

Most enzymes are denatured at temperatures above 60 °C although some can withstand temperatures of up to 100 °C for several minutes. It is the length of time that an enzyme is exposed to a high temperature that is important. Several hours at room temperature may be more harmful than a few seconds at nearly 100 °C.

When enzyme molecules receive heat energy it increases kinetic energy which increases the number of collisions and therefore increases the rate of reaction. Vibrations may cause those chemical bonds which form the secondary, tertiary and quaternary structure of proteins to break, altering the vital shape of active sites, and denaturing the enzyme.

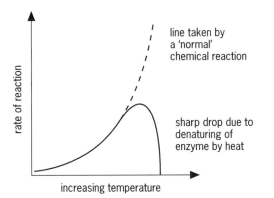

Figure 10.5 Effect of temperature on enzyme activity

The temperature at which an enzyme works most rapidly is known as the **optimum temperature**.

Temperature coefficient (Q$_{10}$)

Between 4 °C and 37 °C (37 °C being the optimum temperature for enzymes in the human body), the rate of enzyme activity doubles for a rise in temperature of 10 °C. This factor of 2 is known as the temperature coefficient, or Q$_{10}$. The equation for this is:

$$Q_{10} = \frac{\text{rate of reaction at } t + 10\,°C}{\text{rate of reaction at } t\,°C}$$

Effect of inhibitors

There are two main groups of inhibitors.

Non-reversible inhibitors

The inhibitors become firmly attached to the active sites of the enzyme, excluding the substrate molecules. Insecticides such as malathion and nerve poisons are examples of non-reversible inhibitors.

Reversible inhibitors

Competitive reversible inhibitors

Competitive inhibitors compete with substrate molecules for the active site on the enzyme. The inhibitor has a shape which the enzyme 'recognises' as a substrate but is then unable to convert it into a product. In Figure 10.6 the inhibitor has a one in four chance of binding to the enzyme's active site.

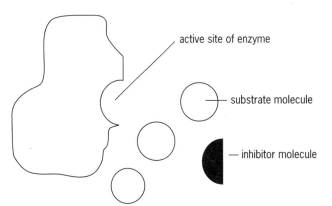

Figure 10.6 Competitive inhibition

Increasing the concentration of the substrate will decrease the inhibition.

Non-competitive reversible inhibitors

Non-competitive reversible inhibitors affect a different site on the enzyme but by doing so they prevent the enzyme's activity. Examples of this type include heavy metals.

FACTORS AFFECTING ENZYME ACTIVITY

1 a State the meaning of the term Q_{10}. (1)
  b How is enzyme activity measured? (1)

2 a Diagram **A** shows an enzyme, and **B** is the substrate of this enzyme.

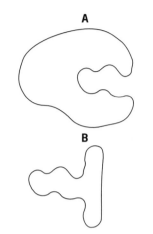

Draw a diagram (based on the above) to show how a competitive inhibitor would affect the activity of the enzyme. (2)

b

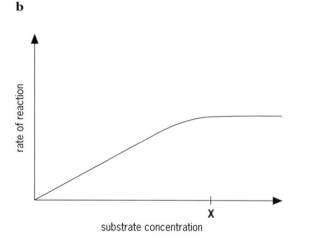

The graph shows the effect of changing the substrate concentration on the rate of an enzyme controlled reaction. Explain why increasing substrate concentration above the value shown as **X** fails to increase the rate of reaction further. (2)

c Explain how adding excess substrate could overcome the effect of a competitive inhibitor. (1)

d Explain what happens to an enzyme molecule when it is denatured by high temperature. (3)
AQA (NEAB), BY01, March 1999

3 Glucose oxidase is an enzyme which catalyses the oxidation of glucose to form gluconic acid and hydrogen peroxide.

An experiment was carried out to investigate the effect of pH on the activity of glucose peroxidase. The activity of this enzyme was determined at a range of pH values. The results are shown in the graph.

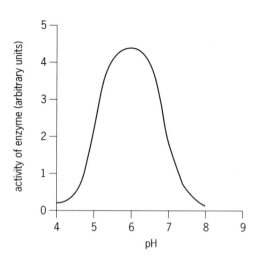

a i State how the different pH values could be obtained in this experiment. (1)
ii Describe the effect of changes in pH on the activity of this enzyme. (2)
iii Explain why changes in pH affect the activity of enzymes. (3)

b Glucose oxidase with another enzyme, peroxidase can be used to measure the concentration of glucose in solutions.

The solution to be tested is first incubated with glucose oxidase, and then with peroxidase, plus an indicator which changes colour when it is oxidised. Peroxidase breaks down the hydrogen peroxide formed by glucose oxidase and simultaneously changes the colour of the indicator.

The intensity of the colour produced is directly proportional to the concentration of the glucose in the solution, as shown in the graph below.

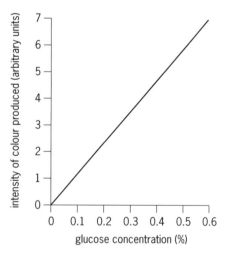

i From the graph, determine the concentration of glucose corresponding to a colour intensity of 6.5 arbitrary units. (1)
ii Describe how this method could be used to compare the concentration of glucose in two samples of fruit juice. (4)
Edexcel (London), Paper B/HB1, January 1999

4 The graph below shows the results of an investigation into the effect of a competitive inhibitor on an enzyme-controlled reaction over a range of substrate concentrations.

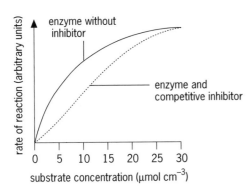

a Give one factor which would need to be kept constant in this investigation. (1)
b i Explain the difference in the rates of reaction at the substrate concentration $10\,\mu mol\,cm^{-3}$. (2)
ii Explain why the rates of reaction are similar at the substrate concentration $30\,\mu mol\,cm^{-3}$. (1)
c The diagram represents a metabolic pathway controlled by enzymes.

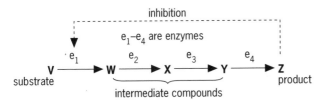

i Name the type of control mechanism which regulates production of compound **Z**. (1)
ii Explain precisely how an excess of compound **Z** will inhibit its further production. (2)

NEAB, BY01, February 1997

5 The monosaccharides glucose and fructose are reducing sugars. Sucrose is a disaccharide which is not a reducing sugar.

Benedict's test is used to detect reducing sugars. When reducing sugars are boiled with Benedict's solution a red precipitate is produced. This precipitate can be filtered from the solution, dried and weighed.

If excess Benedict's solution is used, the mass of precipitate produced is proportional to the concentration of reducing sugar in the solution. The enzyme sucrase is a hydrolase and does not react with Benedict's solution.

a In an experiment, sucrase was added to a solution of sucrose and incubated for 5 minutes. Benedict's test was then carried out on the resulting solution and a red precipitate was produced.
Suggest an explanation for this result. (2)
b A further experiment was carried out to investigate the effect of silver nitrate on the activity of sucrase. The procedure described above was repeated, but different concentrations of silver nitrate were added to the sucrase. The solutions were kept at the same pH for the same time. The mass of precipitate produced by Benedict's test at each concentration was measured. The results are shown in the table below.

Concentration of silver nitrate $(mol\,dm^{-3})$	Mass of precipitate (mg)
0 (control)	50
10^{-6}	37
10^{-5}	27
10^{-4}	10

i Calculate the percentage decrease in the mass of the precipitate produced in the solution containing $10^{-5}\,mol\,dm^{-3}$ silver nitrate compared with the control test. Show your working. (2)
ii Suggest an explanation for the effect of the silver nitrate solution on the activity of the enzyme sucrase. (2)
c i Explain why it is important to maintain constant pH when investigating enzyme activity. (2)
ii State three precautions, other than maintaining constant pH, which should be taken to produce reliable results in the above investigation. (3)

London, Paper B/HB1, January 1997

1 a $Q_{10} = \dfrac{\text{rate of reaction at } t + 10\,°C}{\text{rate of reaction at } t\,°C}$

b Enzyme activity is measured as the amount of substrate converted by a known amount of enzyme in a given time.

2 a The diagram should be based on the one in the question and should show an inhibitor with a complementary shape that would fit into the active site. The inhibitor should be shown in place in the active site.

b The substrate concentration is not limiting/the enzyme concentration is limiting.
All the active sites of the enzyme are full/the enzyme is at its maximum turnover rate.

c One of:
- There is more substrate than inhibitor so formation of an enzyme–substrate complex is more likely.
- The substrate is more likely to enter the active site.

d Correctly named bonds broken/water removed.
The tertiary/globular shape of the enzyme is changed.
The shape of the active site is affected.

3 a i Use a series of buffer solutions.
ii Two of:
- The optimum pH for enzyme activity is pH 6.
- The activity decreases on either side of the optimum.
- There is little enzyme activity at extremes of pH/pH 4/pH 8.

iii Three of:
- Free H^+ or OH^- ions affect the charges on amino acid residues/ionisation of the R groups/affect hydrogen bonding/ionic bonding.
- pH affects the shape of the active site.
- It affects the binding with the substrate.
- Extreme values of pH cause denaturation.

b i 0.56%
ii *In questions such as this you merely need to apply the principles of a fair test to the information given in the question. Do not be put off if you have not heard of this experiment before.*

Four of:
- Use equal volumes of fruit juice.
- Add the same volume/concentration of glucose oxidase to each.
- Incubate for a standard time/stated constant temperature.
- Add the same volume/concentration of peroxidase plus indicator.
- Read the intensity of the colour.
- Use the graph to find the glucose concentrations.

4 a One of:
- temperature
- pH
- quantity of enzyme
- quantity of inhibitor

b i Two of:
- The inhibitor is similar in structure to the substrate.
- The inhibitor attaches to the active site.
- Less substrate attaches to the active site/takes part in reaction.

ii The rates of reaction are similar at high substrate concentrations because few inhibitor molecules can enter the active site since there are so many substrate molecules.

c i negative feedback/end product inhibition
ii Compound **Z** attaches to enzyme e_1 and prevents production of **W**/further reactions leading to **Z**.

5 a Sucrose is hydrolysed to glucose and fructose (monosaccharides).

b i Difference between control and $10^{-5}\,mol\,dm^{-3}$ silver nitrate:

$$= 50 - 27 = 23$$

$$\% \text{ decrease} = \frac{23}{50} \times 100$$

$$= 46\%$$

ii Two of:
- silver nitrate is an inhibitor;
- blocks or affects the shape of the active site;
- so that substrate is no longer able to bind;
- this reduces the rate of the reaction.

c i Two of:
- pH changes will affect the formation of enzyme–substrate complexes.
- It will change the tertiary structure/shape of the active site/enzyme;
- by changing the hydrogen bonding/charges/ionisation.
- There is an optimum pH for the enzyme activity.

ii Three of:
- Same constant temperature.
- Same time of equilibration of enzyme and substrate.
- Equal volume of Benedict's solution.
- Time of heating with Benedict's solution to be the same.
- Same filtration/drying method.
- Dry precipitate to constant mass.
- Equal volume of sucrose.
- Equal volume of sucrase.
- Same concentration of sucrose.
- Same concentration of sucrase.
- Leave enzyme and substrate for the same time to react.
- Equal volume of silver nitrate.
- Use replicates.

Pathways

11 Photosynthesis: Cyclic and non-cyclic photophosphorylation

> Photosynthesis is the activation of chlorophyll by light, resulting in the splitting of water, the release of oxygen and the production of $NADPH_2$ and ATP.

Photosynthesis is a chemically reductive process. Light energy is required and therefore the glucose which is synthesised has a lot of available energy in its molecules.

The equation $6CO_2 + 6H_2O \rightarrow C_6H_{12}O_6 + 6O_2$ is much oversimplified and misleading. *All* the oxygen gas comes from the water. The hydrogen from the water chemically *reduces* the carbon dioxide, leaving its oxygen to be evolved as gas. (Experiments feeding plants with water containing the heavy isotope ^{18}O demonstrate the source of the oxygen gas evolved.)

To provide the hydrogen atoms for the reduction, water must be split. The splitting of water molecules by light energy (absorbed by chlorophyll) is called **photolysis**.

Do not confuse the splitting of water in photosynthesis (photolysis) with the ionisation of water. During ionisation, charged particles (ions) are formed, but when water is *split*, electronically neutral but unstable particles are formed.

In photolysis:

$$\begin{array}{ccc} & \text{both are structurally unstable} & \\ & \downarrow \qquad\qquad \downarrow & \\ H_2O + \text{light} \rightarrow & [OH] & + \quad [H^+ + e^-] \\ \text{energy} & \text{not an ion,} & \text{a hydrogen atom} \\ & \text{no charge} & \text{(a proton with an} \\ & & \text{electron carrying} \\ & & \text{a lot of free energy)} \end{array}$$

Neither the [OH] or the $[H^+ + e^-]$ groups actually exist.

- The [OH] groups combine together to form water and oxygen gas. Remember, all the oxygen comes from the water.
- The $[H^+ + e^-]$ groups find other molecules to react with which become reduced because they gain an electron.

Remember: Reduction is a *gain of electrons* by an atom or ion.

Photosynthesis consists of two main stages – the **light dependent** and the **light independent**. The light-dependent reactions are **photolysis** and **photophosphorylation**. *The chloroplast performs all the reactions of photosynthesis.*

Light-dependent reactions

Photosynthesis uses the energy of sunlight to generate high energy (excited) electrons by splitting water. It is these electrons which are used eventually to reduce carbon dioxide. Note that it is the energy of electrons which reduces the carbon dioxide, *not* the hydrogen ions themselves. It is the *energy path* in photophosphorylation that is important to note.

Chlorophyll molecules absorb light at the red (long wavelength) and blue (short wavelength) ends of the spectrum. That is why chlorophyll looks green. When we measure the amount of energy absorbed by a chlorophyll solution at various wavelengths (by using variously coloured light) we can draw a graph called the *absorption spectrum*.

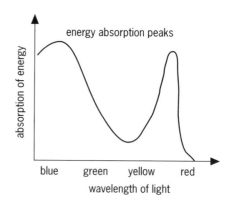

Figure 11.1 Absorption spectrum of chlorophyll

When we measure the rate of photosynthesis that goes on in a leaf at the various wavelengths, we get an *action spectrum*. The shape of the action spectrum graph is virtually identical to that of the absorption spectrum because the rate of photosynthesis is greatest at the blue and red absorption peaks. Measuring the rate of oxygen production is usually the best method of determining the rate of photosynthesis.

The fact that the two graphs are almost identical for most green leaves confirms that light absorption by chlorophyll drives photosynthesis.

Thus, chlorophyll, activated by light, drives an electron stream.

One of the elements in chlorophyll is magnesium (electron configuration 2,8,2). Before illumination the electrons in the outer shell are at ground-state energy level. Light energy causes these two electrons to be temporarily displaced. In this high energy state they are said to be **excited**. A fraction of a second after being displaced, the electrons fall back to ground state. The extra energy that they had is now available for other reactions.

What happens to these excited electrons in a green leaf?

Chlorophyll is found in the grana of chloroplasts (see **13** Photosynthesis: The structure of leaves and factors affecting the rate of photosynthesis). Chloroplasts contain acceptors for the excited electrons and an important series of events now takes place.

Electrons are forced by their energy content into acceptors.

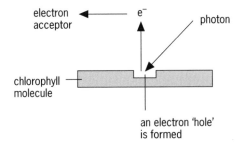

Figure 11.2 A light photon hitting a chlorophyll molecule

The chlorophyll molecules are now in an electron deficient (positively charged, or oxidised) state. *'Holes' in the reaction centre have a positive charge which is powerful enough to split water molecules and produce electrons which replace those lost to the electron acceptor.*

$$2H_2O \rightarrow 4H^+ + 4e^- + O_2$$

The hydrogen protons (H^+) left from the splitting of water molecules associate with NADP to give $NADPH_2$ ($NADPH + H^+$) which is the final step in non-cyclic

photophosphorylation. $NADPH_2$ contains a lot of energy and is a **hydrogen carrier** for the reduction of carbon dioxide.

Non-cyclic photophosphorylation (see Figure 11.3) consists of two systems, PS1 and PS2 which differ in the types of chlorophyll pigments present. Electrons are passed to PS2 and then an electron acceptor. ATP is produced via an electron transport system (ETS), with the electron dropping down to PS1. Then light causes the energy level of the electrons to be raised to a level high enough to be accepted by another electron acceptor. A second ETS leads to the production of $NADPH_2$ with hydrogen coming from the splitting of water. Two photosystems are involved in the reduction of NADP because there is insufficient energy in light to energise the electrons from water straight to NADP. It requires two steps.

Cyclic photophosphorylation is an alternative 'light driven' pathway. Some of the electrons ejected from the chlorophyll molecules of PS1 pass along a series of electron carriers, gradually losing their energy and ending up back in the chlorophyll molecules they started from. As the electrons pass along this ETS some of their energy is trapped in the formation of ATP molecules from ADP and inorganic phosphate. **Only PS1 is involved and only ATP is formed.**

The important products of the light dependent part of photosynthesis are ATP, $NADPH_2$ and oxygen.

The energy from the ATP is needed for the light-independent reactions (see **12** Photosynthesis: The fixation of carbon dioxide).

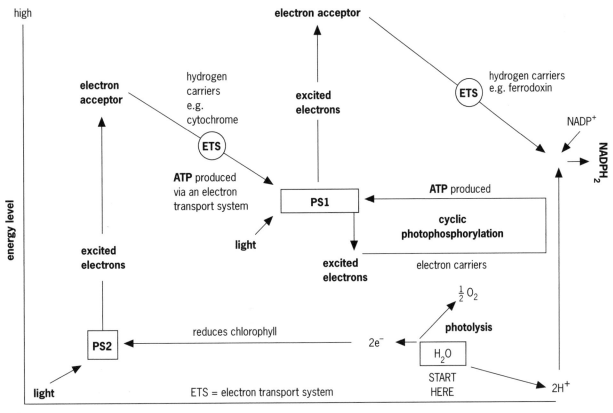

Figure 11.3 Photolysis, cyclic and non-cyclic photophosphorylation

PHOTOSYNTHESIS: CYCLIC AND NON-CYCLIC PHOTOPHOSPHORYLATION

1 a Fill in the spaces to complete the following summary equations for photosynthesis.

$$\underline{\hspace{2cm}} + \text{water} \xrightarrow{\text{chlorophyll}} \text{carbohydrate} + \underline{\hspace{2cm}}$$

$$\underline{\hspace{2cm}} + 6H_2O \xrightarrow{\text{light}} C_6H_{12}O_6 + \underline{\hspace{2cm}}$$

(4)

b Name the following processes which occur in photosynthesis:
i The splitting of water molecules using light energy.
ii The formation of ATP using energy from electrons excited by light.
iii The reactions in photosynthesis consisting of photolysis and photophosphorylation. (3)

2 Read this summary of photosynthesis and then write a list of the most appropriate word(s) to complete the passage.

Photosynthesis involves building up complex organic compounds from _____ (a), _____ (b) compounds. This kind of food production is called _____ (c) nutrition. Light energy is absorbed by _____ (d) molecules, causing _____ (e) to be displaced. The most effective parts of white light are the _____ (f) and _____ (g) colours. The light-dependent reactions occur in the _____ (h) of the chloroplasts. The gas _____ (i) is released when _____ (j) is split. _____ (k) and _____ (l) are produced. These are used in the light-independent reactions to reduce the gas _____ (m). The hexose sugar _____ (n) is produced. (14)

3 Indicate by using the letter A, B, C or D the correct answer to the following.
a The coenzyme used in photosynthesis is:
A – NAD
B – ATP
C – FAD
D – NADP
b Electrons from non-cyclic phosphorylation pass into the dark reaction via:
A – ATP
B – NADPH₂
C – hydroxyl ions
D – H⁺ ions
c Oxygen is evolved during photosynthesis from:
A – the light-dependent stage when water is split
B – the light-dependent stage when carbon dioxide is split
C – the light-independent stage when water is split
D – the light-independent stage when carbon dioxide is split

d The pigment in leaves appears green because:
A – the green wavelengths of light are reflected
B – the green wavelengths of light are absorbed
C – the blue and red wavelengths of light are reflected
D – the blue and red wavelengths of light are absorbed
e The metal element found in chlorophyll is:
A – iron
B – hydrogen
C – manganese
D – magnesium (5)

4 a Define the terms:
i action spectrum (2)
ii absorption spectrum (2)
b Explain what happens to chlorophyll when it is photoactivated. (2)
c Which wavelengths of light are absorbed by chlorophyll? (1)
d Explain why it is advantageous for plants to have more than one type of photosynthetic pigment. (2)
e Explain why the leaves of many plants appear green. (1)

5 The graph shows the percentage of light, of different wavelengths, absorbed by an alcohol solution of chlorophyll a, in the laboratory.

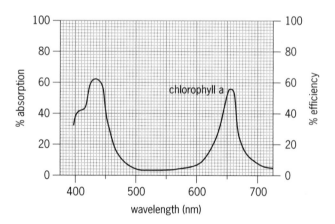

The table below shows how effective these wavelengths are in driving photosynthesis in **intact** plants. The most efficient wavelength is given a value of 100%.

Wavelength (nm)	400	450	500	550	600	650	700
Percentage efficiency	84	96	64	40	58	80	28

a Using the right-hand axis, plot the data on the graph. (2)
b Estimate which of the 50 nm ranges would contain the most effective wavelength. (1)
c What is the name given to the graph you have plotted? (1)

d **i** At what wavelength does the difference between the two plots have the greatest value? (1)

ii Suggest an explanation for this difference. (2)

e Suggest how you might set up an experiment to obtain the data in the table. (3)

f From the information given at the beginning of the question, suggest why a comparison of the two graphs might lead to unreliable biological conclusions. (1)

Welsh, Paper B1, January 1999

6 The part played by chlorophyll in photosynthesis was investigated by Engelmann. He put filaments of green alga on three microscope slides and added a suspension of motile, oxygen-requiring bacteria to each. Each slide was then placed in a small chamber from which air was excluded. Engelmann placed each of these chambers in different light conditions for several hours and then examined them. The results are shown in the diagram.

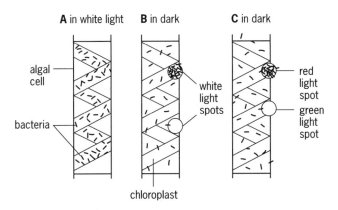

a How does the distribution of bacteria on slide **A** provide evidence that chlorophyll is necessary for photosynthesis? (3)

b One explanation for the results was that bacteria were attracted towards the light. What evidence from the results disproves this idea? (1)

c Suggest an explanation for the result on slide **C**. (2)

NEAB, BY01, June 1998

7 a Research biologists have discovered that some photosynthetic blue–green algae have an important enzyme, nitrogenase which is easily damaged by oxygen. Some types of the micro-organism can protect the enzyme by encasing it in cells which lack enzymes for the photosystem in photosynthesis which splits water. These cells have thick walls.

i Explain how these adaptations will protect the nitrogenase. (2)

ii Which product of the light-dependent reactions will these cells be able to produce? (1)

b Describe how isotopes can be used to show that the oxygen released in photosynthesis originates from the water and not from the carbon dioxide. (4)

8 Only a small percentage of the sunlight falling on a pond is used by the phytoplankton in photosynthesis. Suggest two reasons for this. (2)

9 Explain the part that is played in photosynthesis by:

a water (6)

b cyclic photophosphorylation (4)

10 Describe how a chloroplast is adapted to absorb the energy from light and use it in the light-dependent reaction of photosynthesis. (10)

1 a carbon dioxide + water $\xrightarrow{\text{chlorophyll}}$ carbohydrate + oxygen

$$6CO_2 + 6H_2O \xrightarrow{\text{light}} C_6H_{12}O_6 + 6O_2$$

b i photolysis
ii photophosphorylation
iii light-dependent reactions

2 a simple
b inorganic
c holophytic/autotrophic
d chlorophyll
e electrons
f and **g** blue and red
h grana
i oxygen
j water
k and **l** ATP and NADPH + H$^+$/NADPH$_2$
m carbon dioxide
n glucose

3 a D
b B
c A
d A
e D

4 a i The action spectrum means the effectiveness of the various wavelengths of light; in providing energy for the process (in this case photosynthesis which can be measured by the amount of oxygen evolved).
ii The absorption spectrum means the wavelengths of light used (absorbed); by the various pigments.
b The chlorophyll absorbs light energy and this raises the energy level of the outer shell electrons in the magnesium part of the chlorophyll molecule. These excited electrons are then able to leave the chlorophyll and be passed along a series of electron acceptors.
c Chlorophyll a absorbs the red and blue part of the spectrum/light wavelengths.
d Other chloroplast pigments increase the range of wavelengths from which energy can be obtained. This will result in an increase in photosynthesis.
e Other wavelengths except green are absorbed/green is reflected or transmitted through the leaf.

5 a See graph below. *Deduct one mark for each plotting error (two or more errors scores zero).*

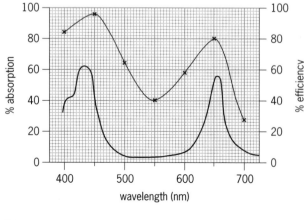

b 400–450 nm
c action spectrum
d i 500 nm
ii Photosynthesis is still occurring (despite the absence of light absorption by chlorophyll)/this is due to the presence of other pigments (e.g. carotene/xanthophyll) in the intact plant which are not present in the chlorophyll extract.
e Three of:
- Source of light with different wavelengths.
- Standard plant material to be illuminated (*note: leaves are too variable so specify same size/surface area*).
- Constant light intensity/temperature/other relevant constant condition.
- Specific method of measuring rate of photosynthesis, e.g. rate of oxygen production/carbon dioxide consumption.

f One graph is obtained from plants, the other from an alcoholic solution of chlorophyll. Chlorophyll might behave differently in intact plants.

6 a Oxygen is produced in photosynthesis. Bacteria move to where oxygen is being produced. Oxygen production is associated with the chloroplast/chlorophyll.
b In **B**, bacteria are attracted to where the light shines on the chloroplast/random distribution in **A**/not attracted to the green light/not attracted to the lower light spot in **B**.
c Red light is used for photosynthesis. Green light is not absorbed by the chloroplast/not used/reflected.

7 a i If the enzymes to split water are missing, no oxygen will be released. The thick cell walls may prevent the entry of oxygen from other cells, which possess both photosystems.
ii ATP
b An aquatic plant is supplied with carbon dioxide containing the normal isotope of oxygen ^{16}O and water containing the heavy isotope ^{18}O. The oxygen evolved can be shown to consist of normal ^{16}O by using a mass spectrometer. When the water supplied has the heavy isotope and the carbon dioxide has normal oxygen; then the oxygen produced is the heavy isotope.

8 Some of the light is reflected or converted to heat. Some of the wavelengths are not absorbed by the chloroplast pigments.

9 a Six of:
- Water is a raw material of photosynthesis which provides electrons;
- these replace those lost from the chlorophyll molecules (non-cyclic);
- $2H_2O \rightarrow 4H^+ + 4e^- + O_2$
- Oxygen is given off as a by-product.
- The H$^+$ reduces NADP to NADPH + H$^+$/NADPH$_2$ and this is used to provide energy;
- eventually the H combines with CO$_2$ forming carbohydrate.

- Water is also a solvent for all the chemical reactions of photosynthesis.
- It provides turgor pressure, allowing the thin-walled mesophyll cells to support the leaf, holding it out to absorb the maximum amount of light energy.

b Light energy (red and blue wavelengths) raises the energy of an electron of a chlorophyll molecule of photosystem 1. This electron can then pass down a series of electron carriers at different energy levels. When the energy released in one of these transfers is sufficient, it is used to build up ATP from ADP. The electron (now with a lower energy level) returns to the chlorophyll molecule.

10 Chloroplasts have groups of chlorophyll molecules/ quantosomes;
arranged on membranes called lamellae/thylokoids.
Grana are piles of these lamellae with a large surface area to absorb light energy.
Chlorophyll absorbs the red and blue wavelengths.
Light striking chlorophyll molecules excites/increases the energy level of electrons.
The excited electrons pass along a chain of electron acceptors situated on the lamella membrane. H^+ collects in lamellar spaces and flows through pores in stalked particles/ATPase.
Photophosphorylation/synthesis of ATP from ADP and inorganic phosphate.
Cyclic – formation of ATP/involves PSI/electrons return to chlorophyll.
Non-cyclic – formation of ATP and NADPH + H^+/involves PS2/electrons from photolysis.

12 Photosynthesis: The fixation of carbon dioxide

Light-independent reactions

The light-independent reactions are sometimes called the dark reactions although they can take place in the light. 'Light independent' is therefore a better term.

The **Calvin cycle** (named after the work of Calvin in the late 1940s) is concerned with the fixing of carbon dioxide into more complex molecules and eventually carbohydrate. Remember, carbon dioxide has to be *reduced* to make carbohydrate, and the energy needed comes from the ATP and $NADPH_2$.

Before reduction can take place, the carbon dioxide has to be picked up by an acceptor molecule.

Six carbon atoms (and therefore six carbon dioxide molecules) are needed to make a hexose molecule (for example, glucose). The reaction is shown in Figure 12.1.

The acceptor, labelled 'A' in Figure 12.1, is a carbohydrate. It is a pentose diphosphate and is called **ribulose biphosphate** or **RuBP**.

Note the following stages in the synthesis of starch.

Stage 1

The first stage is the fixation of the carbon dioxide molecules, each of which combines within a 5-carbon compound (RuBP). Each of the 6-carbon molecules formed (represented by CO_2A in Figure 12.1) immediately splits to produce two molecules of a 3-carbon compound called **phosphoglyceric acid**, or **PGA**.

The equation in terms of the number of carbon atoms is:

$$\begin{array}{lll} C1 & + \; C5 & \rightarrow \; 2 \times C3 \\ \text{carbon} & + \; \text{ribulose} & \rightarrow \; 2 \times PGA \\ \text{dioxide} & \quad \text{biphosphate} & \end{array}$$

Stage 2

The PGA is reduced using the energy from ATP plus $NADPH_2$. Triose phosphate is formed.

All of this is more easily understood if considered just in terms of the number of carbon atoms the substances contain.

In the following equation, six molecules of CO_2 combine with six molecules of RuBP.

$$\begin{array}{llll} & & & \text{energy in} \\ 6\,C1 \; + & 6\,C5 & \rightarrow \; 12\,C3 & \rightarrow \; 12\,C3 \\ CO_2 & RuBP & PGA & \text{triose phosphate} \end{array}$$

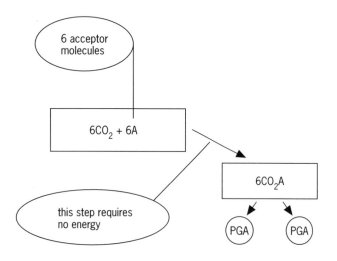

Figure 12.1 Carbon dioxide picked up by an acceptor molecule

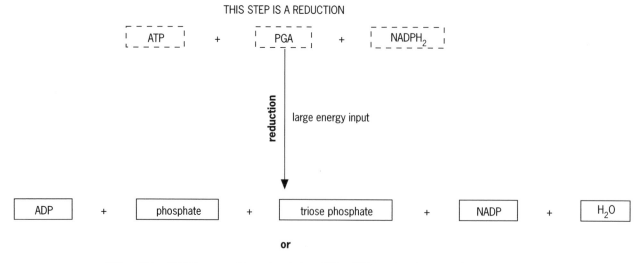

Figure 12.2 Stage 2 in more detail

Stage 3

It is important to realise that for every turn of the cycle one molecule of CO_2 is used, one molecule of RuBP is regenerated, and **one sixth of a molecule of hexose** is stored.

Therefore, it takes six turns of the cycle and six molecules of carbon dioxide to produce one molecule of glucose. In the process, six molecules of RuBP are regenerated.

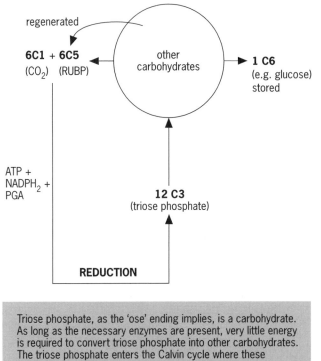

> Triose phosphate, as the 'ose' ending implies, is a carbohydrate. As long as the necessary enzymes are present, very little energy is required to convert triose phosphate into other carbohydrates. The triose phosphate enters the Calvin cycle where these changes take place.

Figure 12.3 The Calvin cycle

Further consideration of the Calvin cycle

A reminder of the total number of carbon atoms involved in the formation of one molecule of glucose:

6C1	+	6C5	=	6C6		=	6C5	+	1C
CO_2		RuBP		a number of carbohydrates (such as the 12 triose phosphates) with a total of 36 carbon atoms			RuBP		hexose stored
6		30		36			30		6

Note that, while the amount of carbohydrate stored increases, the amount of C5 acceptor (RuBP) does not. The 'extra' carbon atoms come from an inexhaustible supply of carbon dioxide in the atmosphere.

The Calvin cycle takes place in the stroma of the chloroplasts (see **13** Photosynthesis: The structure of leaves and factors affecting the rate of photosynthesis).

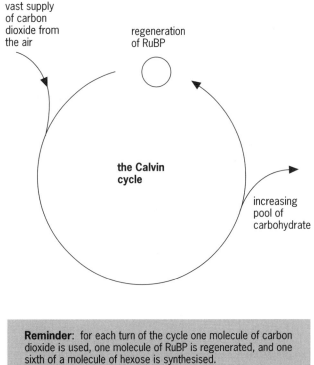

> **Reminder**: for each turn of the cycle one molecule of carbon dioxide is used, one molecule of RuBP is regenerated, and one sixth of a molecule of hexose is synthesised.

Figure 12.4 Regeneration of RuBP

1 a Read the passage about the light-independent stage of photosynthesis and then write a list of the most appropriate word or words to fill the spaces.

In the light-independent stage the molecules of carbon dioxide are accepted by _____. The products _____ and _____ from the light-dependent stage are needed to form _____. This compound can then be converted into a _____ sugar. (5)

b Give the letter of the correct answer to the following:

i The Calvin cycle involves:

A – the splitting of water to release oxygen
B – the reduction of carbon dioxide to form carbohydrate
C – the phosphorylation of ADP to ATP
D – the reduction of NADP to NADPH + H$^+$ (1)

ii The carbon acceptor molecule in the light-independent reaction is:

A – phosphoglyceric acid
B – triose phosphate
C – ribulose biphosphate
D – hexose sugar (1)

c Give the number which is the correct answer to the following.

i How many turns of the Calvin cycle are necessary to produce one molecule of each of the following compounds?

A – ribulose biphosphate
B – glucose (2)

ii Give the number of carbon atoms which one molecule of each of the following contains:

A – carbon dioxide
B – triose phosphate
C – ribulose biphosphate
D – hexose (4)

2 The diagram below shows some of the processes which occur in the light-independent reaction of photosynthesis.

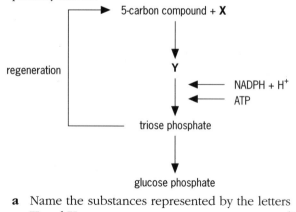

a Name the substances represented by the letters **X** and **Y**. (2)

b State the origin of the NADPH + H$^+$ and the ATP used in the light-independent reaction. (1)

c Where in the chloroplast does the light-independent reaction occur? (1)

London, Paper B2, January 1997

3 The diagram below shows the light-independent (dark) reaction of photosynthesis.

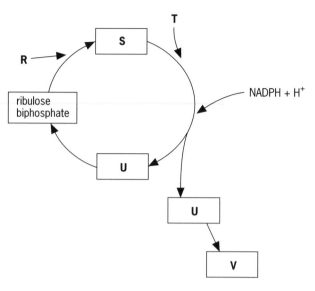

a This cycle of reactions is often named after its discoverer. Give this name. (1)

b Name the molecules which enter the cycle at **R** and **T**. (2)

c Give the full names of the molecules **S** and **U**. (2)

d What is produced at **V**? (1)

e **i** Name the reducing agent which is used up in the cycle. (1)

ii Name the photosynthetic stage which produces this agent. (1)

Welsh, Paper B1, June 1997

4 a Explain the part played in photosynthesis by:

i water (3)
ii ribulose biphosphate (2)
iii ATP in photosynthesis (4)

5

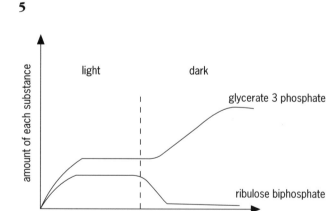

The graph above shows the changes in relative amounts of photosynthetic intermediates found in the chloroplast during periods of light and dark conditions. What information does this give you about the part played by ribulose biphosphate and glycerate 3 phosphate during photosynthesis? (4)

6 Calvin investigated the pathway by which carbon dioxide is converted to organic compounds during photosynthesis. He used the apparatus shown in the diagram below. The apparatus contained cells of a unicellular alga. While the apparatus was in a dark room, Calvin supplied the algal cells with carbon dioxide containing radioactive carbon. The contents of the apparatus were thoroughly mixed, then a light was switched on. At 5 second intervals he released a few of the cells into hot alcohol, which killed the cells very quickly.

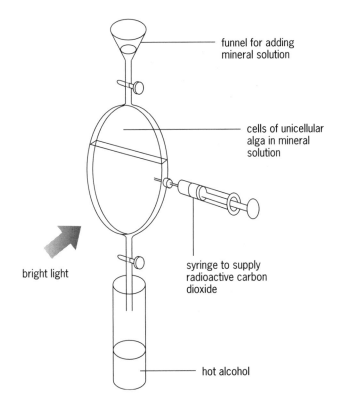

funnel for adding mineral solution

cells of unicellular alga in mineral solution

syringe to supply radioactive carbon dioxide

bright light

hot alcohol

a i Why is the part of the apparatus containing the algal cells thin?
ii How would it be possible to prevent the light source heating the algal cells?
iii Suggest why it was necessary for the algal cells to be killed very quickly. (3)

Calvin homogenised the killed algal cells and carried out two-way paper chromatography. This technique involves running the chromatogram with one solvent, then turning the paper through 90° and running it with a different solvent. By using this technique, Calvin was able to investigate the chromatogram he obtained. The spots are those containing radioactive compounds.

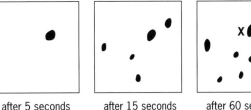

after 5 seconds   after 15 seconds   after 60 seconds

b Suggest how the compound in each spot might be identified. (1)
c i Name the compound first formed in the pathway in which carbon dioxide is converted into an organic compound. (1)
ii Which of the spots **W**, **X**, **Y** or **Z** contained this compound? (1)
iii Explain your answer to **ii**. (1)

NEAB, BY01, March 1998

7 Describe the techniques used to investigate the path of the light-independent reactions of photosynthesis. (12)

It is important that you check the exact requirements of your particular Exam Board as these usually specify how much chemical detail is needed in answers for their questions. Also some chemicals have several names, therefore it is best to use the ones they specify. In these answers, if I have used a question from a specified Exam Board, then I have used the chemical names and detail in their answer scheme.

1 a ribulose biphosphate (5-carbon compound); NADPH + H⁺ (NADPH₂) and ATP (any order); triose phosphate; hexose
b i B **ii** C
c i A = 1, B = 6
ii A = 1, B = 3, C = 5, D = 6

2 a X: carbon dioxide/CO_2
Y: PGA/phosphoglyceric acid/equivalent
b The light (dependent) reaction/stage (of photosynthesis)/grana/thylakoid/non-cyclic photophosphorylation.
c stroma (of chloroplast)

3 a Calvin (this must be spelt correctly)
b R = carbon dioxide
T = ATP
c S = glycerate 3 phosphate (GALP/triose phosphate/PGA)
d glucose/sugar/starch/carbohydrate
e i NADPH + H⁺ / NADPH₂
ii light stage/photosystem 1

4 a i Three of:
• Water is a source of hydrogen.
• Water is a source of electrons.
• The hydrogen reduces NADP to NADPH + H⁺/NADPH₂.
• The electrons replace those lost by chlorophyll.
ii Ribulose biphosphate acts as an acceptor for carbon dioxide; leading to the production of GP.
iii Four of:
• Energy from excited electrons can be used to build up ATP from ADP.
• It acts as a source of energy to drive reactions in the light-independent stage.
• To provide phosphate/for phosphorylation.
• ATP is required in the light-independent state of photosynthesis for the conversion of GP to triose phosphate.
• ATP is used in the formation of ribulose biphosphate in the light-independent phase.

5 During the light-independent stage of photosynthesis, glycerate 3 phosphate is formed;
but if it is to be changed into other metabolites it requires other products of the light-dependent reaction.
The ribulose biphosphate which is present in the light is the carbon acceptor.
The amount present decreases sharply in the dark as it is used up and not regenerated.

6 a i For maximum absorption of light/to supply light to all algae.
ii Insert a glass tank of water between the light and the flask/use fluorescent light.
iii To prevent further reaction from taking place/to stop the enzymes working/to stop photosynthesis/to ensure no more products are formed.
b Run them against known compounds/calculate the R_f value **or** elute the spot and use a valid test such as a mass spectrometer.
c i glycerate 3 phosphate/GP/unstable 6C compound
ii X
iii Because it is the only one on the 5 second chromatogram.

7 The radioactive isotope of carbon is ¹⁴C (compare with normal ¹²C). This was used by Calvin to investigate the light-independent reactions. He used the alga *Chlorella* and supplied it with hydrogen carbonate (carbon dioxide) containing ¹⁴C. The alga was supplied with light. After a short period of time a sample of the photosynthesising alga was dropped into boiling methanol to stop the reaction. Shorter and shorter time periods were used to obtain 'snapshots in time' of the reaction.

A chromatogram of the sample was then produced by homogenising the algal cell contents in a suitable solvent. A concentrated spot was placed on the start line of a piece of chromatography paper. This is achieved by using a fine capillary tube to draw up the extract and placing a tiny spot on the paper. After allowing this to dry, the process was repeated several times. The paper below the spot was placed in contact with a solvent and the chromatogram allowed to run. Turning the chromatogram through 90° and re-running will improve separation. The most soluble substances will be carried furthest along the chromatogram. Radioactive products are identified by placing the chromatogram in contact with X-ray or photographic film. Then the R_f value for a spot is calculated and looked up in standard tables to identify it or that solvent.

R_f = distance moved by the spot/distance moved by the solvent front.

Products which occur after very short time periods occur earlier in the light-independent pathway.

13 Photosynthesis: The structure of leaves and factors affecting the rate of photosynthesis

The functions of green leaves

- To provide surface areas for the gaseous exchanges involved in both photosynthesis and respiration, and for the transpiration of water vapour.
- To provide a large surface area for capturing sunlight.
- To contain the chloroplasts which possess the chlorophyll which absorbs light energy.

The structure of green leaves enables these functions to be performed efficiently. Remember, for photosynthesis to function efficiently, all of the above conditions are necessary.

The leaf is adapted to carry out these functions.

Adaptations of the leaf to obtain light energy (sunlight)

- The leaves are thin and have a large surface area/volume ratio.
- The leaves on a plant arrange themselves so as to reduce overlapping and the likelihood of being in the shade.
- Both the cuticle (which is thicker on the upper surface to reduce water loss by evaporation) and the layer of epidermal cells are transparent to let the light through.
- The palisade mesophyll cells on the upper side of the leaf cells are arranged close together with their long axes perpendicular to the surface. This arrangement (rather than several layers of cells with their long axes horizontal) means that sunlight only has to pass through a minimum number of cellulose cell walls and is therefore brighter when it reaches the chloroplasts. The palisade mesophyll cells, in particular, are packed with chloroplasts.
- The structure of the chloroplasts ensures that the chlorophyll receives the maximum amount of light available.

Adaptations of the leaf for obtaining water and removing sugars in solution

Each leaf has a central midrib with a network of branching veins. This ensures that no cell is very far away from a xylem vessel supplying water or a phloem tube for the transport of sugar solution away from the leaf.

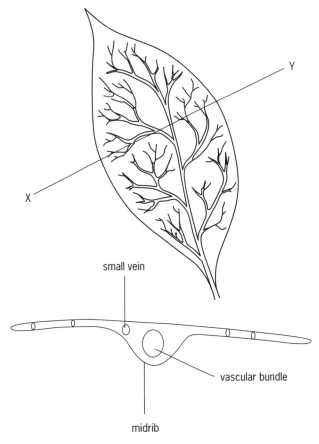

small vein

vascular bundle

midrib

section X–Y

Figure 13.1 Underside of a dicotyledonous leaf showing venation

Adaptations of the leaf for gaseous exchange

- Air, containing carbon dioxide, enters through the stomata when they are open. There may be as many as 45 000 stomata per square centimetre on the lower surface of the leaf. The presence of stomata ensures the exchange of gases without undue water loss. When water is scarce, the stomata close.
- The spongy mesophyll has many air spaces which allow the diffusion of carbon dioxide and oxygen both into and out of the cells. Before carbon dioxide, which is needed for photosynthesis, can diffuse into the cells it dissolves in the surface moisture. The atmosphere in the intercellular spaces is moist due to the water diffusing out of the ends of the xylem vessels.

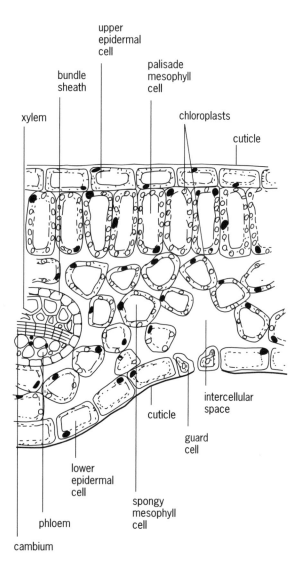

Figure 13.2 Dicotyledonous leaf in vertical section

(The subjects of stomatal movements, transpiration and translocation are dealt with in **29** Transport systems in flowering plants 2.)

Chloroplasts

These are large organelles about 5–10 μm long (1 μm is one thousandth of a millimetre). They are concentrated in the palisade cells of the mesophyll where there is the greatest penetration of light waves, but they are also present in large numbers in the cells of the spongy mesophyll.

Chloroplasts have a double outer membrane and their own DNA. The chlorophyll is contained in the grana where the light-dependent reactions occur. The grana resemble stacks of circular flattened discs called **thylakoids** (or **lamellae**). The chlorophyll molecules are grouped together within granules in the walls of the thylakoids into what are called quantasomes.

The thylakoids are the site of electron carriers and the electron transport chains. Hydrogen ions are actively accumulated into the spaces between the thylakoid membranes and then diffuse down an electrochemical gradient across an ATP synthesase complex. There may be as many as 50 grana, each one typically consisting of about 50 thylakoids. This structure holds the chlorophyll in a position which is efficient for trapping light. It also economises on space and at the same time provides a large surface area. The grana are connected in places by **intergrana**. Surrounding the grana is a matrix called the **stroma**. It is where the light-independent reactions occur.

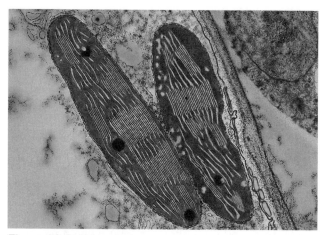

Figure 13.3 Chloroplast

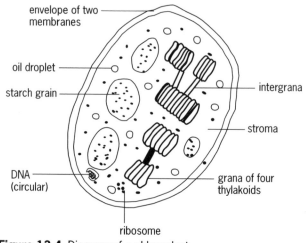

Figure 13.4 Diagram of a chloroplast

Factors affecting photosynthetic rate

> **At a given moment, the rate of a physiological process is limited by the one factor which is in shortest supply, and by that factor alone.**

The limiting factors to be considered are light intensity, temperature and carbon dioxide concentration.

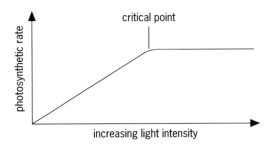

Figure 13.5 Graph showing the effect of light intensity on the rate of photosynthesis

The rate of photosynthesis is directly proportional to the intensity of the light until a critical light intensity is reached. At this point the rate of photosynthesis remains constant unless the supply of carbon dioxide increases, in which case the rate of photosynthesis will further increase until a new critical light intensity is reached.

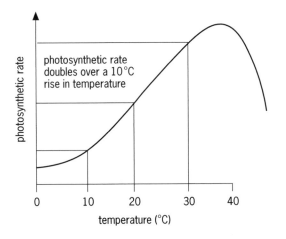

Figure 13.6 Graph showing the effect of temperature on the rate of photosynthesis

The temperature coefficient (Q_{10}) is typically 2. This means that over a 10 degree range in temperature, say, between 20 and 30 degrees Celsius as shown in Figure 13.6, the metabolic rate doubles. At about 40 °C the enzymes start to be denatured and in most cases are completely denatured by 60 °C.

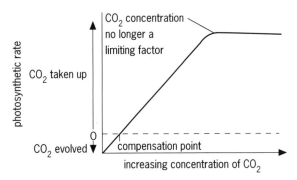

Figure 13.7 Graph showing the effect of carbon dioxide concentration on the rate of photosynthesis

The low concentration of atmospheric carbon dioxide (about 0.04%) is a major limiting factor. The optimum concentration is 0.1%.

At the **compensation point** (see Figure 13.7) the volume of CO_2 given out during respiration equals the amount used in photosynthesis and there is, therefore, no overall gaseous exchange.

1 What are the functions of a plant leaf? (5)

2 a i Use the diagram of an electron micrograph of a chloroplast to name structures **X**, **Y** and **Z**. (3)

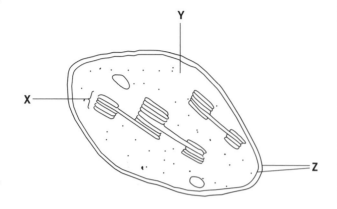

Use the letters on the diagram to identify:
ii Where the light-dependent reactions take place. (1)
iii Where the light-independent reactions occur. (1)
b i Describe the arrangement of chlorophyll within the chloroplast. (2)
ii Suggest why it is arranged in this way. (1)

3 a Below is a diagram of a transverse section through a leaf. Name the cells **A**, **B** and **C**. (3)

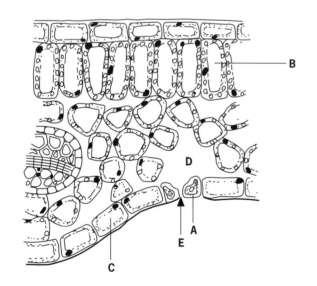

b Green plants can produce their own food. Give the name of this process. (1)
c Name three types of cells in the leaf which contain chloroplasts. (3)
d Name two environmental conditions which may cause **E** to close. (2)
e Describe variations that would occur in oxygen concentrations at point **D** throughout a normal 24-hour period. Explain why these variations occur. (4)

4 The photomicrograph below shows the structure of part of a leaf as seen in transverse section.

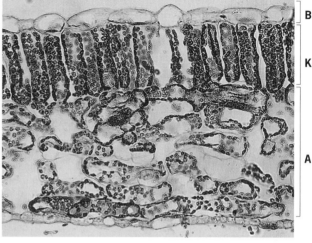

a Name the parts labelled **A** and **B**. (2)
b Explain two ways in which the cells labelled **K** are adapted for their function. (4)

London, Paper B2, June1997

5 The graph below shows how the rate of photosynthesis varies when a plant is grown under different environmental conditions. Use the information on the graph to answer the following questions.

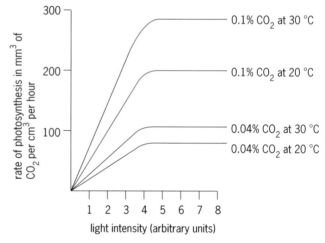

a Describe the way that rate of photosynthesis varies as light intensity increases for a plant grown in an atmosphere containing 0.04% carbon dioxide at 20 °C. (3)
b Describe the effect of increasing the temperature from 20 °C to 30 °C for plants grown in an atmosphere containing 0.04% carbon dioxide. (3)
c If the plant in this experiment was a crop plant grown in a glasshouse, explain the optimum environmental conditions of light and temperature that should be supplied to make it worthwhile to increase the carbon dioxide content of the glasshouse atmosphere. (3)
d Suggest how the carbon dioxide content of the atmosphere and the temperature could be increased at the same time. (1)

6 A shoot of Canadian pond weed was placed in the apparatus shown below.

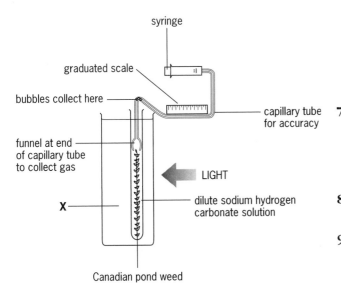

syringe

graduated scale

bubbles collect here

funnel at end of capillary tube to collect gas

X

capillary tube for accuracy

LIGHT

dilute sodium hydrogen carbonate solution

Canadian pond weed

The rate of photosynthesis was measured.
a What is the purpose of **X**? (1)
b Why has sodium hydrogen carbonate been added to the pond water? (1)
c What two precautions would you use to make your results more reliable? (2)

d If you collected a gas bubble of length 9 mm in 10 minutes, how would you calculate the volume of gas produced per hour if conditions remained constant? The diameter of the capillary tube is 2 mm. Show your working. (3)
e What is the gas produced in this experiment? (1)
f Is the gas collected a direct measure of photosynthesis? (1)
g What is the function of the syringe? (1)

7 a State the law of limiting factors. (2)
b Explain what is meant by the compensation point for a plant. (2)
c At what times of day would you expect compensation point to occur? (2)

8 Discuss ways in which the structure of the leaf of a mesophyte plant is adapted for its functions. (10)

9 a Photosynthesis generally takes place in a leaf. Describe how the leaf is adapted to allow this process to occur effectively. (3)
b Explain the roles of water, light and ribulose biphosphate in the process of photosynthesis.
   i water
   ii light
   iii ribulose biphosphate (6)
c Explain why an increase in temperature will increase the rate of photosynthesis. (3)

AQA (NEAB), BY01, March 1999

1 The main function of the leaf is to make food for the plant by photosynthesis. It contains chlorophyll to absorb light energy. It allows rapid diffusion via the stomata of the gases (carbon dioxide and oxygen) which are the raw materials and waste materials of photosynthesis and respiration. It allows transpiration (water loss) from the stomata. This is one of the main forces drawing water and minerals up through the xylem. The carbohydrate sucrose is transported to other parts of the plant in the phloem.

2 a i **X** = granum, **Y** = stroma, **Z** = double membrane/ envelope of chloroplast

   ii The light-dependent reactions occur in X (granum).

   iii The light-independent reactions occur in Y (stroma).

   b i Chlorophyll molecules are grouped together in groups called quantosomes along the thylakoids (lamellae). The thylakoids are stacked together to form the grana.

   ii It provides a large surface area for the most effective absorption of light.

3 a **A** = guard cell, **B** = palisade mesophyll, **C** = lower epidermis

   b photosynthesis

   c palisade mesophyll cells, spongy mesophyll cells and guard cells

   d darkness (lack of light) and lack of water

   e Oxygen is given off as a waste product of photosynthesis. It will therefore start to be produced at dawn, increasing as the light and temperature increases (possibly to a maximum around midday) and then decreasing as dusk approaches. There is no photosynthesis at night, therefore the oxygen concentration will be at a minimum. Respiration occurs at a low level throughout the 24 hour period, using up a small amount of oxygen.

4 a **A** = spongy mesophyll, **B** = (upper/adaxial) epidermis

   b Two of the following pairs:
   • Contain many chloroplasts;
     for efficient photosynthesis.
   • Have thin moist cellulose cell walls;
     for gaseous exchange/light penetration.
   • Longitudinal axis of cells perpendicular to the leaf surface/cells elongated;
     for trapping light energy.
   • Cells fit closely together/rectangular in section;
     for better capture of light.

5 a As the light intensity increases (from 0 to 3.7 arbitrary units) the rate of photosynthesis increases in direct proportion.
   The rate of increase then slows as light intensity increases further.
   The rate of photosynthesis remains constant as the light increases above 4 arbitrary units (at 80 mm^3 of CO_2 per cm^3 per hour).

   b Photosynthesis occurs at a higher rate at 30 °C than 20 °C for all values of light intensity.
   As the light intensity increases from 0 to 4 units the rate of increase is greater at 30 °C.
   After 4.5 light units the rate of photosynthesis remains constant for further increases of light intensity, at

110 mm^3 of CO_2 per cm^3 per hour for 30 °C compared with 80 at 20 °C.

   c At low light intensity, increasing the concentration of carbon dioxide will not significantly increase rate of photosynthesis/yield of the crop plant.
   Whereas when light intensity is higher and not a limiting factor (from 4.5 units upwards), increasing the carbon dioxide concentration will have more effect and will be more profitable.
   From the graph it can be seen that increasing carbon dioxide concentration has far more effect at 30 °C (increases from 110 to 290) than at 20 °C (increases from 80 to 200).

   d Burning a fossil fuel such as oil would increase the temperature and give off the waste gas carbon dioxide.

Note: it is important to refer to exact figures taken from the graph when answering this type of question.

6 a To absorb the heat from the light, minimising temperature fluctuations.

   b To increase the concentration of hydrogen carbonate ions/carbon dioxide available for photosynthesis.

   c Two of:
   • Use the same shoot of Canadian pond weed each time.
   • Maintain the same distance between the lamp and the apparatus.
   • Make sure that all gas bubbles have been expelled before starting the experiment.
   • The experiment should take place in a darkened room.

   d Diameter of tube $(2r) = 2$ mm.

   $$\text{Radius} = 1 \text{ mm}$$
   $$\text{Volume of bubble} = \pi r^2 \times \text{length}$$
   $$= \pi \times 1 \times 9 \text{ mm}^3 \text{ in 10 minutes}$$
   $$= \pi \times 1 \times 9 \times 6 \text{ mm}^3 \text{ in an hour}$$
   $$= 167.56 \text{ mm}^3/\text{hour}$$

   e oxygen-enriched air

   f A small amount of oxygen will be used up in respiration, but this should be fairly constant throughout the experiment and this occurs at a low level compared with oxygen produced by photosynthesis in bright light.

   g The syringe allows any residual air to be expelled at the start of the experiment/it is used to draw the gas bubble from the bend in the tube onto the capillary scale for measuring.

7 a If a process is governed by several factors then the rate of the reaction will depend on the factor which is in the shortest supply.

   b The compensation point is the point at which the rate of photosynthesis and the rate of respiration just balance; so that the oxygen produced in respiration equals the oxygen used up in respiration/the carbon dioxide used up in photosynthesis equals that produced in respiration.

   c Compensation point would occur in conditions of low light intensity; as found shortly after sunrise and shortly before dusk (or appropriate times).

8 *A mesophyte plant is one which grows in an area of generally adequate water supply and so the extreme modifications found in xerophytes do not need to be discussed.*

The main function of the leaf is photosynthesis. It is important for the leaf to receive and absorb maximum light and so leaves have a large surface area and are thin, allowing penetration of light to the cells. The vascular tissues of the leaf veins and also the turgor pressure of the thinner walled cells give support to the leaf so it can be held out to receive light, aided by the petiole which positions it at 90 degrees to the light rays.

The distribution of the tissues in the leaf lamina should be discussed. The palisade mesophyll is under the transparent upper epidermis and well situated to receive light. These cells are packed with chloroplasts which are able to position themselves within the cell so they do not shade one another and can photosynthesise efficiently. The chloroplasts themselves are well adapted in having chlorophyll molecules stacked on lamellae to absorb as much light energy as possible. Palisade mesophyll are thin walled and rectangular in shape with narrow air spaces for gaseous exchange. The spongy mesophyll cells below also have many chloroplasts, but these cells are more rounded and nearer the stomatal pores with the large air spaces to allow diffusion of carbon dioxide into the photosynthesising cells and remove the waste product oxygen. The stomata allow diffusion of gases to and from the atmosphere; they close partially at night to cut down water loss. The leaf has a transparent waxy cuticle on the epidermis to reduce water loss. This is aided by the shape of the leaf which allows the concave lower surface, where most stomata are situated, to trap saturated water vapour.

The water and mineral ions are supplied by xylem vessels of the midrib and lateral veins. The photosynthetic product, sucrose, is transported away by the phloem.

9 a Three of:
- Large surface area to collect solar energy.
- Transparent nature of cuticle to allow light penetration.
- Position of chlorophyll to trap light.
- Stomata to allow exchange of gases.
- Thin/maximum surface area to volume ratio for diffusion of gases.
- Spongy mesophyll/air spaces for carbon dioxide store.
- Xylem for input of water.
- Phloem for removal of end products.

b i Two of:
- Provides hydrogen;
- to reduce NADP.
- Provides electron;
- to stabilise/reduce chlorophyll.

ii Two of:
- Excites oxidises/removes an electron from chlorophyll/photosystem;
- photophosphorylation/ATP produced;
- electron used in reduction of NADP.

iii Carbon dioxide acceptor.
Forms GP.

c Three of:
- Enzymes are involved.
- Extra kinetic energy/molecules move faster.
- Molecules collide more often/more enzyme–substrate complexes formed.
- Increased rate of diffusion of raw materials.

14 Respiration: Glycolysis

> **Respiration is the breakdown of energy-rich substances to build up molecules of ATP from which energy is released for reactions as required in cells.**

The above definition includes respiration in plants and animals, and anaerobic as well as aerobic respiration.

Note the difference between cellular respiration and gaseous exchange, the latter being purely a physical process.

The important points about cellular respiration are the energy paths involved.

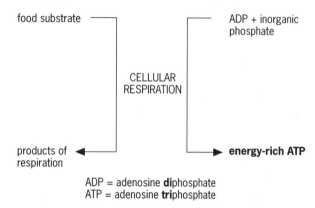

ADP = adenosine **di**phosphate
ATP = adenosine **tri**phosphate

Figure 14.1 Energy paths in cellular respiration

The third phosphate group requires a lot of energy to attach it to the ADP and an energy-rich bond is formed. This energy is released readily when needed and the bond is broken. Consider the analogy of a crossbow. A lot of energy is needed to draw back the spring and set the trigger. A light touch on the trigger and the energy is unleashed. The bow can then be set up once again.

ATP is found in all cells and is therefore the universal energy carrier. ATP provides the energy for:

- the synthesis of materials in living cells
- the transmission of nervous impulses in animals
- the contraction of muscle cells in animals
- active uptake and cation pumps.

Cellular respiration is concerned with how living cells obtain the energy from food substrates such as glucose and transfer it into the energy-rich phosphate bond of ATP.

Phase 1 – Glycolysis

Glycolysis takes place *in the cell cytoplasm*. The necessary enzymes are not found in the mitochondria. It is the first stage of both **aerobic** and **anaerobic** respiration and therefore takes place with or without oxygen.

One molecule of glucose (containing six carbon atoms) is split by a series of reactions into two molecules of pyruvic acid (each containing three carbon atoms).

During these reactions there is a net gain of two ATP molecules plus four hydrogen atoms, the latter being taken up by two molecules of a coenzyme called NAD (nicotinamide adenine dinucleotide).

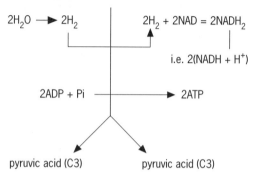

Figure 14.2 Glycolysis

The net gain is two molecules of $NADH_2$ and two molecules of ATP.

The $NADH_2$ is often referred to as **reduced NAD**. The net output of energy at this stage is two molecules of ATP.

If there is no oxygen present the pyruvic acid will be switched to anaerobic pathways.

Anaerobic respiration

Anaerobic respiration is a form of cellular respiration in which energy is released from glucose (and other foods) in the absence of oxygen. The lack of oxygen prevents oxidation via an electron system in the mitochondria. An **anaerobe** is an organism which can respire anaerobically. It may be:

1 **Obligate** – if it cannot survive in oxygen. An example is those bacteria which cause food poisoning such as **botulism**.
2 **Facultative** – when oxygen is present the organism respires aerobically but in the absence of oxygen it respires anaerobically. Most anaerobes are in this group.

Fermentation

A good example is the anaerobic metabolism of sugar by yeast (one of a group of microscopic fungi which are important economically in baking and brewing).

The oxygen supply is quickly used up unless it is replenished by vigorous stirring. After glycolysis, if oxygen is not available the pyruvic acid is converted to ethanol and carbon dioxide. Only two ATPs are formed and the process is only about 2% efficient.

$$\text{efficiency} = \frac{\text{output of energy}}{\text{input of energy (per glucose molecule)}}$$

The ATP provides energy for cellular activities. The ethanol (which still contains much energy) is a by-product.

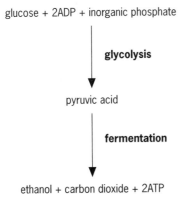

glucose + 2ADP + inorganic phosphate

glycolysis

pyruvic acid

fermentation

ethanol + carbon dioxide + 2ATP

Figure 14.3 The equation for fermentation

It will be appreciated that in the brewing industry it is essential to exclude oxygen from the fermentation process if there is to be a high yield of ethanol. Any air allowed to enter could well carry the spores of other species of micro-organisms which may influence the taste of the brew. Ethanoic (acetic) acid may be produced and form vinegar.

When using yeast in the baking industry, the only consideration is the production of carbon dioxide. The dough is usually kneaded to trap small bubbles of air which assist in the rising of the bread and also supply the yeast with oxygen so that respiration is aerobic and an even greater quantity of carbon dioxide is produced. The small amounts of ethanol which inevitably are produced, evaporate during the baking.

Muscle respiration

When muscle is working too rapidly for its own supply of oxygen, despite an increased breathing rate and the heart beating stronger and faster, it gets its energy by respiring anaerobically. The reaction is similar to that of fermentation (Figure 14.3) but lactic acid is formed instead of ethanol and no carbon dioxide is produced. (Lactic acid has one carbon atom and two oxygen atoms more in its molecule than ethanol.) Once again, only two ATP are produced. The end-product of the reaction (lactic acid), as in fermentation, still contains a lot of energy.

If the lactic acid, which is toxic, builds up to high levels, muscle cramp ensues.

When muscles are forced to respire anaerobically, because there is a shortage of oxygen, an **oxygen debt** builds up. After strenuous exercise there has to be a recovery period when the oxygen is paid back during aerobic respiration, and the lactic acid is converted to carbon dioxide and water with the production of ATP.

The equations for anaerobic respiration

1 Fermentation

$$C_6H_{12}O_6 + 2ADP + 2 \text{ phosphate} = 2CH_3CH_2OH + 2CO_2 + 2ATP + 210\,kJ$$
glucose   ethanol

2 Muscle respiration

$$C_6H_{12}O_6 + 2ADP + 2 \text{ phosphate} = 2CH_3CHOHCOOH + 2ATP + 150\,kJ$$
                        lactic acid

In both fermentation and muscle respiration the two ATPs are those that are produced during glycolysis.

The $NADH_2$ produced during glycolysis is broken down into NAD and hydrogen. The hydrogen is used during fermentation/muscle respiration to reduce pyruvic acid to ethanol or lactic acid. The NAD which is released is needed for glycolysis. Since the amount of NAD in a cell is limited, the process of glycolysis, and the yield of ATP, would quickly come to a halt if anaerobic respiration finished with the production of pyruvate (pyruvic acid) and the NAD is not regenerated.

ATP is used to provide energy for all cellular activities and an active cell will need more than 2 million molecules of ATP per second to provide the energy for its activities.

RESPIRATION: GLYCOLYSIS

1 a Define respiration. (2)
  b i Explain what the term 'anaerobic' means. (1)
    ii Explain what the term 'facultative anaerobe' means. (1)

2 An unfit person went jogging. An investigation showed that during the first 6 minutes of this exercise the concentration of lactic acid in his blood increased.
  a Name the type of respiration occurring in his muscles which produced the lactic acid. (1)
  b What would be the effect if the lactic acid concentration built up to a high level in the muscles? (1)
  c After the exercise the jogger would continue to puff and pant. Explain why it is necessary to obtain large volumes of oxygen after the exercise has ceased. (3)

3 Read the following passage about respiration and write a list of the most appropriate word(s).

The first phase in the respiration of glucose is called _____. This process occurs in the _____ of the cell. The six-carbon glucose molecule is first phosphorylated by _____. The end-products of this first phase are NADH$_2$, _____ and _____. (5)

4 Give the letter which indicates the one correct answer to each of the following.
  i The end product(s) of anaerobic respiration in mammalian muscle tissue are:
  A – ethanol and carbon dioxide
  B – lactate and carbon dioxide
  C – ethanol only
  D – lactate only
  ii Which of the following reactions requires an input of energy?
  A – phosphorylation of glucose
  B – converting pyruvate to lactate
  C – converting phosphoglyceric acid to pyruvate
  D – converting pyruvate to ethanol (2)

5 Anaerobic respiration occurs in yeast cells growing without oxygen. The scheme below shows some of the stages.

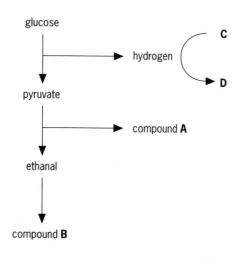

  a Name the compounds **A** and **B**. (2)
  b State which of the compounds on the diagram contains the most energy per gram. (1)
  c The hydrogen atoms are removed by compound **C**. Identify compounds **C** and **D**. (2)
  d What type of enzymes catalyse the removal of these hydrogen atoms? (1)
  e How is compound **C** regenerated? (1)

6 The diagram below shows some of the stages in anaerobic respiration in a muscle.

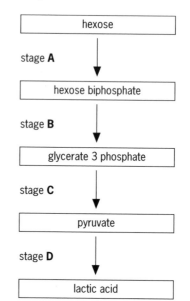

  a i Name the process shown by stages **A** to **C**. (1)
    ii State where in a cell this process occurs. (1)
  b i Give two uses of ATP in cells. (2)
    ii At which of the stages shown on the diagram is ATP used? (1)
  c NADH + H$^+$ is a reduced coenzyme which is involved in anaerobic respiration. At which of the stages shown is NADH + H$^+$ oxidised? (1)
  London, Paper B1, January 1997

7 ATP is 'spent' in the initial stages of glycolysis.
  a Explain why this is necessary. (1)
  b What are the two main roles of ATP in this process? (2)

8 Explain what is meant by glycolysis and describe the way in which it is carried out in the animal cell. List the products of glycolysis, stating how much of each is produced and how they are used by the cell. (10)
  Welsh, Paper B1, June 1997

9 a Give an account of how the following are formed during anaerobic respiration.
    i pyruvate
    ii ATP (6)
  b Describe what happens to the pyruvate in:
    i a plant cell
    ii a muscle cell.
    iii Why is it necessary for the pathway to continue after the formation of pyruvate? (5)

1 a Respiration is the release of energy from the oxidation of organic food substances (for example, glucose) and occurs in the cell.

b i Anaerobic means without oxygen/air.
(Note: whenever 'a' or 'an' precedes a biological word it means not or without.)

ii A facultative anaerobe is an organism which is able to live/respire either in the presence or absence of oxygen.

2 a anaerobic respiration

b muscle cramps

c This is necessary to remove the toxic lactic acid and to obtain energy from it.
This is done by breathing heavily to pay back the oxygen debt. Some of the lactic acid is oxidised to release carbon dioxide and water. Some of the energy is then used to build up some of the lactic acid back to glucose.

3 glycolysis; cytoplasm; ATP; pyruvate and ATP (*or* ATP and pyruvate)

4 i D **ii** A

5 *Take care in this question to distinguish between ethanal (acetaldehyde) and ethanol (an alcohol).*

a compound **A** = carbon dioxide
compound **B** = ethanol

b glucose

c compound **C** = NAD
compound **D** = NADH + H$^+$ (NADH$_2$)

d dehydrogenases

e It is used to reduce ethanal/it is oxidised by ethanal.

6 a i glycolysis

ii In the cytoplasm/cytosol.

b i Two of the following:
- In active transport/ion pumps.
- To phosphorylate a substrate.
- In muscle contraction.
- In the light independent reaction of photosynthesis.
- In cell division/replication of DNA.
- In named anabolic reaction, for example, protein synthesis.

ii stage **A**

c stage **D**

7 a To make the glucose molecule more reactive/to increase the energy level.

b The addition of a phosphate/phosphorylation. To supply energy.

8 Glycolysis is the initial stage in the release of energy from glucose.
The glucose reacts with ATP molecules/becomes phosphorylated to produce a hexose phosphate.
The hexose phosphate breaks into two 3-carbon molecules.
Inorganic phosphate is taken up and combined with ADP and NAD is reduced.
The 3-carbon molecules are converted to pyruvate.
For each molecule of glucose two molecules of ATP are used up and four are produced.
There is a net gain of two ATP;
which is used by the cell as a readily available source of energy.
Two molecules of reduced NAD are produced.
This is a powerful reducing agent/a source of high energy electrons for the electron transfer system.
Two molecules of pyruvate are produced;
which are required by the Krebs cycle for the production of further ATP and NADH + H$^+$.

9 a i Initially two molecules of ATP must be used. One ATP is used to phosphorylate glucose and form glucose phosphate. Another ATP molecule is similarly used producing fructose 1,6-biphosphate. The hexose biphosphate is then split into two molecules of triose phosphate. The coenzyme NAD acts as a hydrogen acceptor, oxidising the triose phosphate to give glycerate 3 phosphate. This is converted into pyruvate.

ii Glycolysis results in a net gain of two ATPs per glucose molecule since four ATPs are formed but two are used up in the initial stages. Inorganic phosphate reacts with the triose phosphate. The two phosphate groups are then used to build up two energy-rich ATPs from two ADPs on the pathway to pyruvate (i.e. a gain of 2ATP per 3C = 4ATP per 6C).

b i The pyruvate has carbon dioxide removed.
Ethanal is formed and this is then reduced to ethanol by NADH + H$^+$.

ii In a muscle cell the pyruvate is reduced to form lactate by the NADH + H$^+$.

iii Converting the pyruvate to either ethanol or lactic acid ensures that the NAD is regenerated from NADH + H$^+$.
NAD is essential if the glycolytic pathway to pyruvate is to continue.

15 Respiration: Aerobic respiration

If oxygen is available after glycolysis then the pyruvic acid which has been formed passes into the **Krebs cycle** (also known as the citric acid cycle).

Phase 2 – The Krebs cycle

At the beginning of this stage most of the energy which originally came from the glucose molecule is locked up in the pyruvic acid (pyruvate). Before the pyruvate enters the Krebs cycle it passes into the *matrix of the mitochondria*. Here it combines with a carrier compound called coenzyme A to form acetylcoenzyme A plus a molecule of carbon dioxide. A pair of hydrogen atoms are also removed and they reduce NAD to $NADH_2$. (This is also known as the **link stage**.)

In Figure 15.1, note the number of carbon atoms in each of the compounds formed in the cycle. The acetylcoenzyme A, which has two carbon atoms, combines with one molecule of a pool of oxaloacetate molecules, each with four carbon atoms, to form a citrate with six carbon atoms in its molecule. In Figure 15.1 the actual Krebs cycle is contained within the dotted square. This, together with the link stage, has a total output of three carbon dioxide molecules, and five pairs of hydrogen atoms.

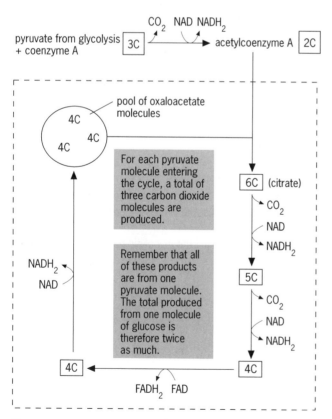

Figure 15.1 The Krebs cycle

Of the five pairs of hydrogen atoms, four combine with NAD and the remaining one combines with a different hydrogen carrier called FAD (flavine adenine dinucleotide).

All of these reactions require the presence of enzymes:

- those that remove carbon dioxide are **decarboxylases**
- those that remove hydrogen are **dehydrogenases**.

The enzymes are *in the matrix which is within the inner membrane of mitochondria*.

The energy is still locked up in the reduced NAD and FAD ($NADH_2$ and $FADH_2$). These carrier molecules can be split and the electrons passed down an electron transfer system. This is the next stage of aerobic respiration and it takes place on the *inner membrane of the mitochondria*.

Phase 3 – Oxidative phosphorylation

The reduced NAD and FAD transfer their hydrogen atoms along a chain of carriers at progressively lower energy levels. The carriers include some iron-containing proteins called **cytochromes** as well as the NAD and FAD.

As the hydrogen atoms (which are actually carried as separate electrons and protons) pass down the chain, the energy which is released is used to build up ATP from ADP plus phosphate. It is this process which is known as oxidative phosphorylation. The final hydrogen acceptor is oxygen which is reduced to water.

The concept of the electron transfer system is easier to grasp if you think in terms of a ball bouncing down a series of steps.

When the ball is at the top it has the greatest gravitational potential energy. It loses this potential as it bounces down each step.

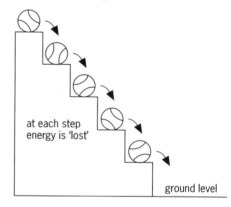

Figure 15.2 Model of an electron transport chain

As the ball in Figure 15.2 bounces down each step, varying amounts of energy are released depending on the vertical height.

At ground level there is no more gravitational potential energy to 'lose'. The formation of water can be considered as 'reaching ground level'. There is no energy in the hydrogen atoms of water as far as living organisms are concerned.

The production of ATP

The ATP is synthesised in the stalked **oxysomes** which are attached to the cristae of the mitochondria (see **1** Cells and tissues). Evidence of this site for ATP synthesis was proposed in the 1960s by Mitchell. The members of the electron transfer system are found within the inner membranes of mitochondria and the electrons pass along them. But what happens to the positively charged hydrogen protons that remain in the space between the inner and outer membranes?

Figure 15.3 illustrates how an electrochemical gradient builds up because electrons move into the inner mitochondrial membrane which is impermeable to hydrogen protons $[H^+]$. A concentration gradient forms which results in a tendency for hydrogen protons to move from the space between inner and outer membranes, across the inner membrane and back into the matrix. The inner membrane is impermeable to hydrogen ions and the only route available for the flow of protons is through the stalked oxysomes. The oxysomes contain the enzyme ATPase which is activated when protons pass through it. Thus, when protons move through the enzyme, ATP is synthesised from ADP and inorganic phosphate. Finally, the hydrogen protons are pumped (active transport) from the matrix, across the inner membrane and back into the inter-membrane space, which thus maintains the gradient.

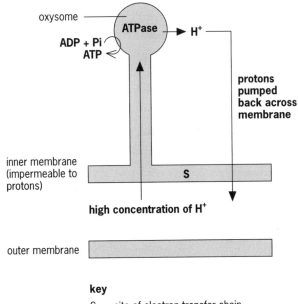

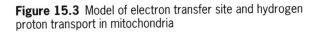

key

S = site of electron transfer chain
H^+ = hydrogen protons

Figure 15.3 Model of electron transfer site and hydrogen proton transport in mitochondria

Altogether, five molecules of reduced NAD/FAD are produced for each molecule of pyruvic acid broken down. Since each molecule of glucose breaks down into *two* molecules of pyruvate the net gain per molecule of glucose is 10 NADH$_2$/FADH$_2$. Add to this the two molecules of NADH$_2$ produced during glycolysis (**14** Respiration: Glycolysis) and we have a total of 12 molecules produced.

The transfer of one pair of hydrogen atoms generates three ATP per molecule of NADH$_2$.

$$NADH_2 + \tfrac{1}{2}O_2 \rightarrow NAD + H_2O$$
$$3\ ADP + 3\ phosphate \qquad\qquad 3\ ATP$$

Twelve molecules of reduced NAD/FAD generate $12 \times 3 = 36$ ATP. Add to this the two ATP that were generated during glycolysis (**14** Respiration: Glycolysis) and we have a grand total of 38 high energy phosphate bonds (ATP) generated from one molecule of glucose. Reduced FAD only produces two ATP but one extra ATP is produced directly in the Krebs cycle.

Remember that reduction is the addition of electrons. It is *electrons* that carry the energy, *not* hydrogen ions.

Summary of ATP generated

- Fermentation generates   2 ATP
- Muscle respiration generates   2 ATP
- Aerobic respiration generates   38 ATP

The respiration of fats

Fats, as well as carbohydrates, can be respired to yield energy. The fat molecules are first broken down into fatty acids and glycerol (**2** Biological molecules). Enzymes then split the carbon atoms from the fatty acid molecules two at a time. Each split is oxidative (energy is released) and some of the energy is used to generate ATP.

Respiratory quotient

Simple experiments can be carried out to show that oxygen is used, and carbon dioxide produced, during respiration. Quantitative comparisons of the amount of oxygen and carbon dioxide in inspired and expired air can be carried out by gas analysis.

Dividing the volume of CO_2 produced by the volume of O_2 consumed gives the RQ (respiratory quotient). Knowledge of the RQ gives information as to the type of food respired, and the kind of metabolism taking place.

The RQ of: carbohydrates = 1.0
fats   = 0.7
proteins   = 0.9

1 a What formula is used to calculate the respiratory quotient? (1)

b Name the group of enzymes which remove carbon dioxide. (1)

c Read the following passage and then write a list of the most appropriate word(s) to fill the spaces.

The pyruvate which has been produced during glycolysis diffuses into the organelles called _____. This second phase of respiration requires _____ conditions. A series of reactions called the _____ cycle then occurs. Hydrogen atoms are removed by _____ enzymes and passed on to _____. Energy is released during many of the reactions and is used to build up energy-rich molecules of _____ from _____ and _____. (8)

2 Write down the letter of the one correct answer to each of the following.

i In respiration the final acceptor for hydrogen ions and electrons is:
A – water
B – glucose
C – oxygen
D – ATP

ii Which of the following is not a definition of oxidation?
A – the addition of electrons
B – the removal of electrons
C – the addition of oxygen
D – the removal of hydrogen

iii The ratio of energy released from a specific quantity of glucose during aerobic respiration compared to anaerobic respiration is:
A – 2:1
B – 19:1
C – 1:2
D – 18:1

iv The link reaction which connects glycolysis to the Krebs cycle results in the formation of:
A – acetylcoenzyme A
B – GALP
C – reduced NAD
D – ATP

v ATP is produced during respiration:
A – only from glycolysis
B – only from the Krebs cycle
C – from glycolysis and the Krebs cycle
D – from glycolysis, the Krebs cycle and electron transfer (5)

3 The diagram below represents a mitochondrion.

a Name **X**, **Y** and **Z**. (3)

b Use the letters on the diagram to locate the part of the mitochondrion:
i Where the electron carrier molecules are located.
ii Where Krebs cycle reactions occur.
iii Where you expect to find the highest concentration of ATP. (3)

c Measure the length of the mitochondrion and calculate its actual length if the magnification is 40 000. Show all your working. (2)

d Mitochondria and chloroplasts have some similarities in their structure. Give two points of similarity. (2)

4 a The inner membrane of the mitochondrion is folded to form cristae. Suggest how this folding is an adaptation to the function of the mitochondrion. (2)

b The equation represents the oxidation of a lipid:

$$C_{57}H_{104}O_6 + 80O_2 \rightarrow 57CO_2 + 52H_2O + energy$$

Use the equation to calculate the respiratory quotient of this lipid. Show your working. (1)

c Suggest an explanation for each of the following:
i The RQ of germinating maize grains is 1. (1)
ii The RQ of a normal healthy person varies over a 24-hour period. (1)

AEB, Paper 1, Summer 1998

5 The diagram below shows some of the reactions of aerobic respiration.

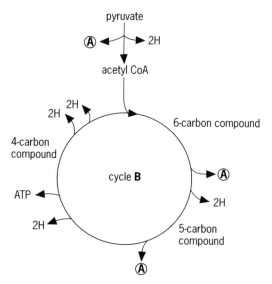

a How many carbon atoms are found in:
i a molecule of pyruvate
ii an acetyl group? (2)

b **i** Identify substance **A**.
ii What is the name given to cycle **B**?
iii What is the name for the process of removing **A**? (3)

c Where exactly in the cell does cycle **B** occur? (1)

d What is the final fate of the H removed from cycle **B**? (1)

6 Substrates other than carbohydrate can be metabolised to release energy. Explain how this occurs in:

a protein

b fat (4)

7 ATP can be considered as a temporary energy store. It supplies energy to cells for a range of processes.

During aerobic respiration ATP is mainly produced in mitochondria by oxidative phosphorylation. In photosynthesis it is produced in chloroplasts during the light-dependent reaction.

a Describe the similarities and differences in the ways in which the ATP is produced in respiration and photosynthesis. (6)

b Describe how ATP is used in processes within cells. (6)

NEAB, BY01, February 1997

8 Give an account of the role played by mitochondria in respiration. (10)

9 Describe the chemical pathway from pyruvate under aerobic conditions and explain how ATP is produced. (9)

10 The figure below represents the formation of ATP in a cell.

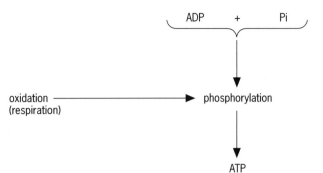

a i Using the information in the figure, explain what is meant by the term phosphorylation. (1)

ii Explain why ATP may be regarded as a nucleotide to which phosphate groups are attached. (2)

b i Name a process in glycolysis in which ATP would be required. (1)

ii Name the 2C compound which enters the Krebs cycle during aerobic respiration. (1)

iii Explain the role of oxygen in the respiratory process occurring in mitochondria. (3)

Cambridge, Module 4802, Winter 1996

1 a $\dfrac{\text{volume of carbon dioxide produced}}{\text{volume of oxygen used}}$

b decarboxylases

c mitochondria; aerobic; Krebs; dehydrogenase; hydrogen carriers/NAD/FAD/coenzymes; ATP; ADP and inorganic phosphate

2 i C **ii** A **iii** B **iv** A **v** D

3 a X = crista; Y = matrix; Z = outer membrane

b i X **ii** Y **iii** X

c Length of mitochondrion = 63 mm
= 63 000 μm

Measurement on diagram/
magnification = 63 000/40 000
Actual size = 1.57 μm

d They are both surrounded by two fluid mosaic membranes. They both have a large internal surface area (cristae in the mitochondrion, lamellae in the chloroplast).

4 a Two of:
- It provides a large surface area.
- For enzymes/carrier molecules.
- Associated with the electron transport system/ATP synthesis.

b Always show your calculations.
RQ = 57/80 = 0.71

c i The respiratory substrate is carbohydrate/named carbohydrate, e.g. glucose/carbon dioxide produced = oxygen consumed.

ii The respiratory substrate changes over the 24-hour period.

5 a i 3 **ii** 2

b i carbon dioxide **ii** Krebs cycle **iii** decarboxylation

c Krebs cycle occurs in the matrix of the mitochondrion.

d The final fate of the hydrogen is it reduces oxygen to become water.

6 a The protein is first hydrolysed into amino acids. These are then deaminated and the residues either enter the respiratory pathway as Krebs cycle intermediates or at the pyruvate stage.

b Fat is hydrolysed and then the fatty acids undergo oxidation into 2C fragments which can enter at the acetyl CoA stage.

7 a Similarities
Two of:
- Both contain electron carriers/electron acceptors/coenzymes/cytochromes.
- The carriers are at decreasing energy levels in both processes.
- Energy is released during electron transfer.
- Both involve reduction/redox/reactions.
- In both phosphorylation of ADP to ATP occurs.
- Any valid suggestion, for example, proton pumps.

Differences
Four of (for paired differences):
- Photophosphorylation uses light.
Oxidative phosphorylation uses organic/named compounds/dehydrogenation.

- Photophosphorylation involves photolysis of water/release of oxygen.
Photophosphorylation involves formation of reduced NADP.
- Oxidative phosphorylation uses NAD/FAD.
Oxidative phosphorylation uses oxygen (as final acceptor).
- Photophosphorylation occurs in the grana.
Oxidative phosphorylation occurs in oxysomes/'stalked granules'/on cristae.
- Respiration produces ATP by substrate level phosphorylation.

b Six of:
- ATP breaks down to ADP + P + energy.
- Use of ATPase enzyme.
- Active transport.
- Example of active transport, e.g. minerals/ions into cell/sodium pumps, etc, against a concentration gradient.
- Reference to carriers/proteins in membrane.
- Synthesis, e.g. proteins/DNA/polysaccharides, etc.
- Reduction of GPA/in light-independent stage/Calvin cycle.
- In glycolysis/first stage of respiration provides activation energy.
- Cell division.
- Cilia/flagella/contractile movement/muscle contraction.
- Any valid point, e.g. bioluminescence.

8 Ten of:
- The first stage of respiration, glycolysis does not occur in the mitochondrion but in the cytoplasm.
- This stage produces the 3-carbon pyruvate which diffuses into the mitochondrion where the enzymes for the rest of the reactions of aerobic respiration are sited.
- The whole of anaerobic respiration, however, occurs outside the mitochondrion in the general cytoplasm.
- The structure of the mitochondrion is adapted to its role, separating the chemical processes within from those occurring outside by means of membranes which are selectively permeable.
- The inner membrane provides a large surface area (cristae) where electron transport and oxidative phosphorylation can occur.
- The matrix contains the enzymes for the Krebs cycle.
- The role of the mitochondrion is to trap the energy released from the breakdown of pyruvate in the form of ATP.
- Also NADH + H$^+$ formed during glycolysis can be used to convert ADP + P$_i$ into ATP.
- The pyruvate entering the mitochondrion is first decarboxylated to form the 2C acetyl CoA which enters the Krebs cycle, combining with a 4C compound to form a 6C compound.
- NAD is reduced at this stage.
- Some ATP is produced directly from the Krebs cycle.
- Dehydrogenation of Krebs cycle intermediates results in reduction of NAD and FAD.

- These then pass on their hydrogen and electrons along the carriers located on the inner membranes.
- Energy released during this process is used to build up ATP from ADP.
- The ATPase is located on the oxysomes and the synthesis is driven by the flow of H^+ ions down the diffusion gradient from within the inter-membrane space.
- The reactions of the mitochondrion require oxygen which acts as the final hydrogen and electron acceptor to form water.
- Cells which require large amounts of energy have large numbers of mitochondria.

9 Under aerobic conditions the 3C pyruvate diffuses into the mitochondrion. It combines with coenzyme A and is decarboxylated to form acetyl CoA. During this process hydrogen is removed by NAD (oxidation). The 2C acetyl CoA then combines with the 4C oxaloacetate, so entering the Krebs cycle. The 6C citrate formed is then decarboxylated, first to a 5C compound and then to a 4C compound which is converted into 4C oxaloacetate again. One molecule of ATP is directly produced during the cycle. In each turn of the cycle four pairs of hydrogen atoms are removed by dehydrogenase enzymes. Three molecules of $NADH + H^+$ are formed and the fourth pair of hydrogen atoms are accepted by FAD. The reduced coenzymes then pass along a system of hydrogen and electron acceptors (cytochromes) coupled with ATP synthesis. The final hydrogen and electron acceptor is oxygen, resulting in the formation of water. Each reduced NAD generates three ATP whereas reduced FAD generates only two ATP. The aerobic oxidation of $2 \times 3C$ pyruvate therefore generates 2×15 ATP. A total of 38 ATP per 6C glucose are produced by aerobic respiration.

10 a i Phosphorylation is the addition of phosphate (molecule/group/inorganic) to ADP.
   ii ATP contains phosphate (groups)/phosphoric acid. It has a pentose sugar/ribose; and adenine/nitrogenous base.
   b i Phosphorylation (of glucose/hexose/fructose phosphate/hexose phosphate); correct conversion, e.g. hexose to hexose phosphate.
   ii acetyl coenzyme A/acetyl CoA/active acetate
   iii Two of:
   - Oxygen acts as the final hydrogen (H) acceptor.
   - Oxygen is reduced to water.
   - Oxygen completes the aerobic process/oxidative phosphorylation which will result in the complete oxidation of glucose/fatty acids/amino acids.
   - Oxygen allows the reuse of carriers/cytochromes.

16 Respiration: Experimentation

Simple respirometers

Respirometers are used to measure respiration rates by measuring the rate of gaseous exchange.

If a plant or animal is respiring aerobically and using glucose as the substrate, the RQ will be 1.0 (**15** Respiration: Aerobic respiration). In this case the volume of carbon dioxide given out will equal the volume of oxygen used.

Avogadro's law states that equal volumes of gases, at the same temperature and pressure, contain an equal number of molecules. This means that there is a direct correlation between an equation and the volumes of gases used or produced.

The following simple equation for respiration demonstrates this fact.

$$\text{glucose} + 6O_2 \rightarrow 6CO_2 + H_2O$$

6 volumes   6 volumes

equal volumes

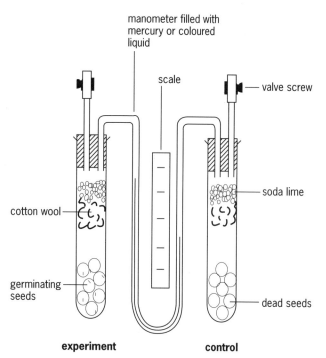

Figure 16.1 A simple respirometer

There are numerous variations on the set-up shown in Figure 16.1.

1 Animal material may be used instead of plant material. In this case it is usually suspended on a gauze platform above the soda lime.
2 If small animals such as blow-fly larvae are used the control chamber is left empty of organic matter.
3 Potassium hydroxide could be used to remove the carbon dioxide. It is much more reactive than soda lime, and will therefore absorb the carbon more efficiently. But it is very caustic and if not used extremely carefully will kill the living material.

Experimental notes

- Valve screws can be used to bring the liquid in the manometer to level positions at the start of the experiment and then for the start of further experiments.
- The mass of germinating seeds and dead seeds need to be the same to ensure a fair control.
- The apparatus can only be used for a short while before renewing the air supply. Once the oxygen is depleted anaerobic respiration will take over.
- You could find the Q_{10} (temperature coefficient) by measuring the rate of respiration at, say 20 °C, and then again at 30 °C.
- The actual volume of gas exchanged could be calculated by measuring the diameter of the capillary tube manometer to calculate the surface area of the hole (πr^2) and then multiplying by the length along the tube that the liquid had moved.
- When using plant material it is essential to ensure that no photosynthesis can take place. Either the experiment must be kept in the dark, which is not very convenient, or else germinating seeds can be used. The use of seeds has the added advantage that their rate of respiration will be faster. However, don't forget that many seeds store oil, and therefore the respiratory quotient will be less than 1.0 and a slightly greater volume of oxygen will be used up than the volume of carbon dioxide produced.
- Animal material will tend to give faster results than plant material because it is usually metabolically more active.

Controlled experiments

Both temperature and pressure can affect the volumes of gases. It is very important, therefore, that experiments using respirometers are carefully controlled. In any controlled experiment only one factor must be different. In the experiment shown in Figure 16.1 the only difference is that the 'experiment proper' contains living seeds while the control contains dead seeds. Any change in gaseous volume in the 'experiment' must be due to the effect of the living material present. The dead seeds are usually disinfected to stop bacterial action.

Redox reactions

A redox reaction is one in which oxidation and reduction occur. The two processes always take place at the same time. In the simple equation in which hydrogen combines with oxygen to form water, the oxygen is reduced by the hydrogen and the hydrogen is oxidised by the oxygen. Remember the definitions:

Reduction is the gain of electrons by an atom or ion.
Oxidation is the loss of electrons by an atom or ion.

In the early 1900s Thunberg discovered that respiring animal tissue contains dehydrogenases which catalyse reactions in which hydrogen is removed from a molecule. Examples include the reactions which take place in the Krebs cycle.

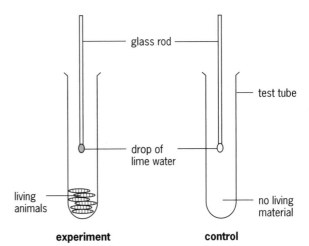

Figure 16.3 Testing for carbon dioxide

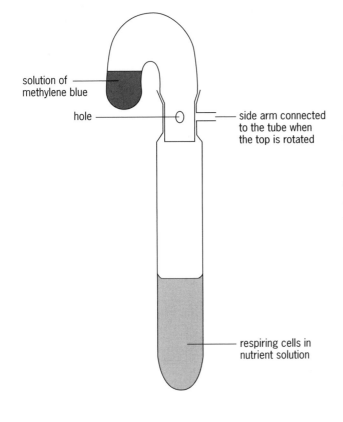

Figure 16.2 Thunberg's experiment

The air is removed from the tube in Figure 16.2 by turning the top until the hole is in line with the side arm and then evacuating the air with a vacuum pump. The top is then turned again so that the hole is sealed and the pump is removed. The contents of the tube are then mixed with the methylene blue which will be slowly reduced and become colourless. If the tube is not evacuated, oxygen in the air will be reduced instead of the methylene blue which will retain its oxidised blue state.

Testing for carbon dioxide

The usual test for the presence of carbon dioxide is that it turns lime water chalky. However, the lime water should be freshly made and, even then, it takes quite a large volume of carbon dioxide to give a positive result. As carbon dioxide is more dense than air it is possible to demonstrate the evolution of carbon dioxide by living material using the technique shown in Figure 16.3. You would need to compare the result by testing another identical sample which contains no living material. The test is not quantitative.

The tubes should be lightly plugged with cotton wool until the time that the glass rods with a drop of lime water are introduced.

Alternatively the gas at the bottom of the tubes can be withdrawn using a syringe with a long needle. The gas can then be expelled through a solution of lime water. Remember that the results will be much clearer if a small amount of lime water is used. Twice as much lime water requires twice as much carbon dioxide to turn it chalky!

Using hydrogen carbonate indicator solution

An accurate way of testing for small amounts of carbon dioxide being produced, or removed, by living material is to use an indicator. Hydrogen carbonate indicator solution can detect small changes in the acidity of solutions caused by either an increase, or decrease, in the concentration of carbon dioxide relative to the 0.04% present in the atmosphere.

- When the indicator is in equilibrium with the carbon dioxide in the atmosphere it is *red* in colour.
- An increase in acidity, as produced by a small amount of carbon dioxide dissolving in the solution, will turn the solution *yellow*.
- A decrease in the original concentration of carbon dioxide present in the solution will cause the solution to turn *red* and eventually *purple*.

Before use, the indicator solution must be aspirated (have air bubbled through it) for at least an hour. It should then be red in colour. Care should be taken when setting up experiments. All apparatus must be clean, and care taken not to breathe over the experiment when setting it up or else the indicator will turn yellow before you are ready. The living material is usually placed on a perforated zinc platform above the indicator solution in a boiling tube. Expected results are:

1. Leaf in *bright* light, photosynthesising. – purple
2. Small animals on a gauze platform suspended over the indicator, respiring. – yellow
3. A *balanced* 'aquarium' of, say, one small tadpole and waterweed. – red

1 a State the colour changes which would occur in the following circumstances:
   i When carbon dioxide is bubbled through lime water. (1)
   ii When carbon dioxide is bubbled through hydrogen carbonate indicator. (1)
   iii When a person breathes through a straw into lime water. (1)
   b Name two chemicals you could use to remove carbon dioxide. (2)
   c Name a chemical you could use to remove oxygen. (1)

2 a State what important precaution must be taken when demonstrating respiration in plant material. (1)
   b Explain why this is done. (1)
   c Explain why faster results would be obtained if a similar mass of animal material were substituted for the plant material. (1)

3 Explain what is meant by a redox reaction. (2)

4 a Write a word equation to summarise:
   i Aerobic respiration of glucose in yeast. (1)
   ii Anaerobic respiration of glucose in yeast. (1)
   b Write a chemical equation to summarise:
   i Aerobic respiration of glucose by yeast. (1)
   ii Anaerobic respiration of glucose by yeast. (1)
   c Calculate the RQ for:
   i Aerobic respiration of glucose. (2)
   ii Anaerobic respiration of glucose. (2)
   In each case show how you arrived at your answer.

5 Three tubes were set up as shown in the diagram.

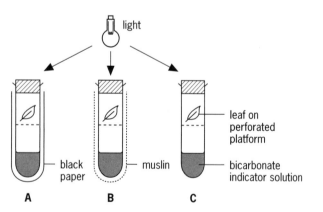

Each tube contained hydrogen carbonate indicator solution which had been aerated.
   a What colour would the indicator solution be at the start of the experiment? (1)
   b What colour change would you expect to find for the indicator in each tube after 1 hour? Explain your reasoning. (6)
   c Describe, giving experimental details, how you would use the bicarbonate indicator to compare the rate of gaseous exchange in the three tubes. (3)

6 Methylene blue is a hydrogen acceptor which is a blue colour in its oxidised state but becomes colourless in its reduced state. It can therefore be used as a redox indicator. Raw milk contains bacteria which are actively respiring. The time for methylene blue to decolorise can be used as a measure of the freshness of the milk.
   a Explain how the methylene blue is decoloured by raw milk. (2)
   b Explain why fresh milk would take longer to decolour methylene blue than sour milk. (2)
   c Explain why opening the tube and then shaking it will result in blue colour being restored. (2)
   d Describe, giving experimental details, how you would use methylene blue to compare the amount of bacterial respiration in two samples of milk. (5)

7 The diagram below shows the apparatus used to compare the carbon dioxide production of different strains of yeast.

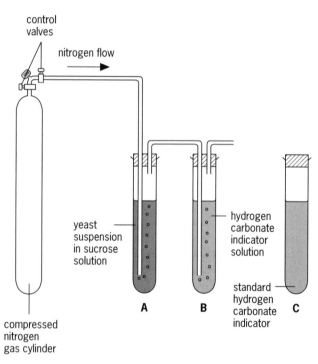

The yeast population to be investigated is suspended in sucrose solution in tube **A**. Nitrogen gas is bubbled through the apparatus during the experiments to ensure that respiration of the yeast is anaerobic.

Tube **C** contains hydrogen carbonate indicator solution through which carbon dioxide has been bubbled. This allows the colour in the tube to develop and this tube is then used as a standard. Hydrogen carbonate indicator is red when neutral, purple in alkaline and yellow in acid conditions.

The time taken for the colour to develop in tube **B** to match the colour in tube **C** is recorded.

a **i** State how the yeast suspension in tube **A** can be maintained at a constant temperature during the experiment. (1)

ii State the colour change which would occur in the hydrogen carbonate indicator solution in the tube during the course of the experiment. (1)

iii Suggest why matching the colour by eye might not be a reliable method of determining the end-point. (1)

b **i** Name one product, other than carbon dioxide, of the reaction occurring in tube **A**. (1)

ii State two changes that occur in tube **A** as the reaction proceeds. (2)

c Describe, giving experimental details, how you would use the apparatus to compare carbon dioxide production in two strains of yeast. (5)

London, Specimen Paper B1, 1996

8 The apparatus in the diagram below was set up to study the rate of oxygen uptake in woodlice.

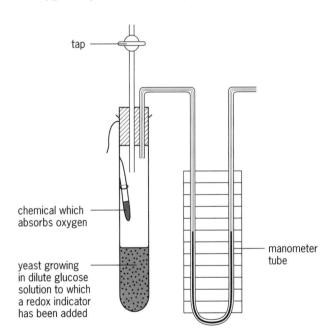

1 cm³ syringe

three-way tap

screw-clip

glass beads

woodlice

perforated platform

soda lime

soda lime

experimental tube

manometer

a Explain why soda lime is placed in the bottom of the tubes. (1)

b Explain why glass beads are placed in the non-experimental tube. (1)

c Explain the purpose of the syringe in this experiment. Describe how you would use it. (3)

d When setting up the experiment, the three-way tap is left open for 10 minutes. Explain why this is done. (1)

e How would you ensure that the temperature was kept constant during the course of the experiment? (1)

f After ensuring that the level of manometer fluid was the same in the two arms, the experiment was allowed to run for 20 minutes at 20 °C. The difference in levels was then found to be 15 mm.

i In which arm would you expect the level to be highest? (1)

ii Assuming the same metabolic rate, calculate the difference in levels after an hour. (1)

iii Predict the difference in levels if the experiment was repeated at 30 °C for 1 hour. Give a reason for your prediction. (2)

g Outline how this apparatus could be used to measure the volume of carbon dioxide produced by the woodlice at 20 °C. (3)

h Predict the difference in levels of carbon dioxide the woodlice would produce in 20 minutes at 20 °C if they were using only carbohydrate as their respiratory substrate. Explain your reasoning. (2)

9 The rate of anaerobic respiration of *Saccharomyces cerevisiae* (yeast) can be measured using the apparatus shown below. The yeast cells are grown in a dilute solution of glucose to which a redox indicator is added. A chemical which absorbs oxygen is placed in the suspended tube.

tap

chemical which absorbs oxygen

manometer tube

yeast growing in dilute glucose solution to which a redox indicator has been added

a Name a suitable chemical which could be used to absorb oxygen. (1)

b Name a suitable chemical which could be used as a redox indicator. (1)

c Describe the colour change you would expect to occur in the tube containing the respiring yeast and the redox indicator named in **b**. (1)

d Explain why this change would occur. (1)

e State the direction in which you would expect the manometer fluid to move during the course of the experiment and explain the reasoning behind your answer. (2)

10 A sample of air was analysed to determine the percentage of carbon dioxide and oxygen it contained. A simple J-shaped capillary tube was used fitted with an airtight screw at one end as a means of drawing in or expelling air or liquid.

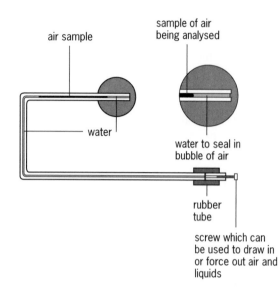

air sample

sample of air being analysed

water

water to seal in bubble of air

rubber tube

screw which can be used to draw in or force out air and liquids

The screw was rotated inwards to its limit. The open end of the tube was then placed into water and a small volume of the water drawn in by turning the screw the other way. It was then removed from the water and a sample of expired air was drawn into the tube. Finally more water was drawn in to enclose an air bubble. The length of this air bubble was then measured.

The air bubble was then brought near to the open end of the tube by using the screw. The end was dipped into potassium hydroxide solution and a small amount drawn into the tube. The bubble was shunted backwards and forwards to mix the air and chemical.

The potassium hydroxide was then pushed out and some pyrogallate solution drawn in and mixed with the air bubble. Again the bubble length was measured.

The results were as follows:

original length of bubble	10.0 cm
length after adding potassium hydroxide solution	9.6 cm
length after adding pyrogallate solution	8.1 cm

a Explain why:
  i potassium hydroxide solution was used
  ii pyrogallate solution was used. (2)
b Assuming that the bore of the capillary tube is uniform, calculate the percentage of carbon dioxide and oxygen in the sample of air. Show your working. (4)
c The J-tube was placed in a sink of water for 10 minutes each time before measuring the bubble length. Explain why. (2)

1 a i colourless to cloudy (white)
  ii red to yellow
  iii colourless to cloudy
 b sodium hydroxide (soda lime); potassium hydroxide (caustic potash)
 c alkaline pyrogallol (pyrogallate)

2 a Light must be excluded from the apparatus.
 b To prevent photosynthesis occurring. This uses up carbon dioxide and produces oxygen.
 c Animal material often has a faster rate of metabolism.

3 A redox reaction is where one substrate is oxidised, resulting in the other being reduced.
 In the respiratory chain for example, hydrogen atoms (or electrons) are transferred from a carrier in the reduced state (which then becomes oxidised) to a carrier in the oxidised state (which then becomes reduced).

4 a i glucose + oxygen → energy + water + carbon dioxide
  ii glucose → energy + ethanol + carbon dioxide
 b i $C_6H_{12}O_6 + 6O_2 \rightarrow energy + 6H_2O + 6CO_2$
  ii $C_6H_{12}O_6 \rightarrow energy + 2C_2H_5OH + 2CO_2$

 c $RQ = \dfrac{\text{volume of carbon dioxide produced}}{\text{volume of oxygen used}}$

  i $RQ = \dfrac{6}{6} = 1$

  ii $RQ = \dfrac{2}{0} = infinity$

5 a red
 b Tube **A** = yellow, because the plant material is respiring and producing the acidic gas carbon dioxide. No photosynthesis will occur in the dark.
   Tube **B** = red, because in the dim light (muslin will only allow a small amount of light through) the processes of respiration and photosynthesis will be balanced (compensation point) so that carbon dioxide output equals carbon dioxide intake.
   Tube **C** = purple because in bright light the rate of photosynthesis will exceed the rate of respiration so there will be a net loss of carbon dioxide. It will become less acidic.
 c Three of:
   • Use equal volumes of bicarbonate indicator solution.
   • Use leaves of equal mass/surface area.
   • Use leaves of the same age and from the same plant.
   • Set the light at the same distance from the tubes.
   • Avoid breathing into the tubes/stopper firmly.

6 a The raw milk acts as a hydrogen donor.
   Respiring bacteria in the milk have dehydrogenase enzymes which transfer hydrogen to methylene blue.
 b Sour milk would contain more bacteria. They would respire more than the fewer bacteria in fresh milk.
   Thus more hydrogen would be transferred to methylene blue which would become reduced more quickly.
 c Shaking the tube will mix the contents with air containing oxygen.
   This will reoxidise the methylene blue.

d It is important to ensure this is a fair test.
  • Use equal volumes of two milk samples. Add an equal volume of methylene blue of the same concentration to each tube.
  • Invert each tube only once to mix.
  • Cover with cling film to exclude additional oxygen entering.
  • The tubes should be placed in a thermostatically controlled water bath to ensure the same temperature.
  • Note the time taken to decolorise the methylene blue (standard end-point) or compare the colour after a set time using a colorimeter.

7 a i Use a water bath (preferably thermostatically controlled).
  ii It would change from red to yellow.
  iii It is difficult to judge the exact shade of a colour; different people have different perceptions of colour.
 b i ethanol/ethyl alcohol
  ii Two of:
   • The sucrose would be converted to glucose and fructose/the sucrose would decrease/there would be less glucose.
   • CO_2 bubbles (froth) would be produced.
   • Yeast cells would die/CO_2 production would decrease/the tube would clear.
   • The pH would fall/become more acidic.
   • Glucose would be changed to an intermediate (named)/the intermediate would be converted to ethanol.
 c Five of:
   • Use constant concentrations/equal amounts of sucrose.
   • Measure equal volumes of both yeast solutions.
   • Use yeast suspensions of the same age/from stock cultures in the same growth phase.
   • Use a constant volume of hydrogen carbonate indicator.
   • Maintain the same nitrogen flow.
   • Keep the temperature constant throughout the experiment.
   • View the colour after a constant time/record the time to reach a standard colour.
   • View the indicator against a white background/using a colorimeter.
   • Use replicates for improved accuracy.

8 a The soda lime absorbs carbon dioxide evolved during respiration. This ensures that the differences in manometer levels are as a result of oxygen uptake.
 b The glass beads act as a control and should be equal in volume to the woodlice.
 c The syringe is used to check that the apparatus is air-tight at the start of the experiment.
   The fluid in the right-hand arm is displaced to the left by depressing the plunger of the syringe.
   The tap should then be closed and the manometer observed to check that the difference in manometer levels does not change.

d This is to allow the air pressure in the apparatus to equilibrate with the atmosphere.

e Use a constant temperature water bath.

f **i** The left-hand arm.

ii 45 mm

iii 90 mm. Enzyme controlled chemical reactions tend to double in rate for a 10 °C rise in temperature (if this is below the optimum)/$Q_{10} = 2$.

g Set up the apparatus with, for example, five woodlice and the soda lime. Equilibrate at 20 °C. Close the screw-clip and the tap and note the fluid level. After a measured period of time (e.g. 20 minutes) read the difference in levels (*x* mm). This represents the oxygen consumption.

Remove the soda lime and repeat the experiment, keeping all the other factors the same.

Note the difference in levels (*y* mm). This represents the difference between the carbon dioxide produced and oxygen used. The difference in levels due to carbon dioxide production is *x* − *y* mm. To convert this to a volume multiply by the cross-section of the tube, πr^2.

h They would produce 15 mm as Avogadro's law can be used to show that the volumes of carbon dioxide produced and oxygen used are equal for aerobic respiration of carbohydrate (see text for equation).

9 a pyrogallol

b, c Two of:
- methylene blue
- blue to colourless
- tetrazolium chloride
- colourless to pink

d Hydrogen and electrons are transferred during the redox reactions of the anaerobic respiration of the yeast to the indicator, which accepts these, becoming reduced.

e The fluid will move to the left because carbon dioxide will be evolved, increasing the pressure in the yeast tube.

10 a **i** Potassium hydroxide absorbs carbon dioxide.

ii Pyrogallate absorbs oxygen.

b Difference due to uptake of carbon dioxide

$$= 10.0 - 9.6$$
$$= 0.4 \, cm$$

% of carbon dioxide in sample $= 0.4$/original bubble length $\times 100$
$$= 0.4/10 \times 100$$
$$= 4\%$$

Difference due to uptake of oxygen $= 9.6 - 8.1 = 1.5 \, cm$

% of oxygen in sample $= 1.5$/original bubble length $\times 100$
$$= 1.5 \times 10 \times 100$$
$$= 15\%$$

c To ensure that the bubble was measured under conditions of constant pressure and constant temperature since both of these factors would affect bubble length.

Body systems

17 Nerves and reflexes

Organisms need to constantly respond to changes in their environment. This increases their chance of survival. All vertebrates have two main co-ordinating systems, the **nervous system** and the **endocrine system**. Although the two systems detect and transmit information in different ways they sometimes work together.

The endocrine system is discussed in **23** and **24**; **17** to **19** are concerned with the nervous system of humans but much of what is discussed here applies to other mammals and, to a lesser extent, other vertebrates.

Functions of the nervous system

- To collect information about the environment, both external and internal.
- To co-ordinate the information gathered and relate it to any relevant previous experience. Decision making may be important at this stage.
- To act on the information gathered. The brain and spinal cord co-ordinate the body's activities.

Changes in external/internal conditions stimulate **nerve endings** in **receptor organs** which in turn send **nerve impulses** along **sensory nerves** to the **central nervous system**. In the central nervous system the nerve impulses are relayed to **motor nerves** which transmit the impulses to **effector organs** which bring about a response.

Behaviour has five components which can be summarised as follows: stimulus – receptor – co-ordinator – effector – response.

The structure and function of nerve cells

Nerve cells are called **neurones**. There are several types of neurone which differ in shape depending on their specific function.

All neurones have a **cell body** containing a nucleus and mitochondria. The cell body has a number of outgrowths, called **dendrites**, which transmit impulses to the cell body.

Impulses leave and travel along the **axon** which may be more than a metre in length. This would be the case with a neurone leaving the base of the spinal cord and connecting with a receptor organ in the foot. Remember that this is just one cell.

The drawing in Figure 17.1 shows the generalised structure of a motor neurone. The arrangement in a sensory neurone is different in some respects. In a sensory neurone the receptor organ may have nerve endings enclosed in a capsule (for example, touch-sensitive nerve endings) or it may consist of free dendrites (for example, cold-sensitive nerve endings).

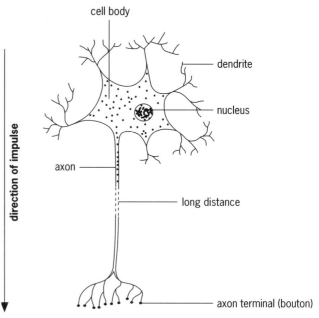

Figure 17.1 Generalised structure of a motor neurone

Note the position of the cell bodies in the sensory neurones shown in Figure 17.2. When we look at the structure of the spinal cord in transverse section, the reason for the different organisation will become apparent.

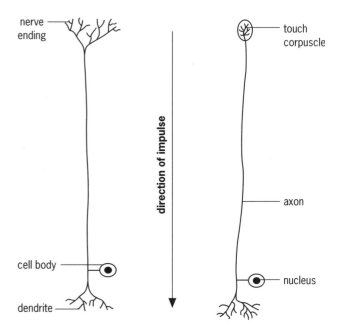

Figure 17.2 Two types of sensory neurone

A nerve consists of a bundle of nerve fibres, or neurones. A nerve will usually have both **afferent** neurones (sensory neurones carrying impulses *to* the central nervous system) and **efferent** neurones (motor neurones carrying impulses *away* from the central nervous system to effector organs such as muscles).

Some of the larger nerve fibres are covered with a fatty **myelin sheath** formed by **Schwann cells**. The sheath acts as an electrical insulator and makes possible a greater current flow. An **action potential** (**18** Action potential and the synapse) cannot form in the axon where it is covered with myelin. Action potentials can form at the **nodes of Ranvier** (see Figure 17.3) and so they jump from node to node; this increases the speed with which they are transmitted.

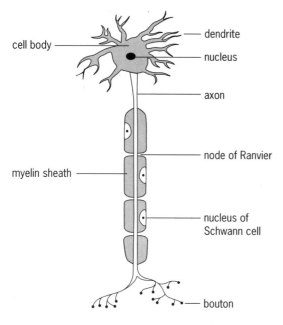

Figure 17.3 Myelinated motor neurone

Grey and white matter

The material of the central nervous system (brain and spinal cord) consists of grey and white matter. Grey matter contains the nerve cell bodies and is where co-ordination occurs. In the spinal cord it is in the shape of a letter H. The white matter lies outside the grey matter in the spinal cord but inside the grey matter in the brain. It is the fatty myelinated sheaths which give the white matter its appearance.

Spinal reflex actions

A reflex action is an automatic and invariable response to a specific sensory stimulus.

In some of the simplest reflex actions three nerves are involved:

1 A sensory neurone, stimulated, for example, by a pain stimulus in the hand.
2 A connector (relaying) neurone, in the spinal cord which can pass the stimulus on to many other neurones.
3 A motor neurone, which stimulates an effector organ, such as a muscle fibre.

Connector neurones (Figure 17.4) are not always involved. An example of a reflex action involving only a sensory and a motor neurone is the *knee jerk* – a stretch reflex.

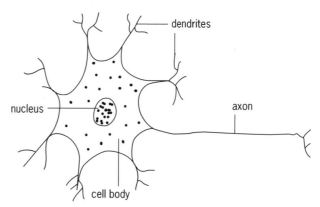

Figure 17.4 A connector neurone

The brain is not necessary for a reflex action, but relaying neurones make connections with sensory fibres running to the brain which may make one aware that the reflex has occurred.

Other examples of reflexes in the human include coughing, pupil dilation/contraction, and the ejaculation of sperm.

Figure 17.5 shows a **reflex arc**, which is the path taken by the stimuli during a reflex action. Do not confuse 'reflex arc' with 'reflex action'.

Conditioned reflexes

A conditioned reflex is one in which the actual response made to a stimulus is conditioned by the consequences. A simple example is that if you pick up something which is very hot you drop it. However, if that 'something' is a best dinner plate out of the oven then you are more likely to put it down as soon as possible but without dropping it onto a hard surface such as the floor.

Conditioned reflexes are discussed further in **20** Behaviour.

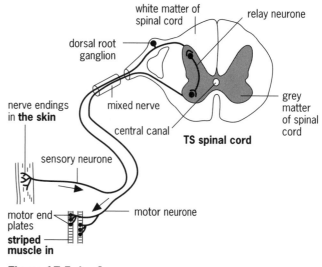

Figure 17.5 A reflex arc

1   The diagram shows a mammalian neurone.

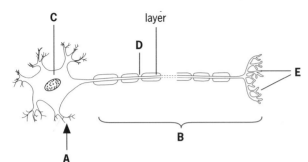

a   What type of neurone is shown? (1)
b   Name the parts labelled **A**, **B**, **C** and **D**. (4)
c   **i** Suggest two reasons why the fibre is surrounded by a fatty layer. (2)
    ii Name the fatty layer. (1)
d   **i** The part labelled **E** is branched. Suggest reasons for this. (2)
    ii Name the swellings at the branched ends of **E**. (1)
e   The pressure receptor is found in the skin of the hand. Where would the other end of the neurone be located? (1)

2   An example of a reflex action occurs when you burn your finger. This would cause you to contract the biceps, thus pulling the finger away from the source of heat. Such a reflex action involves three neurones.
a   **i** Name the effector in this action. (1)
    ii Explain the meaning of the term 'reflex action'. (2)
    iii Draw a diagram to show such a spinal reflex. Label the neurones. (3)
    iv Give two advantages of this response being a reflex action. (2)
    v Give two other examples of reflex actions. (2)
b   When the detective Hercule Poirot was thinking he often referred to his 'little grey cells'. Explain what these consisted of and where in the body they would be found. (We would call it grey matter.) (2)
c   What gives the white appearance to 'white matter'? (1)
d   Name the two main co-ordinating systems. (2)

3   Read through the following and then list the most suitable word(s) to fill the gaps in this passage about the nervous system in humans.

There are three main types of nerve cells (_____). Electrical impulses travel along _____ from the receptor to the central nervous system. The message may then pass to an intermediate nerve cell and then be carried along a _____ to the muscle or gland. These bring about the response and are known as _____. (4)

4   Pavlov observed that dogs salivated when a man came to feed them. A bell was rung just before feeding time. He then noticed that the dogs salivated when the bell was rung even if the food was not brought.
a   What is the original stimulus which results in salivation? (1)
b   What was the new stimulus which brought about the response? (1)
c   **i** What term can be used to describe this second type of behaviour? (1)
    ii Give another example of this type of behaviour. (1)

5   Some of the early research into conduction of impulses was carried out on the giant axons of molluscs such as squid and octopus. These do not have a myelin sheath around their axons. The table below shows the speed of conduction in the axons of cat and octopus.

	Speed of conduction (m/s)	Diameter of axon (μm)
cat (myelinated axon)	25	3.5
cat (non-myelinated axon)	15	3.5
octopus (non-myelinated axon)	25	600

a   Describe how the rate of conduction is related to the structure of the axon in cats. (1)
b   Describe how the rate of conduction is related to the diameter of non-myelinated axons. (1)
c   Explain how a myelin sheath affects the rate of transmission of impulses along the axon. (3)
d   Suggest why giant axons may have developed in squid and octopus rather than snails. (2)

6   Compare the structure and functions of sensory neurones and motor neurones. (10)

1 **a** motor neurone
   b **A** = dendrite; **B** = axon; **C** = cell body/cytoplasm with Nissl's granules; **D** = node of Ranvier
   c **i** The fatty layer electrically insulates the nerve fibre. It speeds up the transmission of the impulse.
      ii myelin
   d **i** This provides a greater surface in contact with the muscle or gland.
      A larger area of the muscle will be able to contract/a larger region of the gland will be stimulated.
      ii motor end plates
   e In the grey matter of the spinal cord.

2 **a** **i** biceps muscle
      ii A reflex action is an automatic (involuntary) action; in which the stimulus results in a rapid response.
      iii

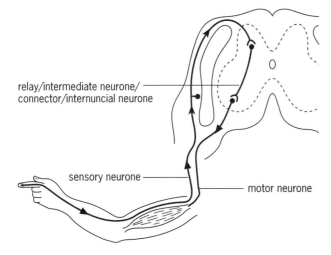

relay/intermediate neurone/
connector/internuncial neurone

sensory neurone

motor neurone

      iv Two of:
      • The response will be automatic.
      • The response will be faster.
      • This will help to increase survival and minimise harmful effects.
      v Any two reflex actions, for example, knee jerk reflex; pupil reflex (smaller in bright light), etc.
   b Grey matter consists of nerve cell bodies; located in the inner part of the spinal cord or outer part of the brain.
   c The myelin sheaths give a white appearance to nervous tissue.
   d The nervous system and the endocrine system.

3 neurones; sensory neurones/nerve cells; motor neurone/nerve cell; effectors

4 **a** The sight, smell or taste of food.
   b The sound of the bell.
   c **i** A conditioned reflex.
      ii Any example where a simple reflex arc has had another stimulus or response associated with it, creating a new automatic pathway (see example in text).

5 **a** Possession of a myelin sheath increases the rate of conduction of impulses.
   b As the diameter of the axon increases, the rate of conduction of impulses increases.
   c The myelin acts as an electrical insulator.
      The action potential jumps from one unmyelinated node of Ranvier to the next;
      so increasing the rate of conduction.
   d *Either* Molluscs such as snails are well protected from predators by their shells; whereas squid and octopus are not and might need a faster speed of conduction to escape.
      or Many snails are herbivores and do not need fast responses; whereas squid and octopus are predators and need to react more quickly, etc.

6 *In each case you need to stress similarities or differences between the two types of neurone.*

   • The sensory neurone usually has a longer dendron than the axon/or they are equal in length. The motor neurone has a long axon whereas the dendrons are short.
   • The cell body is in the middle of the sensory neurone/located in the dorsal root ganglion.
   • The cell body of the motor neurone is located in the brain or spinal cord.
   • The function of the sensory neurone is to carry impulses to the central nervous system (brain/spinal cord).
   • The motor neurone carries impulses from the central nervous system.
   • The sensory neurone carries impulses from a receptor.
   • The motor neurone carries impulses to the effector/muscle/gland.
   • The sensory neurone passes the signal across synapses to the motor neurone, often via an intermediate neurone.
   • Impulses are carried to the cell body by dendrons, whereas axons carry the impulses from the cell body.
   • Some neurones are myelinated to speed transmission and electrically insulate the axon.
   • At junctions between neurones, one-way synapses control the passage of information from one neurone to the next.

18 Action potential and the synapse

All living cells maintain a potential difference across their membranes by creating a difference in the electrical charges between the cell cytoplasm and the extracellular tissue fluids. This is brought about by the distribution of anions and cations.

Unlike other cells, neurones are able to alter this potential difference. When there is no nervous impulse the potential difference is approximately −70 mV (1000 millivolts = 1 volt). This is known as the resting potential and the membrane is said to be **polarised**. A minus sign is used here to show that the charge on the inside of the membrane is negative compared with the outside.

Resting potential

Energy from ATP has to be used to achieve and maintain the potential difference. The **sodium–potassium exchange pump** is an important membrane 'pump'. It pumps sodium ions (Na^+) out of the cell and into the extracellular fluid, and potassium ions (K^+) into the cell. The ions are exchanged in equal numbers.

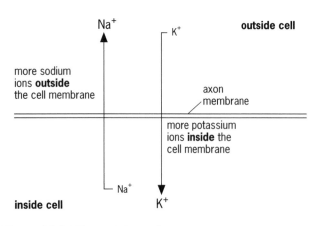

Figure 18.1 Effects of the sodium–potassium exchange pump

The sodium–potassium exchange pump thus establishes concentration gradients down which the sodium and potassium ions will diffuse. The sodium ions slowly diffuse back through the cell membrane and into the cell, but the potassium ions diffuse out of the cell at a much faster rate because the cell membrane is more permeable to potassium ions than to sodium ions.

Both ions have a positive charge but because they diffuse at different rates there will be net gain of positively charged ions on the outside of the cell. The membrane, therefore, will be negatively charged on the inside compared with the outside. The difference in the concentration of ions on either side of the neurone membrane is maintained by the sodium–potassium exchange pump which actively transports the ions against diffusion gradients.

Depolarisation

When a neurone is stimulated, the charges can be reversed at the site of the stimulus. The negative charge of −70 mV on the inside of the membrane changes to a positive charge of 40 mV. This is known as an action potential and the membrane is said to be depolarised.

This condition lasts for about 2 milliseconds (2 thousandths of a second), after which the same portion of membrane is **repolarised** and returns to the resting potential.

What causes depolarisation?

Depolarisation occurs when there is a sudden increase in the permeability of the membrane to sodium ions, but this only happens if the stimulus is strong enough and reaches a certain **threshold**. If the stimulus is not strong enough to reach the threshold value the action potential is not produced and there is no impulse. This is called the **all or nothing response**. The action potential is either generated or it is not.

The increased permeability of the membrane to sodium ions at threshold occurs when pores, or channels, called **sodium gates**, open and allow the sodium ions to flood into the cell. There is a high concentration of sodium ions outside the cell membrane at this point due to the sodium pumps which have been actively pumping them out of the cell.

The entry of sodium ions depolarises the membrane. At the same time the depolarisation makes the membrane more permeable to sodium ions which leads to further depolarisation. This is an example of **positive feedback** where the products of a process cause further activation.

When sufficient sodium ions have entered to make the charge inside the membrane positive, the permeability of the membrane to sodium ions starts to decrease. As the sodium ions move inwards, so the potassium ions move outwards along a diffusion gradient but at a much slower rate. This continues until the membrane is repolarised and the resting potential is restored.

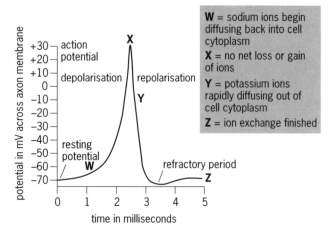

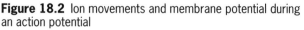

Figure 18.2 Ion movements and membrane potential during an action potential

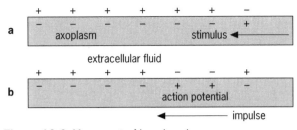

Figure 18.3 Movement of impulse along an axon

In Figure 18.3a the stimulus sets up an influx of sodium ions into the cytoplasm which causes a flow of current across the membrane just in front of the point of stimulus.

The current itself is a stimulus and so the reaction proceeds along the axon as shown in Figure 18.3b and creates an impulse. An impulse may travel as fast as 100 m/s.

Two factors will increase the speed of transmission of an impulse:
- the larger diameter of the axon
- myelination of the nerve.

Refractory period

This is the recovery time (for repolarisation) needed between one impulse and the next. It lasts about 1 millisecond, during which time a fresh action potential cannot be generated.

The synapse is the point where the axon of a neurone joins with the cell body or dendrites of another neurone. The synaptic cleft is a small gap of about 15 nm between the junctions of neurones. (One nanometre = one thousand millionth of a metre or 10^{-9} m.) The junctions between neurones are similar to the junctions between the last neurone in a pathway and an effector organ such as a muscle. Axons terminate in bulb-like structures called synaptic knobs or boutons. They contain small vesicles containing transmitter substances and many mitochondria (work has to be done!).

When an impulse arrives at a bouton, calcium ions enter and cause some of the vesicles, which contain transmitter substances, to fuse with the **pre**synaptic membrane. They release their contents onto the **post**synaptic membrane where there are specific sites to which they attach. The postsynaptic membrane is depolarised and this generates an EPSP (excitatory postsynaptic potential). The impulse crosses the synaptic cleft and then passes along the postsynaptic neurone in the same manner as it passed along the axon of the presynaptic neurone.

There are a number of transmitter substances including **acetylcholine** and **noradrenaline**.

Transmitter substances are rapidly degraded by enzymes after they have been released. For example, acetylcholine is degraded by cholinesterase which breaks it down into choline and ethanoic acid. These products are then absorbed and resynthesised into acetylcholine. The energy comes from ATP produced by the mitochondria in the boutons. The temporary effects of the transmitter substances ensure that the postsynaptic neurone is not over stimulated.

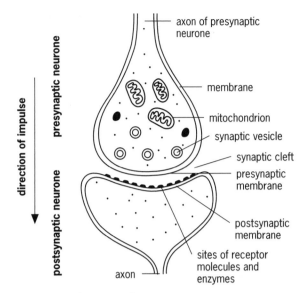

Figure 18.4 Structure of a synapse

The effects of drugs on synapses

There are two types of drugs.

Inhibitory

These *stop* synaptic transmission by:

- Preventing the release of the transmitter; for example, Botulinus toxin which stops the release of acetylcholine. Botulism is a form of food poisoning caused by a bacterium which is an obligate anaerobe (see **14** Respiration: Glycolysis) and which grows well in unsterilised airtight containers.
- Blocking the action of the transmitter with the receptor molecules on the postsynaptic membrane; for example, propranolol which is a β-blocker, and organophosphates which are used as weedkillers and insecticides.
- Stopping depolarisation of the postsynaptic membrane; for example, curare which is used as a poison for arrows.

Excitatory

These *stimulate* synaptic transmission by:
- Copying the action of the transmitter; for example, nicotine which stimulates the nervous system, and amphetamines which mimic the action of noradrenaline.
- Stimulation which results in the release of more of the transmitter; for example, caffeine which increases the rate of cell metabolism.

Summation

If two or more synaptic knobs are needed in order to release enough transmitter; to set up an action potential, this is known as **spatial summation**. If one knob does not release enough transmitter for an action potential but a second impulse is received quickly after the first, then an action potential may still arise. This is known as **temporal summation**.

1 a Read the following passage about synapses carefully and then write a list of the most suitable words to fill the gaps.

The junction between the axon of one neurone and the _____ of the next neurone is called a synapse. Here the axon terminal or synaptic _____ as it is called, contains many _____ to supply energy. The synaptic _____ are bounded by membranes and contain a transmitter substance called _____. This diffuses across the synaptic _____ to bind with _____ on the _____ membrane. (8)

b Name the following:
i The ions actively pumped across the membrane of the neurone. (2)
ii The term for the critical value of the stimulus which must be reached before depolarisation of the neurone membrane will occur. (1)

c State whether the following drugs have an inhibitory or excitatory effect on synaptic transmission.
i caffeine
ii nicotine
iii organophosphates
iv β-blockers (4)

2 Explain the meaning of the following terms:
a transmitter substance
b resting potential
c impulse
d positive feedback (8)

3 The diagram shows some of the events which occur in a synapse after the arrival of an impulse at the presynaptic membrane.

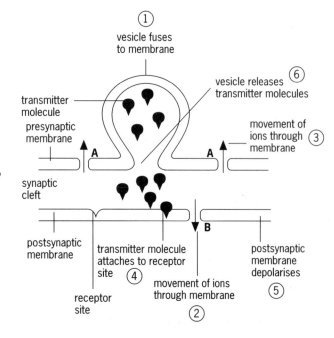

a i Put the events 1–6 on the diagram in the correct sequence. (1)
ii Name the ions labelled **A** and **B**. (2)
iii By what process do transmitter molecules move across the synaptic cleft? (1)
iv Name one transmitter molecule released by synaptic vesicles. (1)

b One impulse arriving at the presynaptic membrane does not produce an action potential in the postsynaptic neurone, but several impulses arriving in close succession do.
i Explain this observation. (2)
ii What name is given to the process described? (1)

NEAB, BY04, June 1997

4 The membrane potential of a neurone leading from a stretch receptor in a muscle was measured before, during and after a period in which the muscle was stretched. The results are shown in the graph.

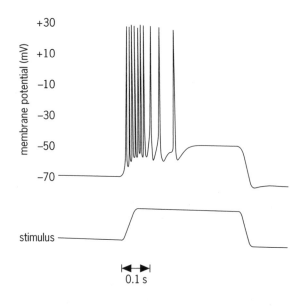

a Explain what causes the initial increase in membrane potential. (2)
b How do these results illustrate the 'all-for-nothing' principle? (1)
c i For how long was the muscle stretched? (1)
ii Describe and explain the effect on the neurone of keeping the muscle stretched for this period. (2)
iii Suggest **one** biological advantage of this effect. (1)

NEAB, BY04, March 1999

5 Explain how an action potential is transmitted along a myelinated neurone. (10)

6 The graph shows the electrical events associated with a nerve impulse.

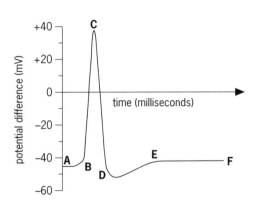

a What is the value of the resting potential? (1)
b Which part of the graph corresponds to:
  i repolarisation of the axon membrane (1)
  ii the refractory period? (1)
c Explain why there is a change in the potential difference across the axon membrane between points **B** and **C** on the graph. (2)

AEB, Specimen Module 4, 1997

7 The diagram below shows a synapse as seen with an electron microscope.

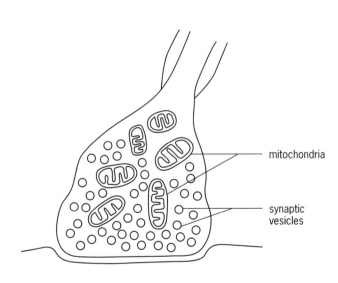

a Explain the function of each of the following:
  i the mitochondria (2)
  ii the synaptic vesicles (2)
b The graphs show the effect of adding acetylcholine to skeletal muscle and heart muscle.

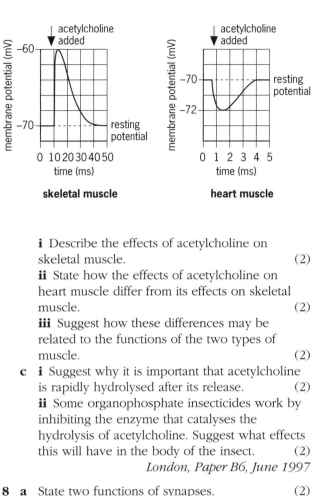

i Describe the effects of acetylcholine on skeletal muscle. (2)
ii State how the effects of acetylcholine on heart muscle differ from its effects on skeletal muscle. (2)
iii Suggest how these differences may be related to the functions of the two types of muscle. (2)
c i Suggest why it is important that acetylcholine is rapidly hydrolysed after its release. (2)
ii Some organophosphate insecticides work by inhibiting the enzyme that catalyses the hydrolysis of acetylcholine. Suggest what effects this will have in the body of the insect. (2)

London, Paper B6, June 1997

8 a State two functions of synapses. (2)
b Calcium ions enter neurones through presynaptic membranes when action potentials arrive.
  i Explain how these ions enter the neurones. (2)
  ii Describe the events which follow the entry of these ions until the depolarisation of the postsynaptic membrane. (5)
  iii Explain the importance of cholinesterase at synapses. (2)
c The functioning of a synapse can be affected by the presence of drugs as shown in the table below.

Drug	Effect on synapse
curare	blocks the receptors at the postsynaptic membrane of a motor neurone–muscle junction
morphine	activates inhibitory receptors in the presynaptic membrane of sensory neurones

Suggest the consequences, on the action of the synapse, of:
  i curare (2)
  ii morphine (2)

UCLES, Biology Foundation Paper, June 1998

1 **a** cell body or dendrite; knob; mitochondria; vesicles; acetylcholine or noradrenaline; cleft; receptors; postsynaptic
   b **i** sodium and potassium ions
      ii threshold
   c **i** excitatory
      ii excitatory
      iii inhibitory
      iv inhibitory

2 **a** A transmitter substance is a chemical (acetylcholine or noradrenaline) which is released from vesicles into the synaptic cleft when an impulse arrives at the synaptic knob. It diffuses across the cleft and depolarises the postsynaptic neurone.
   b A resting potential is the potential difference occurring across the membrane of a polarised neurone. The outside of the membrane is positively charged (excess Na^+ ions) with respect to the inside.
   c An impulse is a wave of depolarisation which passes along the axon. This momentary reversal of the resting potential is called the action potential.
   d Positive feedback is when the products of a process cause further activation.

3 **a** **i** 3 1 6 4 2 5.
      ii **A** = calcium; **B** = sodium
      iii diffusion
      iv noradrenaline/acetylcholine/any named brain transmitter
   b **i** Two of:
      • The one impulse releases insufficient transmitter substance to produce an action potential.
      • Several impulses release more transmitter which results in the threshold being reached to produce an action potential.
      • The charge on the postsynaptic membrane (excitatory postsynaptic potential) builds up as more transmitter substance diffuses across until the membrane is sufficiently depolarised to exceed the threshold value.
      ii summation

4 **a** The initial increase is caused by the increased permeability to ions/an influx of sodium ions. Excess positive ions inside the axon make the inside positively charged/change to $+30\,mV$.
   b All the action potentials are the same size.
   c **i** 0.44 s (accept 0.4–0.47 s)
      ii The action potentials become less frequent, then cease after 0.2 s.
      Adaptation occurs (NOT adaptation of the synapse)/the stimulus no longer generates an impulse.
      iii It avoids a response to 'background noise'/harmless stimuli can be ignored/it prevents overloading of the CNS.

5 In its normal state the membrane of the neurone is polarised, having an excess of positively charged sodium ions outside. This excess of sodium ions makes the inside more negatively charged in comparison to the outside. Such a condition is known as the resting potential. Stimulation depolarises the membrane. Channels in the membrane of the neurone open, changing the permeability of the neurone to sodium and potassium ions. This allows sodium ions to pass through the membrane by diffusion since there is a much higher concentration of sodium outside the membrane. The sodium channels then close. The potassium channels then open and potassium ions diffuse out. If sufficient sodium ions enter to change the charge above a threshold value then an action potential will be generated. Depolarisation causes local electrical circuits to set up alongside the area of depolarisation and this triggers an action potential in the adjoining regions. In this way a wave of depolarisation (action potentials) travels along the axon.

Action potentials can only be generated at points where the myelin sheath is absent/there is no electrical insulation at the nodes of Ranvier. Local electrical circuits set up between adjacent nodes and the action potential jumps from node to node (saltatory conduction) increasing the speed of transmission. Another factor increasing the speed of transmission is the diameter of the neurone.

An action potential is always the same size. A strong stimulus will produce a high frequency of action potentials. If depolarisation of the membrane is not great enough to reach the threshold value and so generate an action potential then an impulse is not produced. This is called the 'all or nothing law'.

Neurones are usually stimulated at one end and the impulses travel along them in one direction. Synapses will only allow transmission in one direction.

The resting potential must be restored before a second impulse can be generated. This involves repolarisation of the membrane. Its permeability alters, allowing potassium ions to diffuse out of the neurone. The sodium–potassium pump then actively restores the ionic balance by pumping out sodium. The time taken to recover the resting potential is called the refractory period.

6 **a** $-45\,mV$
   b **i** CD **ii** DE
   c Movement of sodium ions into the axon.

7 **a** **i** Two of:
      • The mitochondria are responsible for the production of ATP/for the release of energy (not the production of energy!).
      • This occurs during aerobic respiration.
      • The energy is used in active transport/movement of vesicles/synthesis of the transmitter substance.
      ii Two of:
      • The vesicles contain the transmitter substance, for example, acetylcholine.
      • They fuse with the presynaptic membrane;
      • releasing the transmitter substance into the synaptic cleft.

b i (*Note that the scale is very different on the two graphs!*)

The acetylcholine makes the membrane potential less negative/there is a rapid depolarisation.

There is a slower return to the resting potential.

ii Two of:

- Acetylcholine causes the hyperpolarisation of heart muscle (membrane potential becomes more negative).
- The membrane potential changes much less in heart muscle than in skeletal muscle.
- Acetylcholine has an inhibitory effect on heart muscle whereas it has an excitatory effect on skeletal muscle.

iii Acetylcholine makes skeletal muscle contract.

It regulates/slows the rate of activity of heart muscle.

c i Two of:

- Acetylcholine must be rapidly hydrolysed to remove it from the synapse/neuromuscular junction.
- If it is not removed it would accumulate and maintain the postsynaptic potential.
- This would cause repeated impulses/stop new impulses.
- This would cause loss of muscle control.

ii Two of:

- In the insect this would cause uncontrolled muscle contraction.
- Affect flying/feeding/ventilation movements (make this specific to insects).
- Leads to death.

8 *Note: the correct terms should be used throughout the question. Imprecise terms such as 'messages/signals/ amplification' will not gain marks.*

a Two of:

- The action potential/impulse can pass to the neurone/muscle.
- Allows transmission in one direction only.
- Allows impulses to travel to many neurones from one neurone/to one neurone from many neurones (*summation is allowable*).
- Allows inhibition.
- Allows integration.
- Allows association/learning/memory.
- Allows varied responses.
- Filters out weak impulses.
- Prevents overstimulation (for example, fatigue/adaptation/accommodation).

b i By diffusion; through channels in the membrane (which open when the impulse arrives).

(*Note: gates is an acceptable term but not pumps and carriers*)

ii Five of:

- Vesicles move towards the presynaptic membrane.
- Fuse with it.
- Release the transmitter substance/acetylcholine, into the synaptic cleft/synapse.
- The transmitter substance diffuses across the cleft.
- Binds onto the postsynaptic membrane.
- The channel/proteins open.
- Allowing sodium ions to enter.

iii It breaks down acetylcholine/into acetate and choline. (*Inactivates/deactivates are not acceptable here.*)

and one of:

- Breaks down for recycling/to be reused/returns to the presynaptic membrane.
- Breaks down to prevent constant stimulation of postsynaptic membrane/if not broken down there will be constant stimulation.

c i Two of:

- It prevents the transmitter substance/acetylcholine binding to the membrane/receptors.
- It prevents the action potential/impulse from crossing the synapse/reaching the muscle.
- It prevents an influx of Na$^+$.
- There will be no muscle contraction/paralysis/no effector response.

ii Two of:

- It prevents the action potential from reaching the synaptic bulb/presynaptic membrane/intermediate neurone/relay neurone/motor neurone.
- The impulse does not cross the synapse.
- It prevents an influx of calcium ions.
- It prevents the release of neurotransmitter.
- The impulse does not reach the CNS/brain.
- It reduces the frequency of impulses.
- Refer to hyperpolarisation.

19 The peripheral and central nervous systems and the brain

The brain and spinal cord are hollow and filled with a tissue fluid called **cerebrospinal fluid** (CSF). The CSF also fills the space between two of the three membranes which completely surround the brain and spinal cord. All of these spaces, both inside and outside of the brain, are interconnected and continuous with the blood plasma in blood capillaries.

Functions of the cerebrospinal fluid

- A constant pressure is maintained around the CNS (central nervous system).
- The fluid pressure supports and protects the CNS by acting as a shock absorber between the nerve tissue and the protective bones of the skull and vertebral column.
- It keeps the CNS moist which makes the diffusion of substances into and out of the nerve cells possible.

The hypothalamus and the thalamus will be dealt with in **23** Endocrine system: General principles and ADH.

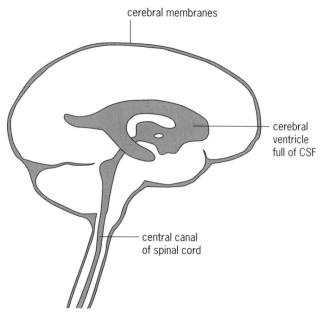

Figure 19.1 Ventricles of the brain

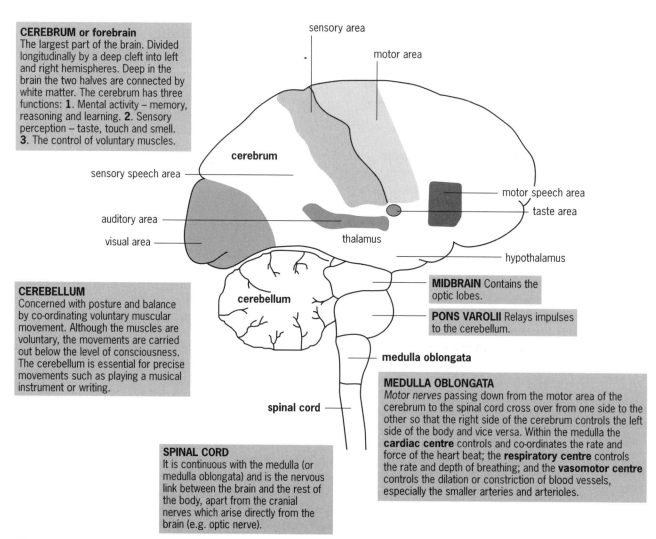

CEREBRUM or forebrain
The largest part of the brain. Divided longitudinally by a deep cleft into left and right hemispheres. Deep in the brain the two halves are connected by white matter. The cerebrum has three functions: **1.** Mental activity – memory, reasoning and learning. **2.** Sensory perception – taste, touch and smell. **3.** The control of voluntary muscles.

CEREBELLUM
Concerned with posture and balance by co-ordinating voluntary muscular movement. Although the muscles are voluntary, the movements are carried out below the level of consciousness. The cerebellum is essential for precise movements such as playing a musical instrument or writing.

SPINAL CORD
It is continuous with the medulla (or medulla oblongata) and is the nervous link between the brain and the rest of the body, apart from the cranial nerves which arise directly from the brain (e.g. optic nerve).

MIDBRAIN Contains the optic lobes.

PONS VAROLII Relays impulses to the cerebellum.

MEDULLA OBLONGATA
Motor nerves passing down from the motor area of the cerebrum to the spinal cord cross over from one side to the other so that the right side of the cerebrum controls the left side of the body and vice versa. Within the medulla the **cardiac centre** controls and co-ordinates the rate and force of the heart beat; the **respiratory centre** controls the rate and depth of breathing; and the **vasomotor centre** controls the dilation or constriction of blood vessels, especially the smaller arteries and arterioles.

Figure 19.2 The main parts of the central nervous system in humans – lateral view

The peripheral nervous system

This is the rest of the nervous system, that is the whole of the nervous system apart from the central nervous system (brain and spinal cord).

It consists of three parts:

- The cranial nerves which originate from the brain.
- The spinal nerves which originate in pairs along the spinal cord and emerge from between vertebrae.
- The autonomic nervous system which controls the *involuntary activities* of the body.

The autonomic nervous system

The autonomic system is divided into two parts: the **sympathetic** and the **parasympathetic** nervous systems.

The two systems work together inasmuch as they send impulses to the same organs, but the effects that the two systems produce oppose each other. One nerve stimulates and the other inhibits. It is rather like the accelerator and brake pedals of a car. They work in opposition but both are needed for proper control. The two nervous systems regulate very accurately the involuntary activities of both glands and organs. It is possible to exact a measure of control over certain activities by training. Examples of these are training leading to bladder and anal control.

Some examples of the opposing effects of the sympathetic and parasympathetic nervous systems are given in the table.

Organ	Parasympathetic	Sympathetic
heart	inhibits pacemaker	stimulates pacemaker
intestines	stimulates peristalsis	inhibits peristalsis
iris	stimulates contraction	inhibits contraction
kidney	urine increased	urine decreased

The opposing effects of the two systems are due to different chemical transmitters being secreted at the synapses between the neurone and the effector organ (Figure 19.3).

As well as the opposing functions of the two systems, there are also distinct structural differences between the two sets of nerves. These are outlined in Figure 19.4.

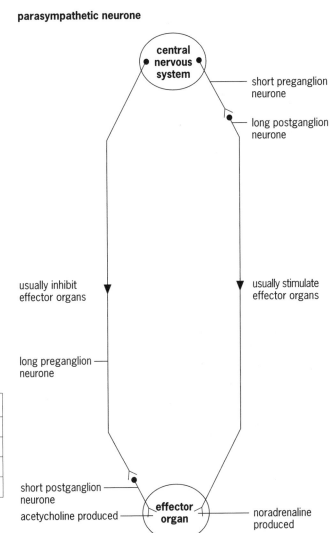

Figure 19.4 Differences between sympathetic and parasympathetic neurones

In the parasympathetic nervous system, the preganglion fibres from the central nervous system are long, and the postganglion fibres, which are embedded in the wall of the effector organ, are short. In the sympathetic nervous system, the preganglion fibres from the central nervous system are short because the synapses are alongside the vertebrae.

A complex of cell bodies and synapses is called a ganglion.

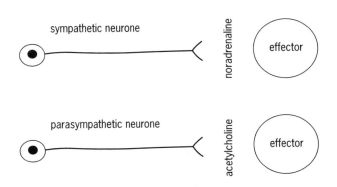

Figure 19.3 Sympathetic and parasympathetic neurones

1 a Name the following:
i The fluid found between the membranes surrounding the brain and spinal cord. (1)
ii The largest part of the brain. (1)
iii The two parts of the autonomic nervous system. (2)
iv A swelling containing synapses and cell bodies. (1)
b Complete the table below by writing the name of the part of the brain which has the function described. (3)

Function	Region of the brain
regulates the rate of breathing	
muscular co-ordination and the control of posture	
monitoring the composition of the blood	

2 The diagram shows the human brain as seen from below.

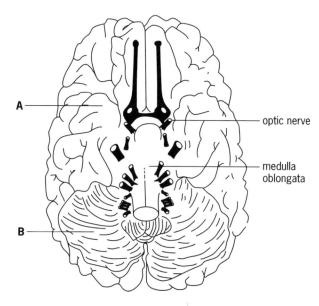

(Adapted from: Simkins J., Advanced Human Biology. Harper Collins Publishers, 1997.)

a Name the regions of the brain labelled **A** and **B**. (2)
b Give one of the functions of the medulla oblongata. (1)
c The optic nerve transmits impulses to the brain.
i Which part of the brain receives these impulses? (2)
ii Describe how the brain might process the information contained in the impulses. (2)
NEAB, BY04, February 1997

3 a Name the region of the brain which controls the rate of beating of the heart. (1)
b Describe the part played by the autonomic nervous system in the control of the rate of heart beat. (3)

4 a Describe the differences between the parasympathetic nervous system and the sympathetic nervous system. (4)
b Compare the effects of the parasympathetic and sympathetic nervous systems on:
i bronchial dilation
ii pupil dilation
iii ventilation rate of the lungs
iv blood pressure. (4)

5 The diagram shows a vertical section through a human brain.

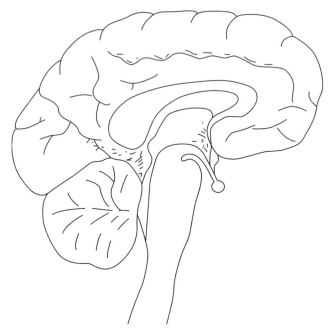

a **i** A person drinks a glass of lemonade. Name the parts of the brain to which the following descriptions would apply:
P – receives impulses from touch receptors in the lips.
Q – enables the person to co-ordinate the movements necessary to drink from the glass.
R – contains receptors that respond to a change in the concentration of the blood plasma. (3)
ii Use the appropriate letter and a guideline to show the position of each of the parts **P**, **Q** and **R** on the diagram of the brain. (3)
b When the lemonade touches receptors in the throat, a swallowing reflex occurs. What are the effectors in this reflex action? (1)
AQA (NEAB), BY04, March 1999

6 a Many pills for relieving the effects of sea-sickness contain drugs which have an inhibitory effect on the parasympathetic nervous system. Suggest why patients taking such pills may experience difficulty with their vision and a feeling of dryness in their mouths. (4)
b State which part of the brain is responsible for regulating the following functions:
i control of temperature
ii speech (2)

1 a i cerebrospinal fluid
ii the cerebrum/forebrain
iii the sympathetic and parasympathetic nervous system
iv a ganglion

b

Function	Region of the brain
regulates the rate of breathing	medulla oblongata
muscular co-ordination and the control of posture	cerebellum
monitoring the composition of the blood	hypothalamus

2 a A = cerebrum/cerebral hemispheres/cerebral cortex
B = cerebellum
b Controls heart rate/breathing rate/dilation or constriction of blood vessels.
c i Sensory/association areas of forebrain; cerebral hemispheres/visual cortex/occipital lobe.
ii Activates visual association areas.
Interprets impulses in terms of shape/colour/brightness/movement/image.
Recognises objects by reference to past experience.

3 a medulla oblongata
b The cardiac centre co-ordinates the rate of heart beat. Receptors detect changes and send impulses along the appropriate autonomic nerves. The message to increase the rate travels along the sympathetic neurones which release noradrenaline at the junction with the pacemaker, causing it to increase the rate at which cardiac impulses are generated. The parasympathetic nerve decreases the heart rate by activating the pacemaker by means of the neurotransmitter acetylcholine.

4 a Parasympathetic NS in general inhibits effectors, whereas the sympathetic NS in general stimulates effectors.
Parasympathetic NS produces acetylcholine at the effector, whereas the sympathetic NS produces noradrenaline.
The synapse between the two neurones is situated nearer the effector organ in the parasympathetic NS, whereas in the sympathetic NS it is situated nearer to the spinal cord.
The parasympathetic preganglion neurone is short and the postganglion neurone is long, whereas the sympathetic postganglion neurone is short and the preganglion neurone is long.
b i Parasympathetic constricts/sympathetic dilates bronchioles.
ii Parasympathetic constricts/sympathetic dilates the pupil.

iii Parasympathetic decreases/sympathetic increases the rate of ventilation of the lungs.
iv Parasympathetic decreases/sympathetic increases the pressure of the blood.

5 a i P – cerebral hemisphere/cerebrum/cerebral cortex
(Note: sensory area is not an acceptable answer)
Q – cerebellum
R – hypothalamus

ii *(Note: each label should be attached by a guideline to the area named below. Attaching the same letter to two different areas will result in no marks.)*

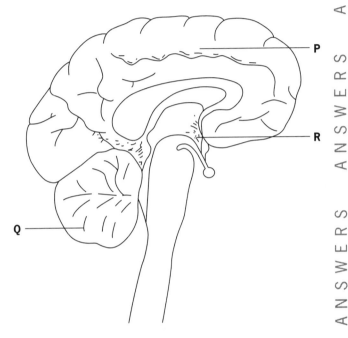

P – anywhere in the cerebral hemisphere
Q – anywhere in the cerebellum
R – anywhere in the area *above* the pituitary gland
b Muscles of the gullet/throat/oesophagus, or circular/peristaltic muscles *(no marks are gained from the unqualified answer muscles/muscles of the mouth).*

6 a The parasympathetic system has the effect of constricting the pupil of the eye.
If inhibited by drugs this would result in lack of ability to regulate the amount of light entering and lack of focusing ability.
The parasympathetic system stimulates the secretion of saliva.
If inhibited by drugs then saliva production would cease causing dryness of the mouth.
b i hypothalamus
ii cerebral cortex

20 Behaviour

> Behaviour is instinctive (innate), or learned and modified as a result of experience.

Innate behaviour (or species-characteristic behaviour)

Innate behaviour is characteristic of a particular species and is the result of specific nerve pathways determined by the DNA which is inherited. It is, therefore, genetically controlled.

The response to a particular stimulus will always be the same and is the same for all members of the species. It is the result of selection over many generations. The response is selected for its survival value.

There is a huge variety of responses ranging from simple reflex actions which avoid danger, to the complex social behaviour shown by courting birds and the behaviour of social insects such as ants, bees and wasps.

There is no intelligence involved and no appreciation of the purpose that a particular response serves. The more that a particular species shows innate behaviour, the less learned behaviour, or intelligence, there will be.

Innate behaviour is sometimes the result of a series of reflex actions, each one, in turn, acting as the stimulus for the next.

There are a number of types of innate behaviour: taxes, simple reflexes and kinesis – details of each follow.

Taxes

Taxes involve the whole animal moving towards (positive taxis) or away from (negative taxis) a particular stimulus. Examples of taxes are given below.

Taxis	Stimulus	Example
phototaxis	light	woodlice move away from light
chemotaxis	chemicals	*Planaria* move towards food
geotaxis	gravity	fruit flies move upwards
rheotaxis	resistance to movement	moths fly into the wind

Simple reflexes

A rapid response is made to a stimulus such as the presence of food or a stimulus which is potentially dangerous. Invertebrates have an escape reflex. An earthworm, for example, withdraws down its burrow as soon as vibrations on the ground are felt.

Kinesis

This is a response which involves the whole animal moving but the response is not directional. An example is the movement of woodlice towards damp areas. If a woodlouse is placed in a dry environment it moves faster and keeps changing its direction randomly until a damp area is found. The rate of movement depends on the intensity of the stimulus rather than its direction.

Invertebrates tend to be relatively short-lived and therefore do not have the time to learn behaviour patterns on a trial and error basis. Instead, they rely on instinctive behaviour.

Although vertebrates usually live longer and have time to learn behaviour patterns, instinctive behaviour is still important. The young of vertebrates rely on instinctive behaviour until they have had time to learn through experience. Mammals show the most advanced patterns of learned behaviour and yet observation of a baby mammal will soon make one aware of the extent to which instinct contributes towards development and survival.

Behaviour and communication between animals have obvious links. Chemical signals play a part, largely unrecognised, in the lives of humans. We sometimes talk of humans having a smell of fear. This chemical signal is recognised by bees which will then sting. Bee-keepers, who show little fear, tend to be stung much less frequently. Humans also give out chemicals in response to their sex or their mood. We respond to these messages without realising it.

Insects produce pheromones (chemical substances used to attract other insects of the same species). Male emperor moths can detect a female of the same species several kilometres away. Some of the lower plant groups also produce pheromones. They are secreted into water by female gametes to attract male gametes.

Many male mammals mark their territory by spraying the boundary with urine and/or faeces. This serves as a warning to rival males to keep away. Male Siamese fighting fish will respond aggressively to the image of another male fighting fish. Birds use a combination of sounds and visual signals to communicate effectively, especially during courtship and when caring for their young. Examples of this kind of behaviour are endless.

Learned behaviour (or individual-characteristic behaviour)

Experience enables behaviour to be modified. The early environment and experiences of an organism may have pronounced effects on its later behaviour. Learned behaviour takes place in most animal groups to a lesser or greater extent. It is not confined to mammals or even vertebrates.

There are a number of different types of learned behaviour.

Habituation

This is particularly important in young animals. It occurs when a stimulus is continually repeated but nothing happens. The animal learns to ignore the stimulus because there is neither punishment nor reward. This is the simplest form of learned behaviour. Animals learn to ignore the unimportant signals in the environment which cause them no harm and offer no reward. Birds learn to ignore scarecrows, and animals in general, ignore background noises such as the wind.

Imprinting

This is a specialised type of learning in which there is a rapid development of a response to a stimulus at an early stage of development. Konrad Lorenz did some classic work on imprinting with young greylag geese. He demonstrated that on hatching, the young birds 'recognised' the first thing that they saw as a mother figure, even if it was an animal of a different species. The geese then follow this imprinted object, and relate to similar objects, for the rest of their lives. Lorenz imprinted young greylags on himself. Imprinting also occurs with bird song when young and inexperienced birds have adult bird songs imprinted on them.

Associative learning

This involves the association of two or more stimuli. There are two types.

Conditioned reflexes

A conditioned reflex is a reflex action which is modified by experience. The classic experiments were performed by Ivan Pavlov on dogs. The sight, smell and taste of food caused the dogs to salivate. Pavlov then rang a bell immediately before giving the dogs food and repeated this a number of times. In the end the dogs would salivate on the sound of the bell without there being any food produced.

The following are the features of a conditioned reflex:

- Two stimuli are presented together and are associated.
- The condition is temporary.
- The behaviour is involuntary.
- The response is reinforced by repetition.

Operant conditioning

This is sometimes referred to as 'trial and error learning'.

In this type of associative learning the stimulus *follows* the action and is not simultaneous with it (compare with the first point of 'conditioned reflexes' above).

Animals learn to adopt a particular pattern of behaviour when errors are followed by a stimulus which is unpleasant, and correct responses are followed by a stimulus which is pleasant. The training of animals is an example of operant conditioning. If a small reward of food is given when a dog 'obeys' a command, and a small slap or the use of a stern voice applied when the command is not obeyed, then the dog soon learns to associate responding correctly with a pleasant outcome and its behaviour is changed. Repetition improves the response.

Courtship

Courtship consists of behaviour patterns involving displays and rituals which are a necessary precursor to copulation. Hormones and pheromones influence courtship behaviour.

Courtship behaviour only develops in sexually mature individuals and therefore futile matings, involving sexually immature individuals, are avoided. Often, sites for nesting and raising offspring are at a premium and would be wasted if occupied by immature individuals.

Mating rituals and differences in appearance between male and female members of a species make them attractive to the opposite sex, ensuring that time and energy are not wasted on courting members of the same sex.

The courtship behaviour of frogs is well known. The male mounts the female and remains there for several days before the eggs are laid. It is essential that the sperm are released immediately the eggs are released and before the surrounding jelly layer absorbs water and swells, making fertilisation impossible.

Sex hormones are released into the bloodstream in response to visual, tactile and chemical stimuli (for example, pheromones). In humans, for example, such stimuli result in an increased blood supply to the genitalia which causes erection of the penis in the male, and the labia and clitoris becoming swollen with blood in the female. The cells lining the wall of the vagina produce a fluid which makes penetration of the penis more easy and also neutralises the normally acid conditions of the vagina which would kill the sperm.

Semen contains hormones called **prostaglandins** which cause muscular contractions of the uterus and oviducts, thus helping the sperm to reach an ovum.

The oestrous cycle

Adult female mammals have a reproductive cycle, called the oestrous cycle, controlled by the pituitary gland which produces gonadotrophic hormones (this is dealt with in greater detail in **24** The endocrine system).

Depending on the species of mammal, the oestrous cycle lasts from 5 to 60 days or until pregnancy occurs. Many mammals only mate at certain times of the year, often coinciding with adequate supplies of food and water, also to allow sufficient time for development before the onset of bad weather. In order that fertilisation may take place as early as possible in the spring, the horseshoe bat mates in the autumn and the female stores the sperm in a plug of mucus. In spring the plug dissolves and the sperm are released. This is called **delayed fertilisation**.

Ovulation takes place in the early stage of the oestrous cycle and it is only then that the female of most species will accept the male, and copulation takes place. In animal husbandry this is very useful as it means that a male of the species can be taken to the female when she is *on heat* at the beginning of the cycle. In this way one can be sure of the parentage and selective breeding is made much easier.

1 Name the following types of behaviour:
  a Responding to food chemicals by moving towards the stimulus.
  b A randomly directed movement of the organism in response to a stimulus.
  c Animals learn not to respond to unimportant signals in the environment.
  d A series of patterns of behaviour which may lead to mating.
  e A rapid development of a response to a stimulus at an early stage in life. (5)

2 Cockroaches are insects that live in cracks in walls, emerging to feed at night. In its natural environment the behaviour of the cockroach is influenced by contact between its body and objects under which it moves, and by the presence of a pheromone.

  In order to investigate the relative importance of these two stimuli a specially designed choice chamber was used. This was made of glass plates separated by spaces. The distance between the plates on side **A** was just sufficient to allow cockroaches to move between them. The diagram shows the apparatus with the sides removed to show the arrangement of glass plates. The space between the top plates was sealed after the cockroaches had been added.

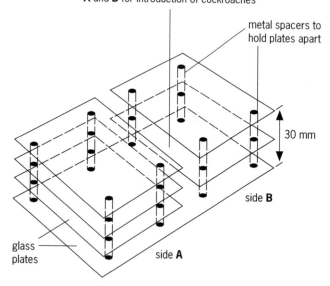

space between the top plates of sides
A and **B** for introduction of cockroaches

metal spacers to hold plates apart

30 mm

side **B**

glass plates

side **A**

A series of pheromone dilutions was made to treat the surfaces of side **B** of the chamber. For each dilution 60 cockroaches were put into the chamber. The cockroaches were then left for 24 hours in total darkness before their distribution was recorded.

Concentration of pheromone (arbitrary units)	Number of cockroaches recorded in side B after 24 hours
0	3
1	4
2	5
3	6
4	7
5	10
6	15
7	30
8	50

  a i What is a pheromone? (1)
  ii Describe the general role of pheromones in influencing behaviour in nature. (3)
  b i What is the relationship between the concentration of pheromone and the number of cockroaches in side **B** after 24 hours? (1)
  ii Give one conclusion that can be made about the relative importance of body contact and pheromone concentration in determining the response of cockroaches. (1)
  iii Suggest one way in which the behaviour shown by cockroaches in this experiment may help cockroaches survive in their natural environment. (1)

NEAB, BY04, February 1997

3 The diagram shows a chamber set up for an investigation into the movement of woodlice in response to humidity.

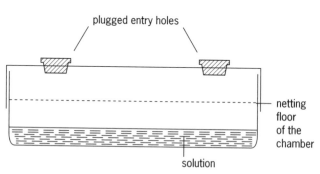

plugged entry holes

netting floor of the chamber

solution

Eleven chambers were set up, each with a different relative humidity obtained by using different concentrations of a solution in the base of the chamber. The rate of movement of the woodlice was recorded. This was repeated ten times for each of the chambers using different woodlice each time, and the means were plotted on a graph.

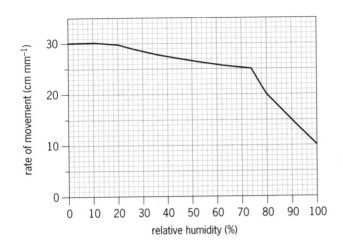

a Explain how the response shown increases the chance of survival of the woodlice in natural conditions. (2)

b **i** Suggest why the woodlice were kept in a dry environment before the investigation was carried out. (1)
ii Suggest why different woodlice were used each time. (1)

c **i** Name the type of behaviour observed in this investigation. (1)
ii Give a reason for your answer. (1)

NEAB, BY04, June 1997

4 A petri dish was divided in two. Half was illuminated and the other half kept in the dark. The drawing shows the path of a free-living aquatic flatworm plotted for 5 minutes in each condition.

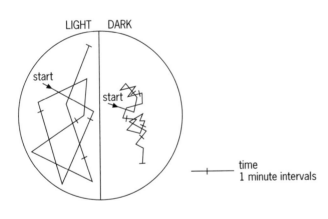

a Describe two differences between the movement of the flatworm in the light and in the dark. (2)

b **i** What type of response is shown by the flatworm in this investigation? (1)
ii Explain the importance of this response in the life of the flatworm. (2)

AEB, Specimen Module 4 (Section A), 1997

5 Robins have individual territories during the autumn and early winter, but from early in the new year pairs of birds begin to share the same territory, which they maintain throughout the spring and summer.

a Describe and explain the advantages of territorial behaviour, with reference to the behaviour of the robin throughout the year. (7)

b The territories are usually defended by song and displays, which often involve exhibiting the red breast as much as possible to any intruding robin. Fighting is sometimes involved, especially when the territory is being established. Suggest why robins usually defend their territories by song and display rather than fighting. (3)

c A robin shows aggressive behaviour towards models of adult robins and bunches of red feathers placed within its territory. However, models of young robins which do not have a red breast, are not threatened.

What can be deduced from these observations about the factors controlling the aggressive behaviour of robins? (2)

NEAB, BY04, June 1997

6 Contrast innate behaviour with learned behaviour. (8)

AEB, Paper 4, Summer 1997

1 **a** positive chemotaxis
   b kinesis
   c habituation
   d courtship
   e imprinting

2 **a** **i** A pheromone is a scent or chemical produced by one animal which affects the behaviour of another.
     ii Three of:
- pheromones are sex attractants
- indicates sexual maturity
- stimulates courtship behaviour
- territory markers
- dominance signal
- co-ordination of group response.

   b **i** As the concentration of pheromone increases more cockroaches move to side **B**.
     ii When pheromone concentration is low (or zero) contact is more important.
     iii One of:
- Contact is likely to reduce dessication in natural conditions.
- Cockroaches are less likely to be eaten by predators in crevices.
- Cockroaches are attracted to high pheromone concentrations as they are likely to aggregate where there is food or safety.

3 **a** Woodlice move faster in an environment in which they are likely to dehydrate.
Fast movement increases their chances of finding a suitable environment/slow movement increases their chances of remaining in a favourable environment.
   b **i** One of:
- They were kept in a dry environment to make them more active at the start of the experiment.
- To ensure that all the animals were in the same state of hydration at the start of the experiment.

     ii Woodlice will show natural variation in their response/using a large sample enables a 'typical' response to be found.
   c **i** kinesis
     ii The rate of movement is related to the intensity of the stimulus.

4 **a** In the light the flatworm moves faster/covers greater distance.
The flatworm turns less/less area covered.
   b **i** kinesis
     ii Enables organism to remain in dark areas.
This would give more protection from predators/allow the flatworms to remain in a more favourable area.
     or
It is an advantage to cover a greater area in the light areas; since there is a greater chance of encountering food, etc.

5 **a** During autumn and early winter it is necessary to have a large enough territory for each robin to find sufficient food and shelter to survive when resources are scarce, so the territory is defended from other robins.
The behaviour must change to allow access to the opposite sex in spring and summer so that courtship, mating and pair-bonding can occur. This will allow the parents to retain the mate and share nest-building, incubation of eggs and feeding and care of each other and the young. This will help to ensure survival and successful rearing of the next generation. This behaviour helps ensure a lower chance of predation, and less transmission of disease since the robins are not in contact with many members of the species. Only the 'fittest' birds obtain territories in the competition to survive (natural selection).

   b Three of:
- Fighting involves the possibility of harming the combatants. This is not a strategy which will result in survival of the species and so will be selected against.
- Fighting will use up much valuable energy/food which could be used for activities which help the robin/robin species survive.
- Fighting is used when two individuals both have a good chance of obtaining a territory.
- In established territories the intruder is submissive, withdrawing so that fighting is not necessary.
- Song and display can be used in courtship to advertise 'fitness' and attract mates.

   c These observations indicate that it is not the shape of a robin which is the stimulus for the aggressive innate behaviour; but instead the sign stimulus/release is the colour red. (*This is important for the survival of the young robins which must not be driven away before they are independent.*)

6 *Note: contrast learned behaviour with instinctive behaviour for each point.*

Instinctive behaviour is inherent in an organism from an early age; it is not acquired during life like learned behaviour.

The basic forms of innate behaviour are similar in many types of organism, for example, taxes, kinesis, etc, whereas learned behaviour can show a huge range of modifications.

Members of a species will display similarities in their instinctive behaviour, whereas learned behaviour can be very varied even between closely related animals.

Instinctive behaviour is important for survival, particularly at an early stage in the life of vertebrates, before they have time to acquire learned behaviour patterns. A certain pattern of instinctive behaviour is often a permanent feature throughout the life of the individual whereas learned behaviour can easily be modified in response to new circumstances or because of previous experiences. Learned behaviour of a particular type may only be used for a short period of time.

Innate behaviour may consist of one reflex response acting as a stimulus for the next reflex action etc, whereas learned behaviour can vary in sequence so that it is better modified to the occasion and need not follow a particular pattern.

Instinctive behaviour is genetically determined to a far greater extent than learned behaviour. Instinctive behaviour does not involve intelligence or an appreciation of the purpose of the response whereas learned behaviour may be intelligent, involving a knowledge of the value of a particular response in a given circumstance.

21 Human health and disease

> A disease is a harmful abnormality in the form or function of an organism.

However, if the abnormality is minimal then the **disease** will not be serious.

Before diseases are considered it is necessary to understand what is meant by being healthy. This can be defined theoretically by certain statistical measurements. We can measure, for example, body temperature, height, pulse rate, breathing rate, sensitivity of hearing and visual acuity. Most people will fall within a range which is regarded as 'normal' but some individuals will fall outside the range and yet be healthy and must be considered as examples of biological variation.

Old age is an example of the difficulty in defining health adequately by using theoretical norms. Elderly people have bones which are more brittle and break more easily, their muscular strength is diminished, their sight is less sharp and their hearing becomes diminished.

A better definition of **health** is *an ability to function normally within one's environment*.

Illness is not the same as disease. It is possible to have a disease for years without being aware of it (for example, Huntington's chorea – a genetic disease which may not manifest itself until the person is about 35 years old). A diabetic who is being satisfactorily treated with insulin has the disease (sugar diabetes) but is not ill.

Physical fitness is the ability to do physical work at a constant rate of heart beat – maybe running a marathon! Some experts choose to divide the range of health and disease into three categories: health; absence of disease; and disease.

Diseases may be:

- pandemic – occurring over a wide part of the globe, for example, malaria
- epidemic – many cases of the disease within an area, for example, an epidemic of influenza
- endemic – limited to a particular geographical area, for example, endemic goitre, caused by a lack of iodine in the water supply which is essential for the secretion of the hormone thyroxine by the thyroid gland.

Classifying diseases is important when compiling statistics on the causes of illness and deaths in an area or country. For example, in the middle of the twentieth century lung cancer was discovered to be the most serious form of cancer in males. Previously it had been rare. Further research led to the conclusion that smoking and lung cancer were closely connected.

Smoking and disease

In 1999 about 120 000 people in the UK died of smoking-related diseases. This annually costs the taxpayer and the National Health Service as much as £500 million for serious cases – time, money and often hospital beds that could be used elsewhere. A direct link between smoking and cancer was established in 1950, and in 1991 studies in the UK established that passive smoking (breathing in other people's cigarette smoke) causes lung cancer. Children whose parents smoke are more likely to suffer from asthma and respiratory diseases. These facts and figures exist despite the government's anti-smoking advertising campaigns and the use of cigarette filters and milder tobaccos.

Smoking-related diseases are caused by carbon monoxide, tar and nicotine (a highly toxic poison) and include the following.

- **Chronic bronchitis.** Develops after about 15 years of smoking. The number and size of mucus glands in the bronchioles increase and cause blobs of mucus to be coughed up ('smoker's cough'). The smaller bronchioles may become inflamed and blocked and this leads to a fall in arterial oxygen tension and a rise in carbon dioxide tension (see **33** Oxygen and carbon dioxide transport in mammals). Ventilation of the lungs becomes seriously affected with accompanying 'shortage of breath'.

- **Emphysema** (chronic obstructive pulmonary disease). A disease associated with chronic bronchitis and characterised by a loss of elastic tissue and blood capillaries plus the degeneration of the walls of the alveoli. The lungs become filled with pockets of air; the affected person wheezes and feels a tightness in the chest; expiration becomes difficult. There may be loss of weight and the skin sometimes turns bluish due to insufficient oxygen in the blood. Lung tissue destroyed by emphysema cannot be repaired.

- **Lung cancer.** A relatively rare disease until World War II, but by the end of the twentieth century it had become the leading cause of death from cancer. **Epidemiological** studies (those concerned with the incidence and distribution of diseases within a population) show a definite link between lung cancer and smoking. The following facts have emerged:
  - 80% to 90% of lung cancer cases are due to smoking.
  - The risk is greater for those who start smoking when young.
  - Heavy smokers run a greater risk than light smokers.
  - Passive smoking causes a large number of deaths (as much as 2% of lung cancer deaths).
  - The survival rate for those with lung cancer has shown little improvement in the last 40 years.
  - Lung cancer incidence among women is rising.

At first it was thought that the increase in the number of cases of lung cancer was due to better diagnosis of the disease, but by the 1960s epidemiological studies disproved this.

The early stages of lung cancer may show no symptoms. Later stages may include coughing, lack of breath, blood in the phlegm, chest pains, attacks of pneumonia, loss of weight, and loss of appetite.

Treatment for lung cancer is most effective in the early stages but most diagnosis occurs when the cancer is extensive. There are three types of treatment:

- surgery
- radiation
- chemotherapy.

If you are a smoker you need to face the facts: most die within 1 year of diagnosis and only about 10% of sufferers survive for as long as 5 years.

Infectious diseases

> **An infectious disease is one which is caused by an organism (pathogen), usually a micro-organism, that adversely affects a person's health.**

Some infectious diseases have a worldwide importance in terms of their economic, social and biological effects.

Cholera

India, Pakistan and Bangladesh, together with other countries in south-east Asia, are particularly subject to epidemics of cholera. During the last two decades (1980–99) cholera spread through the refugee camps of the Sudan and Ethiopia. The disease is caused when the bacterium *Vibrio cholerae* enters the body, via the mouth, usually in contaminated water. Toxins from the bacteria combine with a substance in the cells of the intestine wall which causes the body to rapidly excrete fluids. This makes the sick person thirsty and he or she is then likely to drink more contaminated water which will exacerbate the problem.

Outbreaks of cholera can be prevented by better sanitation and clean supplies of water. Immunisation has a very limited effect for about 6 months and although it may help to protect an individual it does not prevent the spread of infection.

Malaria

This is the most prevalent of serious infectious diseases, occurring especially throughout the subtropical and tropical areas of the world which are inhabited by malarial mosquitoes (*Anopheles* sp.). The disease is caused by the malarial parasite (*Plasmodium* sp.), a protoctist (see **41** Natural selection and classification) which feeds on the red blood cells of animals, including humans. The blood cells eventually burst and 16 asexually produced offspring, together with their toxic wastes, are released into the blood plasma. Large numbers of blood cells rupture simultaneously and it is this stage which causes the fever.

A female mosquito requires a meal of blood in order to produce fertile eggs. When she thrusts her **stylet** (a mouthpart like a hypodermic syringe) into her victim, she first injects saliva which contains an **anticoagulant**. If the mosquito is infected with the malarial parasite, then some of the parasites will be transferred to the animal host where they migrate firstly to the liver and then, after multiplying asexually, into the bloodstream. Similarly, if the host is infected, parasites will be drawn into the mosquito's stomach. Sexual reproduction takes place and a motile stage of the parasite develops in the mosquito's salivary glands.

Most *Plasmodium* strains became resistant in the late twentieth century to the drugs that had previously been effective in treating malaria, which is once more on the increase. There are probably as many as 150 million cases of malaria in the world with as many as 1.5 million deaths per year. The socio-economic consequences of the disease are enormous.

Prevention is better than cure and this can be accomplished by:

- Draining marshes and swamps which are the breeding grounds of the *Anopheles* mosquito.
- Spraying breeding grounds with oil.
- Ensuring that anything which will hold rain is not left upturned or uncovered.
- Using window screens and mosquito netting, especially at night.
- Using insect repellents, mosquito coils and insecticides that are not non-persistent.

Tuberculosis (TB)

This is caused by *Mycobacterium tuberculosis*. The bacterium is hardy and can resist dryness for months as well as mild disinfectants.

The disease occurs worldwide, particularly in developing countries where homes and workplaces may be overcrowded, the conditions humid and the diet poor. It is spread by droplets of moisture when an infected person coughs or sneezes (one of the symptoms of the disease is violent coughing) and by drinking unpasteurised milk from infected cows.

Tuberculosis is increasing all over the world and is a significant cause of death in Asia, Africa and Latin America.

Tuberculosis is best treated by good hygiene and a nutritious diet plus the early treatment of infected persons with a cocktail of antibiotics such as streptomycin which must be taken over a long period. The BCG vaccine provides **immunity** and has helped to control the disease among children in the developing world. Multi-resistant strains have developed and AIDS sufferers are at high risk of contracting TB.

AIDS (acquired immunodeficiency syndrome)

The HIV virus destroys the immune system which is the body's natural defence against disease. AIDS is the final stage during which the body is vulnerable to the infections that finally cause death. There is a gap of about 10 years between infection with HIV and, what has become to be known as 'full blown AIDS'. However, during that period of 10 years there are a number of stages with serious symptoms and infections. To date, although there are some drugs which temporarily block the development of AIDS, there is no known cure.

AIDS is pandemic with more than 90% of cases occurring in developing countries. Because of the size of the problem, up-to-date figures are impossible to obtain but by 2000 there were:

- 5.6 million people newly infected with HIV during 1999; this includes 570 000 children under 15 years
- 33.6 million people living throughout the world infected with HIV/AIDS, including 1.2 million children under 15 years
- 16.3 million deaths from AIDS since the beginning of the epidemic, including 3.6 million children under 15 years.

(Source: UNAIDS Joint United Nations Programme on HIV/AIDS epidemic update, May 2000.)

The estimated total global costs of AIDS and HIV is estimated for the year 2000 to be $500 billion.

Like other viruses HIV must enter a cell in order to multiply. The HIV virus is mainly transmitted by exposure to blood, semen, other genital secretions and breast milk. The main modes of transmission are:

- **Heterosexual** intercourse when one of the partners is already infected.
- **Homosexual** activity with an infected partner.
- **Drug abusers** sharing the same hypodermic needle contaminated with infected blood.
- **Contact with infected blood** via an open wound or by contamination with infected blood during transfusions before blood was screened in the late 1980s.
- **From mother to child** – either in breast milk or across the placenta.

Retroviruses

Viruses are the smallest organisms and are about 0.1 μm in size.

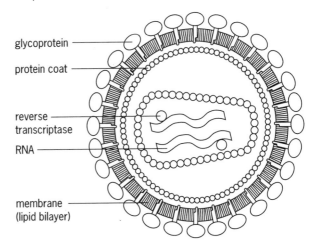

Figure 21.1 Human immunodeficiency virus

The HIV virus is a **retrovirus** which is one of a group of viruses having RNA and *not* DNA for their genetic blueprint. Apart from HIV, retroviruses are responsible for one form of human cancer.

Retroviruses contain a special enzyme called **reverse transcriptase** which uses RNA to make DNA. DNA normally codes for RNA – the other way round (see **6 The genetic code and protein synthesis**). The DNA thus synthesised becomes permanently included in the DNA **genome** (all the genetic material) of an infected cell. The synthesised DNA may remain inactive for some time but if the cell divides, new viral RNA will be produced. This explains why those who are HIV positive do not immediately show AIDS symptoms.

How can HIV be prevented?

- By changing sexual behaviour, i.e. monogamy and not changing sexual partners.
- By not sharing hypodermic needles at any time.
- By using condoms and other 'safe sex' methods – *not* caps.

The immune system

> **Immunity is the ability of the body to recognise and resist foreign antigens such as viruses and bacteria.**

- An antigen is a substance which, when introduced into the body, is not 'recognised' and produces an **immune response**, i.e. the production of antibodies. Antigens may be micro-organisms, toxins or molecules on the surface of cells, for example, the A/B antigen on red blood cells.
- Antibodies are immunoglobulins (types of protein) produced by lymphocytes. They react with specific antigens and are part of the body's immune system.

Two types of white blood cells (**leucocytes**) help to defend the body against antigens:

1 **Phagocytes**, which are motile (as well as being transported by the blood and lymphatic tissues) and ingest certain bacteria by forming a cup-shaped depression and then encapsulating them within a cytoplasmic vacuole. Phagocytes are formed in the white marrow of the long bones (e.g. femur).

2 **Lymphocytes**, which are not motile (but are also transported by the blood and especially the lymphatic tissues). They are also formed in the marrow of the long bones and each lymphocyte is able to recognise one specific antigen. For this reason each individual possesses a whole range of different lymphocytes.

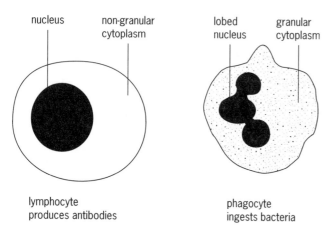

nucleus   non-granular cytoplasm   lobed nucleus   granular cytoplasm

lymphocyte produces antibodies   phagocyte ingests bacteria

Figure 21.2 Two types of leucocyte (white blood cells)

The body's reaction to the presence of antigens is known as the **immune response**. There are two types of lymphocyte (T-cells and B-cells) which differentiate in the bone marrow when they are formed. Each type responds differently:

• **T-cells** are produced in the bone marrow but mature in, and are activated by, the thymus gland which is situated near to the heart. Mature T-cells circulate in the blood or move to lymph nodes. T-cells:
  – Move to areas of infection. They have receptor proteins on their cell surfaces which bind onto invading complementary antigens (lock and key mechanism) and destroy them with enzymes produced by the T-cell's **lysosomes** (cytoplasmic organelles). An antibody–antigen complex is formed.
  – Produce substances which stimulate the phagocytes to ingest the antibody–antigen complex.
  – Produce **interferon** which prevents viruses replicating.
  – Divide repeatedly as a result of the binding onto the antigen and a batch of identical clone cells are produced. Each clone cell is capable of destroying the same antigen. All the above are examples of **natural active immunity**.

• **B-cells** are produced in the bone marrow and circulate in the blood. They have receptor proteins on their cell surfaces which bind onto antigens but the result is different to that of T-cells. When the B-cells replicate, two different types of cell are formed:
  – **Plasma cell clones** which produce large numbers of antibodies specific to the invading antigen.
  – **Memory cells** which remain in the lymphoid tissue. They replicate and protect the body against infection by the same antigen at a later date. Thus the body acquires long term immunity.

Types of immunity

Natural active immunity
This is dealt with under 'T-cells' above.

Natural passive immunity
Preformed antibodies are passed from mother to baby, either before birth across the placenta, or after birth in the mother's milk, to provide temporary immunity until the baby's immune system is active.

Acquired passive immunity
Antibodies formed in one individual are extracted and then injected into a different individual. A good example is in the treatment of tetanus when antibodies from the blood of horses are extracted and then injected into the patient.

Induced immunity
This occurs when small quantities of weakened, killed or detoxified microbe or antigen (the **vaccine**) are introduced so as to cause the body to produce the required antibodies. Vaccines do not act quickly but give the body time to build up its own supply of antibodies. Vaccination has virtually eradicated smallpox.

Allergies

Immune responses can be triggered by any foreign molecule entering the body. The immune system cannot distinguish between microbes and other relatively harmless materials.

Pollen, especially from wind-pollinated plants such as grasses, causes **hay fever** in some people. One third of those with hay fever also suffer from **asthma**.

Asthmatic sufferers may be allergic to pollen, dust, animal fur or feathers, mould spores or certain foodstuffs, etc. During an asthmatic attack, the muscles of the bronchial tubes contract and excess mucus is produced. Prolonged or frequent attacks can become dangerous if too little oxygen is taken in or if emphysema develops. Treatment involves:

• using drugs (e.g. adrenaline) which widen the bronchial tubes
• determining which substances cause the allergy and then trying to prevent exposure to them.

Note: it is most important to be specific and use the correct terms. It is particularly important to distinguish between the pathogen (disease-causing organism) and the disease (a set of symptoms, which may be caused by a pathogen or environmental factors, diet deficiencies, genetic factors, etc).

1 a Define the following terms:
   i antigen (2)
   ii B-cell (2)
   iii pandemic (2)
b Explain the term infectious disease. (2)

2 a Fill in the table to show two differences between artificial and natural immunity.

Artificial immunity	Natural immunity

(4)

b Describe the part played by T-helper-cells in the immune response. (2)

3 The diagram shows the structure of the human immunodeficiency virus (HIV).

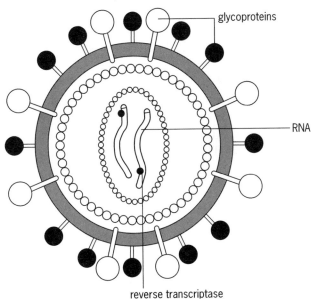

glycoproteins

RNA

reverse transcriptase

a What is the function of the following?
   i RNA (1)
   ii reverse transcriptase (1)
b One form of pneumonia and some types of cancer are normally very rare. They are much more common in people who are infected by HIV. Explain why. (2)
c Viruses such as the influenza virus and HIV have very high mutation rates. Suggest why this makes it difficult to produce a vaccine against them. (2)

AQA, Specification A Specimen Paper 3, for 2001/2002 exam

4 Fitness can be defined as the capacity to do physical work at a particular constant heart rate. Describe and explain how each of the following affects fitness.
a emphysema (6)
b regular exercise (6)

AQA (NEAB), BY08, March 1999

5 a i Name the organism that causes tuberculosis. (1)
   ii State one way in which this organism is transmitted from person to person. (1)
b The table below shows (to the nearest 1000) the number of deaths per year from tuberculosis in England and Wales at 10 year intervals from 1905 to 1995.

Year	Number of deaths per year
1905	66 000
1915	54 000
1925	43 000
1935	32 000
1945	26 000
1955	9 000
1965	4 000
1975	3 000
1985	1 000
1995	3 000

(Adapted from: Denning D. and Kavanagh S., Tuberculosis, Biological Sciences Review, Vol. 8, No. 2. Philip Allan Publishers, 1995.)

i Calculate the mean reduction per year in annual death rate from 1905 to 1945. Show your working. (2)
ii Antibiotics and immunisation against tuberculosis were introduced during the 1940s. What evidence is there in the data that this affected the annual death rate from tuberculosis? (2)
c Suggest two reasons why the annual death rate from tuberculosis was falling before antibiotics and effective immunisation were introduced. (2)
d The data for 1995 suggest a rise in annual death rate from tuberculosis. This may be linked to the increase in HIV infection.
i Suggest why HIV infection might increase the risk of developing tuberculosis. (2)
ii Suggest one reason, other than HIV infection, which might lead to an increase in deaths from tuberculosis. (1)

Edexcel (London), Module 4 (Section A), June 1999

6 Fill in the following table to show:
 a The name of the pathogen.
 b The group of organisms to which the pathogen belongs.
 c A preventative measure.

Disease	Pathogen	Group to which pathogen belongs	Preventative measures (other than vaccination)
malaria			
AIDS			
cholera			

(9)

7 A vaccine has recently been developed against malaria. A trial of this vaccine was carried out in South America. The graph shows some of the results of this trial.

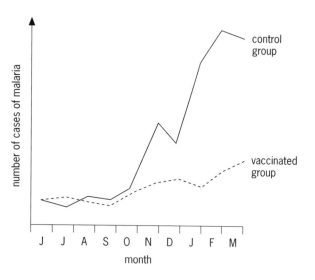

 a i Suggest how the control group should have been treated. (1)
 ii Explain why it was necessary to have a control group. (1)
 b Describe and explain the evidence from the graph which suggests that there was a period of heavy rain from October to March. (3)
 The table shows some more data collected during this trial. It shows the number and percentage of people in different age groups who caught malaria during the first year of the trial.

Age group (years)	Vaccinated group		Control group	
	Total number	Percentage	Total number	Percentage
1–4	3	0.07	13	0.32
5–9	32	0.44	43	0.58
10–14	36	0.57	58	0.75
15–44	68	0.62	83	0.57

 c Explain the advantage of giving the percentage of people who caught malaria as well as the total number. (2)
 d From the data concerning the percentage of people catching malaria, the researchers concluded that the vaccine was most effective with people 1–4 years old.
 i Explain the evidence from the table that supports this conclusion. (1)
 ii Suggest why the vaccine was most effective with people of this age group. (2)
 e Explain how B-lymphocytes, plasma cells and memory cells help to protect the body from disease. (5)

AQA, Specification A Specimen Paper 3, for 2001/2002 exam

8 a In the 1960s many doctors became so convinced that there was a link between smoking and the incidence of lung cancer that they started to recommend that their patients should give up smoking. Explain what evidence there is for such advice. (6)
 b Name two other diseases associated with smoking. (2)
 c The babies of mothers who smoke heavily in pregnancy often have a smaller birth weight.
 i Name a gas produced by smoking which may cause this effect. (1)
 ii Suggest how this gas may affect birth weight. (3)

9 The simplified diagram shows the way in which the body responds to invasion by *Vibrio cholerae*.

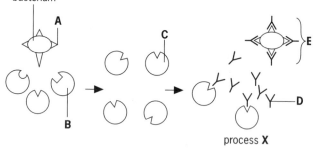

 a State two terms which could be used to describe the type of immunity shown in the diagram. (2)
 b Which of the components labelled **A, B, C, D** and **E** represents:
 i an antibody
 ii an antigen? (2)
 c Use the information in the diagram and your knowledge to explain fully what would happen if this bacterium entered the body again some months after the process labelled **X** had taken place. (3)

Welsh, Module B2, June 1996

1 a i An antigen is a foreign protein.
It causes antibody production/stimulates a response of the immune system.
ii A B-cell is a type of lymphocyte, produced in the bone marrow.
It responds to antigens by producing specific antibodies/humoral immunity.
iii In a pandemic there are many cases of the disease; occurring in many parts of the world.
b A set of symptoms/abnormality in form or function/ damaged tissues;
caused by a pathogen/pathogenic micro-organism.

2 a Two of the following pairs of points:

Artificial immunity	Natural immunity
pathogen is less virulent/altered/ attenuated	pathogen is in the normal/natural state
antigen is injected/deliberately introduced into the body	antigen enters body naturally
antibodies are injected into the blood	maternal antibodies travel through the placenta/milk conferring immunity

b T-helper-lymphocytes activate macrophages for phagocytosis of pathogens.
The specific T-helper-cells instruct B-cells to divide/secrete specific antibodies.

3 a i RNA forms the genetic material/codes for virus proteins.
ii Reverse transcriptase makes a DNA copy of the virus RNA.
b Normally the immune system/T-cells help to control these.
People with HIV have damage to their immune systems/ T-cells.
c Two of:
- Such viruses will always be producing different proteins (because of mutation).
- The proteins act as antigens.
- Antibodies will only work against specific antigens.

4 a Six of:
- Shortness of breath/difficulty in breathing.
- Inability to carry out any strenuous exercise;
- due to breakdown of alveolar walls.
- Enzymes are released from macrophages/phagocytes.
- Smaller surface area for gas exchange.
- Less oxygen in the blood/carried to the muscles.
- Loss of elasticity of alveolar walls.
- Lungs permanently inflated.

b Six of:
- Sustained exercise is possible;
- due to increased lung capacity.
- Lower resting heart rate.
- Heart stronger/more cardiac muscle tissue.
- Increased cardiac output/stroke volume.
- Decreased blood pressure.
- Increased muscle size/strength of skeletal muscles.
- Increased blood flow to muscles.

5 a i *Mycobacterium tuberculosis*
ii breathing in of droplet infection/coughs/sneezes/spitting

b i *Divide the reduction in deaths by the number of years between 1905 and 1945.*

$$66\,000 - 26\,000 = 40\,000$$

$$\frac{40\,000}{40} = \text{mean reduction of 1000 deaths per year.}$$

ii After 1945 there is a sudden, sustained decrease in deaths.
Work out a mathematical comparison, for example:
- Between 1945 and 1955 there was a mean reduction of $17\,000/10 = 1700$ per year; compared with the mean of 1000 per year previously.
- The reduction in deaths per year was $1700/1000 \times 100\%$ greater than the mean reduction for the previous 40 years $= 170\%$ reduction.

c Improved health and hygiene.
Improved living conditions/people living in less cramped conditions.
d i HIV affects the helper T-cells/immune system;
leading to an impaired immune response.
ii One of:
- Mutation leading to strains of the bacterium which are resistant to many antibiotics/multi-resistant strains.
- Selection of multi-resistant strains due to overuse of antibiotics in animal feeds/animal rearing.
- Production of multi-resistant strains due to transformation by other bacteria in the body.
- Selection of multi-resistant strains due to overuse of antibiotics in medicine to treat non-serious infections.

6

Disease	Pathogen	Group to which pathogen belongs	Preventative measures (other than vaccination)
malaria	*Plasmodium vivax*	protoctists	one of: drain breeding swamps; insect repellents/ insecticides; netting/ screening, etc
AIDS	HIV (human immuno- deficiency virus)	viruses	one of: use of condoms; restrict sexual partners; not sharing needles
cholera	*Vibrio cholerae*	bacteria	one of: improved sanitation; boiled/sterile water/good food hygiene

7 a i The control group could be injected with water/saline/something that did not include the antigen.
ii The results could be compared with the control group/to make sure that nothing else in the treatment produced the effects.

b The increase in malaria; associated with an increase in the number of mosquitoes.
Mosquitoes breed in wet weather.

c This allows a comparison to be made; as different numbers of people might have been treated.

d i The largest difference between the vaccinated group and control group was found in this age group.
ii They have not been exposed to as much malaria/fewer have had malaria.
They have no natural immunity.

e Five of:
- B-lymphocytes respond to a specific antigen.
- They divide rapidly/form a clone.
- Plasma cells form.
- The plasma cells secrete antibodies.
- Some form memory cells which become active on the second exposure to the antigen;
- and produce the antibodies faster.

8 a Six of:
- Animal experiments have shown that exposure to smoke increases tumours.
- Tar can trigger tumour formation in animals.
- Tar contains carcinogens.
- Statistics show a relationship between the number of cigarettes smoked and lung cancer deaths.
- A relationship between inhaling and increased lung cancer.
- An increased risk of lung cancer for passive smokers/non-smoking members of families of smokers.
- An increased risk of lung cancer as years of smoking increase.
- Historical records show lung cancer was less common before tobacco was commonly used.

b emphysema; bronchitis

c i carbon monoxide
ii Haemoglobin in red blood cells has a greater affinity for carbon monoxide than oxygen.
Less oxygen passes across the placenta to the fetus.
Less oxygen available for respiration/release of energy/growth.

9 a Two of:
- Active immunity.
- Acquired immunity.
- Specific immunity.
- Adaptive immunity.

b i D
ii A

c Some of C/memory cells would remain in the body.
and two of:
- If the body is reinvaded they divide/clone;
- secreting antibodies;
- more quickly which combine with the antigen/bacterium;
- labelling it for destruction by phagocytic cells/lysis.

The eye is a receptor organ

The eye is a delicate organ which is essential for the survival of an animal in the wild. It is protected by the bony socket which is part of the skull. It has a tough, white, outer coat (the sclera); eyebrows to shield the eyes from sweat; eyelashes to protect the eyes from particles of dust and dirt; eyelids which wipe the eye in conjunction with an isotonic tear fluid containing an enzyme which attacks bacteria.

The function of the eye is to produce a focused image of near or distant objects on the retina which contains numerous nerve endings sensitive to light waves. Impulses are transmitted from the retina to the brain via the optic nerve.

Controlling the amount of light entering the eye

The iris (or iris diaphragm) is the coloured part of the eye. It consists of both radially arranged muscles and circular muscles.

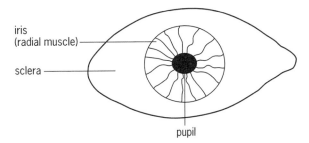

Figure 22.2 Front view of right eye

The two sets of muscles work antagonistically:

Light conditions	Radial muscles	Circular muscles	Pupil size
bright	relax	contract	small
dim	contract	relax	large

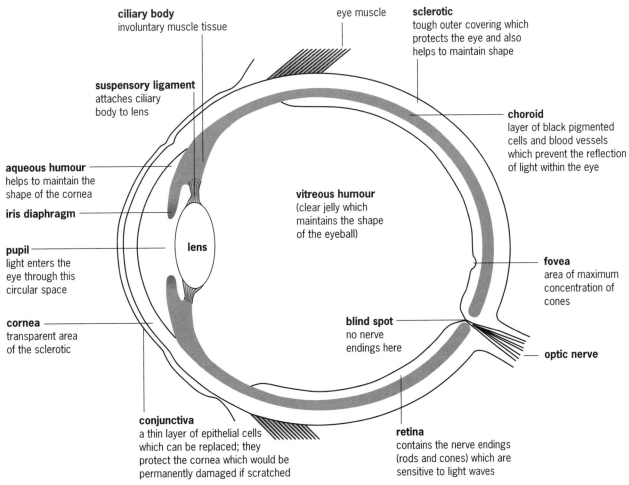

ciliary body
involuntary muscle tissue

eye muscle

sclerotic
tough outer covering which protects the eye and also helps to maintain shape

suspensory ligament
attaches ciliary body to lens

choroid
layer of black pigmented cells and blood vessels which prevent the reflection of light within the eye

aqueous humour
helps to maintain the shape of the cornea

iris diaphragm

vitreous humour
(clear jelly which maintains the shape of the eyeball)

lens

pupil
light enters the eye through this circular space

fovea
area of maximum concentration of cones

cornea
transparent area of the sclerotic

blind spot
no nerve endings here

optic nerve

conjunctiva
a thin layer of epithelial cells which can be replaced; they protect the cornea which would be permanently damaged if scratched

retina
contains the nerve endings (rods and cones) which are sensitive to light waves

Figure 22.1 Diagrammatic transverse section of the human right eye as seen from above

Focusing the image

When we look at an object, light rays from that object are diverging (spreading out) in all directions (see Figure 22.3). Light rays from close-up objects enter the pupil in a cone shape while the light rays from objects far off appear almost parallel to each other.

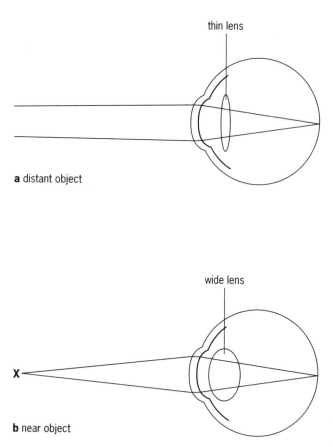

a distant object

b near object

Figure 22.3 Light rays focusing on the fovea

When using a magnifying glass, objects are brought into focus by moving the lens nearer or further away from the object. The eye solves the same problem by changing the shape of the lens. The lens takes on a nearly-spherical shape when the muscles of the ciliary body contract and the suspensory ligaments which connect the ciliary body to the lens are no longer pulling on the lens. When the circular muscle relaxes, the lens is pulled into a flattened shape. This explains why the eye muscles are more relaxed when one is focusing on distant objects, and 'close' work is more tiring on the eye muscles. This ability of the eye to change the shape of the lens when focusing on an object is termed **accommodation**. In order to understand what is happening you need to remember that the ciliary body is a circular muscle. The usual sectional view of the eye as is seen in Figure 22.1 gives the impression that the lens would be pulled into a flattened shape when the muscle contracts. *This is not the case because the ciliary muscle is circular,* see Figure 22.4.

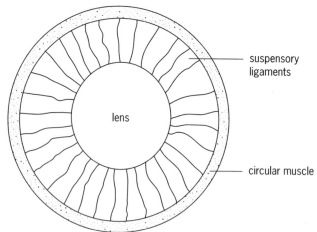

Figure 22.4 Front view of eye showing ciliary muscles

If you focus on part of this page without moving your head or eyes you might be surprised to discover how very few words are in sharp focus at one time – probably a length of no more than ten letters at the most. Reading involves constant moving of the head and eyes across the page. The same principle applies when looking at objects that are far away.

An object comes into sharp focus when the light rays from that object are refracted (redirected) so that they come to a sharp point at the fovea (yellow spot) which is an area of the retina where cones (nerve endings sensitive to coloured light) are concentrated, and rods (not sensitive to colour and with poor sharpness of vision) are absent.

The rest of the image, which is not in sharp focus, is also projected onto the retina at the back of the eye. The only part of the image that is focused onto the fovea is that part which is in sharp focus. As can be seen in Figure 22.5 the image is actually upside down. The light-sensitive nerve endings in the retina are stimulated to send impulses to the brain which then resolves the impulses of about 100 million neurones into an image. At the same time the image is inverted so that we see objects the 'right way up'.

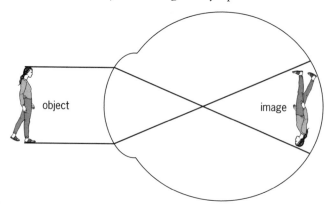

Figure 22.5 The inverted image on the retina

For simplicity, of the countless number of lightwaves reaching the retina from the object, only the outermost are shown in diagrams such as Figure 22.5.

Photosensitive pigments

Nerve impulses are transmitted to the brain when photoreceptors in the retina, acting as transducers, convert light energy into electrical energy. There are more than 100 million of these receptors which are of two types:

- **Rods**, which cannot discriminate between colours and are more effective in poor light conditions. They occur in large numbers in nocturnal animals which sacrifice colour vision for good sight in dim light. The light-sensitive pigment in rods is called **rhodopsin**.
- **Cones**, which are concerned with colour vision and are present in the eyes of most vertebrates. The light-sensitive pigment is called **iodopsin**.

Note the different types of connection to receptor neurones, and the different shapes of the cells, as seen in rods and cones (Figure 22.6). Rods and cones contain mitochondria.

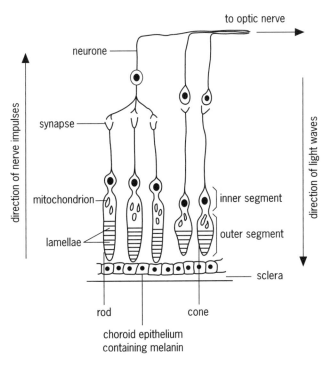

Figure 22.6 Structure of the retina (simplified)

When exposed to light the rhodopsin molecules split into **opsin** (a protein) and **retinene** (formed from vitamin A). This is termed **bleaching**. When rhodopsin is bleached a generator potential is set up in the rod. If the potential is large enough, or a number of rods are stimulated at the same time producing **summation** (see **18** Action potential and the synapse), an action potential is set up. In dim light summation makes it possible for action potentials to occur. An action potential produces an impulse which is transmitted from the rod cell to a receptor neurone. The rhodopsin is then resynthesised using energy from ATP. Note the presence of many mitochondria in rods and cones.

Iodopsin is not so sensitive as rhodopsin to changes in light intensity and cones are of little use in dim light. We cannot see colours in dim light. There are three types of iodopsin. Each one is sensitive to light of a different wavelength: green, red and blue. Different cones contain different forms of iodopsin and colours are seen when the appropriate cones are stimulated in varying proportions. For example, the equal stimulation of red and green cones would be seen as yellow. This is called the **trichromatic theory of colour vision**.

Sharpness of vision is known as **visual acuity** and is due largely to the density of packing of the receptors on the retina.

Eye defects

If the curvature of the lens is too great the light rays that enter the eye will be refracted too much and the focal point will be in front of the retina. This gives rise to a condition known as myopia (short-sightedness). A thin lens, with a curvature which is too little, refracts the light rays so that the focal point would be behind the retina. This condition causes hypermetropia (long-sightedness).

Both defects can be corrected by the appropriate lens (Figure 22.7).

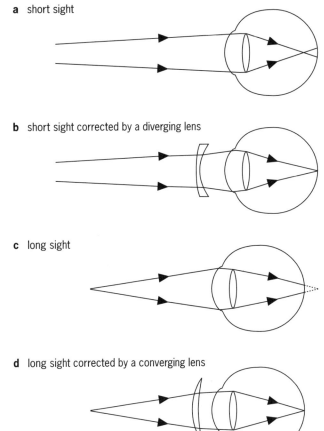

a short sight

b short sight corrected by a diverging lens

c long sight

d long sight corrected by a converging lens

Figure 22.7 Long- and short-sightedness

THE MAMMALIAN EYE

1 a The diagram shows a section of the eye.

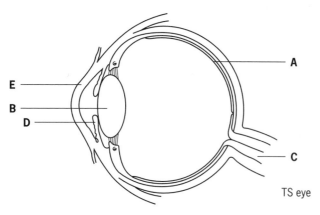

TS eye

Name the parts of the eye labelled **A**, **B**, **C**, **D** and **E**. (5)

b Name three interfaces of the eye which refract light. (3)

c Name the muscles which contract to make the lens more convex. (1)

d Name the part of the eye containing receptors. (1)

2 a Draw a diagram to show the appearance of an eye from the front in dim light. (2)

b Name the automatic adjustment to regulate the amount of light entering the eye. (1)

c Describe the muscular changes that bring about the changes in pupil size associated with entering a brightly lit room. (3)

d Name:
i the receptor
ii the effector
involved in the changes described in **c**. (2)

3 a Draw a simple diagram to show how an image from a distant object can be focused on the back of the eye. (3)

b Name the part of the eye where this image is focused. (1)

4 The diagram shows a single rod cell from the retina of the human.

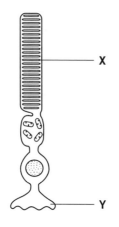

a Name the regions labelled **X** and **Y** on the diagram. (2)

b State the functions of:
i X
ii Y (2)

c State where most of the rod cells are located in the human eye. (1)

d i Name the light-sensitive pigment which is present in rod cells. (1)
ii Describe where this pigment is located in rod cells. (1)
iii Describe what happens to this pigment when affected by light. (1)

5 The diagram shows the arrangement of rod cells and their associated neurones found in the retina.

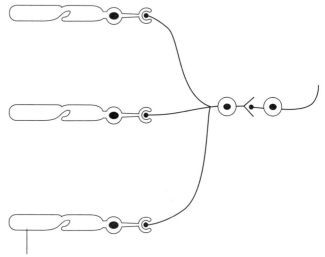

rod cell

a Describe how a generator potential is produced by the action of light on rod cells. (5)

b Many mitochondria are located in the inner segment of a rod cell. Explain their function. (2)

c Many plants are sensitive to the length of day, responding by flowering at the appropriate time of year. Compare the ways light is detected by a rod cell in the human eye and these flowering plants. (3)

6 Mammals active in the daytime have a layer of cells containing the black pigment melanin below the rods and cones.

a Explain the function of this layer of cells. (1)

b Nocturnal animals often have a reflective tapetal layer in the retina. Explain the function of this layer. (1)

Day-active mammals may have a high proportion of cones in their retinas whereas night-active mammals will have fewer cones and more rods.

c Explain how:
i the relative number of rods and cones (2)
ii the distribution of rods and cones (2)
affects the vision of objects.

7 A person was instructed to close the left eye and stare with the right eye at a cross drawn on a plain white board. Different objects were then moved into the field of view from one side. The points where the objects were first seen and where their colours were first identified were recorded. The results are shown below.

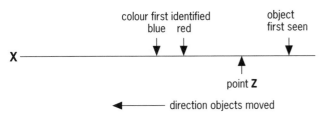

a On what part of the right eye would the cross be focused? (1)

b Explain why an object might be seen at point **Z** but its colour would not be identifiable. (1)

c **i** Explain how possession of different sorts of cone cell allows the colour purple to be identified. (2)

ii Draw an arrow below the line on the diagram showing where you would expect the colour purple to be first identified. (1)

AQA Specification A Specimen Paper 6, for 2002 exam

8 The diagram shows the distribution of rods and cones across part of the retina of a human eye.

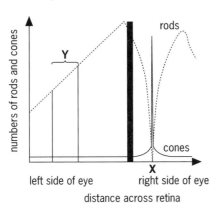

a **i** Which part of the retina does point **X** represent? (1)

ii Give a reason for your answer. (1)

b **i** Give one difference between the connections that a rod cell and a cone cell make with other neurones in the retina. (1)

ii Explain why the part of the retina in region **Y** is very sensitive to dim light. (2)

AEB, Paper 4, Summer 1998

9 **a** The diagram shows part of the structure of the retina.

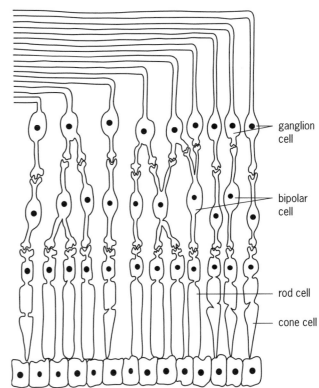

Use the diagram to help you explain how the structure of the retina and its neuronal connections enable a person to have:

i A high level of visual sensitivity in low light levels. (3)

ii A high degree of visual acuity. (3)

b A student was asked to stare intently at a brightly illuminated blue cross for several minutes. After this time the teacher replaced the cross with a plain white screen at which the student continued to stare. The student then saw an image of a yellow cross on the white paper. Explain why the student saw:

i The cross as blue in colour.

ii An image of a yellow cross when staring at the plain white screen. (6)

AQA (NEAB), BY10, March 1999

1 a **A** – retina; **B** – lens; **C** – optic nerve; **D** – iris (diaphragm); **E** – cornea
  b Three of:
  • Between air and cornea.
  • Between cornea and aqueous humour.
  • Between aqueous humour and lens.
  • Between lens and vitreous humour.
  c ciliary muscle
  d retina

2 a See Figure 22.2
  b the iris reflex
  c Circular muscles contract; radial muscles relax; the pupil becomes smaller (admitting less light).
  d **i** Receptor = rods and cones of the retina.
  ii Effector = muscles of the iris.

3 a The image must be inverted (upside down).
  The image must be smaller.
  Parallel rays of light must come from a distant object.

  b retina

4 a **X** = outer segment; **Y** = inner segment
  b **i** **X** – absorbs light energy/generates action potential
  ii **Y** – contains mitochondria to produce ATP to re-synthesise rhodopsin
  c peripheral regions of the retina
  d **i** rhodopsin (visual purple)
  ii on the lamellae
  iii It becomes bleached/converted to a different form.

5 a The light energy is absorbed by rhodopsin.
  This energy changes the rhodopsin into its isomer.
  This then dissociates into opsin and retinal/retinene.
  This causes depolarisation of the membrane of the rod cells, allowing the entry of sodium ions as its permeability alters.
  This gives rise to a generator potential.
  b The mitochondria produce ATP as a product of respiration.
  The ATP is used to provide energy to build up rhodopsin from retinene and opsin.
  c In each case a pigment detects the light.
  The pigment absorbs a particular wavelength of light.
  The pigment then changes to another form or substance.

6 a The black pigment melanin absorbs the light rays, preventing the internal reflection of light and the formation of hazy images.

b The reflective tapetum enables the maximum use of the dim night light.
  c **i** A high proportion of rods which are more sensitive to dim light will result in better night vision/vision in dim light.
  A high proportion of cones, which consist of three types responding to different wavelengths, will result in good colour perception in brighter conditions.
  ii Cones have a higher visual acuity than rods and are packed together in the fovea, which is the part of the retina where the clearest image is perceived. No cones are found in the periphery.
  Rods are scattered throughout the retina but many are in the periphery so that poorly lit objects can be perceived if focused there.

7 a fovea
  b The image falls on the rod cells/does not fall on the cones.
  c **i** Different cones are sensitive to red and blue light. Purple stimulates both these types of cone.
  ii Arrow drawn opposite the point where blue is first identified.

8 a **i** fovea/yellow spot
  ii There are no rods/only cones are present.
  b **i** The rods show convergence/several rods connect with one neurone/a single cone connects with one neurone.
  ii Two of:
  • There is a high density of rods/many rods.
  • Rhodopsin/rod pigment is changed at low light intensity.
  • There is synaptic convergence/summation.
  • The rods have a lower threshold.

9 a **i** Three of:
  • Rod cells are responsible for sensitivity in dim light.
  • Several rod cells are connected to each bipolar cell.
  • The additive effect of a small amount of light striking several rod cells;
  • creates a large enough depolarisation to generate an action potential.
  ii Three of:
  • Cone cells are responsible for acuity.
  • Each cone cell is connected to an individual neurone.
  • Light strikes each individual cone cell to generate a separate action potential/impulse.
  • A very small area of the retina is stimulated, and this results in very accurate vision.

  b **i** The blue-sensitive cones are stimulated;
  by blue light reflected from the cross.
  ii The blue-sensitive pigment has become bleached/the blue-sensitive cells are unable to respond/fatigued.
  White light stimulates the remaining types of cone cell.
  The red- and green-sensitive cells.
  The combination produces the yellow image.

23 The endocrine system: General principles and ADH

The co-ordination of the body is under the control of the nervous system (topics **17** to **19**) and the endocrine system (topics **23** and **24**).

The endocrine glands secrete **hormones** into the blood when the body requires tissues to be stimulated for a much longer period of time than is the case with nervous impulses.

Hormones:

- can reach the entire body because they are carried in the blood plasma
- are effective in minute quantities and act either on specific target organs or else have an effect over a wide area of the body
- do not act in isolation from the nervous system, but rather the two systems complement each other.

> **Hormones are proteins or steroids which are secreted by endocrine (ductless) glands directly into the bloodstream. They bring about responses from specific organs or tissues.**

The facts in Table 23.1 are very commonly a part of examination questions.

Table 23.1 Differences in control between the nervous and endocrine systems

Endocrine system	Nervous system
the response usually lasts over a period of time	the response is usually immediate and short-lived
the response is relatively slow compared with nerves	the response is very fast
the effects are usually widespread	the effects are very specific because impulses are transmitted to effector organs

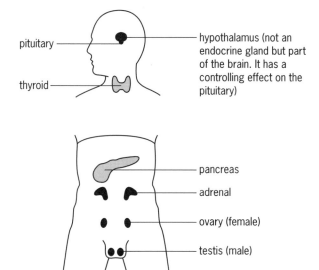

Figure 23.1 The major endocrine glands in the human

Some of the glands have more than one function. The pancreas is both an endocrine and an exocrine gland. (Exocrine glands secrete via a duct, e.g. digestive glands.) The pancreas produces both a hormone (insulin) and digestive enzymes. The ovaries and testes produce sex cells as well as sex hormones.

The glands have many capillaries so that the hormones which they produce can pass easily into the blood system.

How hormones work

Hormones only affect target cells which contain special chemicals called receptor molecules on their surfaces. The receptor molecule and the hormone have molecular shapes which are complementary to each other. This is another example of the 'lock and key' mechanism which was discussed in **9** How enzymes work. When the hormone has locked onto the receptor molecule, a 'messenger' molecule which is located inside the cell membrane is activated, moves away from the membrane, and stimulates a specific chemical change to take place within the cell.

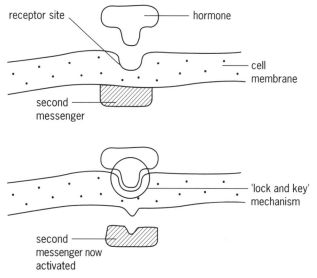

Figure 23.2 Mechanism of hormone action

Such a change may be one of:

- an increase in membrane permeability
- the secretion of products produced by the cell
- protein synthesis
- enzyme activation.

Note that steroid hormones (e.g. sex hormones and corticosteroids) have a different chemical structure and, because of this, they are able to *pass through cell membranes* and enter a target cell where they combine with an intracellular receptor protein. The hormone–receptor complex then enters the nucleus and stimulates the activity of specific genes.

The mechanism of hormone release

Hormones will be released due to one of the following three conditions.

- Due to the direct stimulation of the endocrine gland by a nervous impulse, for example, the release of **adrenaline** by the **adrenal glands** as a direct response to stimulation by the sympathetic nervous system.
- Due to the presence of another hormone which has already been secreted into the blood, for example, **luteinising hormone (LH)**, produced by the **anterior pituitary gland**, further stimulates the **corpus luteum** in females to produce **progesterone**, or the **testes** in males to produce **testosterone**.
- Due to increased levels of chemicals in the blood, for example, the **islets of Langerhans** in the pancreas secrete **insulin** when the level of blood glucose rises.

The release of a hormone in response to another hormone, or chemical, being present in the blood, is controlled by a **negative feedback loop** whereby a decrease in the amount of metabolite results in the secretion of less hormone and vice versa. Figure 23.3 shows how a negative feedback loop works in controlling blood glucose levels.

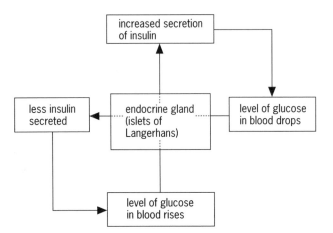

Figure 23.3 Example of a negative feedback loop

The adjustments in a feedback loop are continuous and so a very fine level of control is achieved.

The pituitary glands and the hypothalamus

The hypothalamus is part of the brain and is linked directly to the pituitary glands. The hypothalamus controls body temperature, sleeping and wakefulness, drinking, feeding and osmoregulation.

The anterior pituitary is connected to the hypothalamus by blood vessels and the posterior pituitary is connected by nerves.

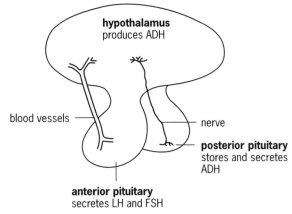

Figure 23.4 The hypothalamus and pituitary

The hypothalamus only weighs 4 grams but it performs a number of vital functions. It is the link between the endocrine and nervous systems.

ADH (antidiuretic hormone)

ADH increases the permeability to water of the walls of the distal convoluted tubules, and collecting ducts of the kidneys. (See also **35** Structure and function of the kidneys: Homeostasis.)

Increased permeability of the tubule wall results in more water being absorbed from the glomerular filtrate into the surrounding blood capillaries which makes the osmotic potential of the blood less negative. The amount of ADH that is secreted is determined by the osmotic potential of the blood circulating to **osmoreceptors** in the hypothalamus. Here we have another example of negative feedback control.

> **Negative feedback is a mechanism by which an increase in the products of a biological process causes the products themselves to act as regulators by shutting down further production until the concentration of product is reduced as necessary.**

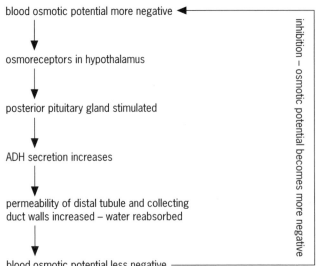

Figure 23.5 Negative feedback mechanism for the control of ADH secretion

1 a Name the two systems which help to co-ordinate the body. (2)
  b List three differences between the action of these two types of systems. (3)
  c Name the part of the pancreas that produces insulin. (1)
  d Give a general term used to describe a receptor which responds to the concentration of solutes in the blood. (1)

2 a Read through the following passage about hormones and then write a list of the most appropriate words to fill the gaps.

  Hormones are produced by _____ glands which therefore must be well provided with blood _____. Hormones are transported around the body in the blood _____ but affect only _____ organs. Here the hormone attaches to a _____ molecule located on the cell _____. This type of mechanism is often described as _____ and _____. (8)

  b Name two events which might trigger the release of a hormone. (2)
  c What effect would an increase in the concentration of antidiuretic hormone have on the concentration of urea in the urine? (1)
  d Explain how an increase in the concentration of antidiuretic hormone in the blood results in this change. (3)
  e Explain why some cells are affected by hormones and not others. (3)

3 The control of body temperature is summarised in the diagram.

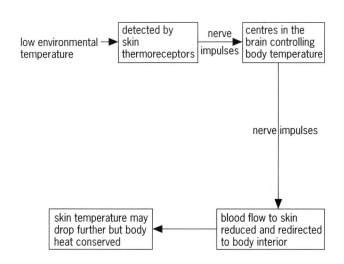

  a Name the part of the brain which co-ordinates body temperature control. (1)
  b What other functions are controlled by this organ? (3)
  c Describe the principles of homeostatic control. (4)

4 a Explain what is meant by *homeostasis*. (2)

  The diagram shows part of a generalised negative feedback system.

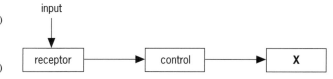

  b With reference to the diagram:
  i State what **X** represents. (1)
  ii On the diagram draw an arrow to show where the negative feedback takes place. (1)
  c With reference to the diagram below, explain why this part of the system does not show negative feedback. (3)

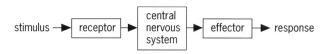

  Homeostatic mechanisms are responsible for the control of blood carbon dioxide concentration. At high altitudes, an increase in breathing rate occurs which decreases the carbon dioxide concentration of the blood, and leads to an increased urine production.
  d Explain the effect of high altitude on urine production. (3)
  Cambridge, Module 4804, June 1977

5 a The diagram shows some important features of homeostatic mechanisms in the mammalian body.

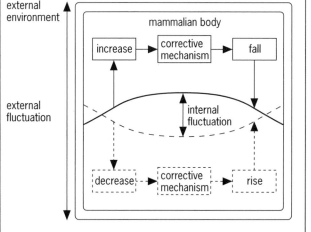

  Use the information in the diagram to help explain the importance of a mammal maintaining a constant internal temperature. (4)
  b Explain the role of the hypothalamus and nervous system in regulation of body temperature. (6)
  AQA, Specification A Specimen Paper 6, for 2002 exam

6 **a** State the response you would expect in a negative feedback system if the level of a metabolite or condition decreased. (1)

b Explain the system controlling the water balance in mammals. (8)

7 In order to maintain the pressure of the blood in the arterial system of mammals a homeostatic mechanism operates. Stretch receptors in the wall of the heart and pulmonary and aortic blood vessels detect a rise in blood pressure and send nerve impulses to the cardiac centre in the brain. Impulses are sent to change the rate of cardiac activity.

a What change in heart rate would you expect in this example? (1)

b Explain what part negative feedback plays in this process. (2)

8 The diagram shows the hypothalamus.

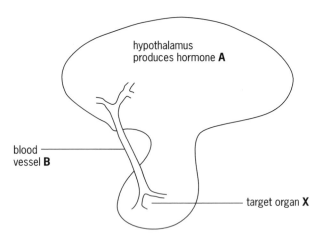

a Hormone **A** is transported to its target organ **X** in blood vessel **B**. How does this differ from the usual transport of hormones to their target organs? (2)

b Suggest two advantages of this method of delivery of hormone **A**. (2)

1 a The nervous system and the endocrine system.
   b See Table 23.1. *Note: you must contrast the differences to obtain the mark.*
   c The islets of Langerhans (beta cells). *Note: some boards require this detail.*
   d osmoreceptors

2 a endocrine; capillaries; plasma; target; receptor; membrane; lock; key
   b Two of:
      • nervous stimulation
      • hormonal stimulation
      • increase in the level of a metabolite.
   c It would lead to an increase in urea concentration.
   d The ADH increases the permeability to water; of the collecting duct/distal tubule. Water is therefore reabsorbed into the blood and this results in a higher concentration of urea in the urine.
   e Hormones have their own specific shape. This fits onto a complementary receptor site; located on the membranes of the target cells.

3 a hypothalamus
   b Three of:
      • sleeping and wakefulness   • feeding
      • drinking   • osmoregulation.
   c In order to maintain the internal environment at optimal (constant) levels. It is necessary to monitor body levels of metabolites/conditions using receptors. Negative feedback mechanisms bring about a response by the action of effectors (glands or muscles); via a co-ordinator.

4 a Two of:
      • Homeostasis is the maintenance/control of the internal environment.
      • Keeps conditions at a set point/constant/stable;
      • despite external changes.
   b i effector/muscle/gland/output
      ii The arrow should go from **X** to receptor/to the arrow between the receptor and control/to the control/to the arrow between the control and **X**/to input.
   c Three of:
      • The input of low environmental temperature;
      • is not reduced or changed.
      • The skin temperature may drop further (positive feedback);
      • it is not returned to the set point.
      • The internal environment is not monitored.
   d Three of:
      • (The decrease in blood carbon dioxide) leads to an increase in blood pH.
      • There is a low concentration of H^+ ions.
      • There is an increased excretion of HCO_3^-.
      • Alkaline urine is produced.

5 *The answer to this question requires continuous prose. Quality of language is important. In order to gain credit, answers must be expressed logically in clear scientific terms.*

   a Four of:
      • Body temperature does not fluctuate as much as that of the environment.
      • Temperature control allows humans to live in different places/climates.
      • A temperature too low would result in enzyme reactions/metabolism too low.
      • A temperature too high would denature proteins/enzymes;
      • and upset the balance of substances produced in metabolism.
   b *First you must explain the general principle.*
      This system of control involves receptors, the hypothalamus (control) and effectors to bring about the response of returning the body temperature to normal. *Then you must give the detail.*
      Five of the following:
      • Receptors in the hypothalamus.
      • Measure the blood temperature.
      • Skin receptors.
      • The hypothalamus has heat gain and heat loss centres.
      • Give details of the responses of the effectors, e.g. increased production of sweat if blood temperature rises.
      • Explain the physical effect this has on body temperature, e.g. as sweat evaporates it takes the latent heat of vaporisation from the body, thus lowering the body temperature.
      • Mention specific behavioural responses such as moving into the shade, etc.

6 a An increase in the metabolite or condition.
   b Eight of the following:
      • When the osmotic/solute potential of the blood becomes more negative;
      • osmoreceptors in the hypothalamus;
      • stimulate the posterior pituitary gland;
      • to release antidiuretic hormone (ADH);
      • this is carried in the blood plasma;
      • to the kidney.
      • The permeability of the collecting duct walls and distal convoluted tubule increases;
      • so more water is reabsorbed into the blood which then has a less negative osmotic/solute potential (see Figure 23.5).
      • A small volume of concentrated urine is produced.
      • The sensation of thirst may be produced by stimulation of the thirst centre in the brain.

7 a As the blood pressure increases the cardiac activity will decrease.
   b The stimulus (increase in pressure) results in a response which returns conditions to normal, i.e. decrease in force of contraction which produces a decrease in pressure.

8 a This is more direct/delivery directly to the target organ by blood vessel **B**; whereas usually hormones travel all round the body in the blood plasma, affecting only target organs.
   b This would probably require less of hormone **A** than the usual method of transport. It might result in a quicker response.

24 The endocrine system: Insulin and glucagon; sex hormones

Insulin and glucagon

Within the pancreas there are groups of cells called the islets of Langerhans. There are two types of cells.

- Large **α cells** which respond to a *drop* in blood glucose levels by producing a hormone called **glucagon**. Liver cells have receptors for glucagon and when the glucagon molecules arrive in the liver and bind to the membranes of liver cells, the level of glucose is increased in one of *two* ways:
  - the rate of conversion of stored glycogen into glucose is increased
  - the rate at which excess amino acids are converted into glucose is increased.
- Smaller **β cells** in the islets of Langerhans respond to a *rise* in blood glucose levels by producing a hormone called **insulin**. Most body cells have receptors for insulin. Insulin lowers the blood glucose level in one of *four* ways:
  - it promotes an increase in the respiration rate so more glucose is used
  - the rate of conversion of glucose to glycogen is increased
  - the rate of conversion of glucose to fat is increased
  - the rate of absorption of glucose by body cells, muscle cells in particular, is increased.

Both the above include examples of negative feedback loops. The loop for glucagon is shown in Figure 24.1.

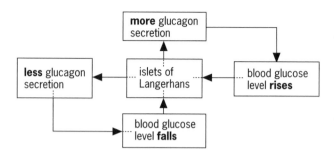

Figure 24.1 Negative feedback loop of glucagon secretion

The feedback loop for insulin was given in Figure 23.3 (**23** Endocrine system: General principles and ADH).

Insufficient insulin production causes **diabetes mellitus** which is fatal if left untreated. Raised blood sugar levels are known as **hyperglycaemia** while low levels are called **hypoglycaemia**. The two hormones, insulin and glucagon, interact to give a very sensitive control of blood sugar level.

The sex hormones

The sex hormones stimulate the development of the sex organs as well as secondary sexual characteristics, such as the growth of pubic hair. The main sex hormones are testosterone, progesterone and oestrogen. Their secretion is controlled by two hormones secreted by the pituitary gland (see Figure 24.2).

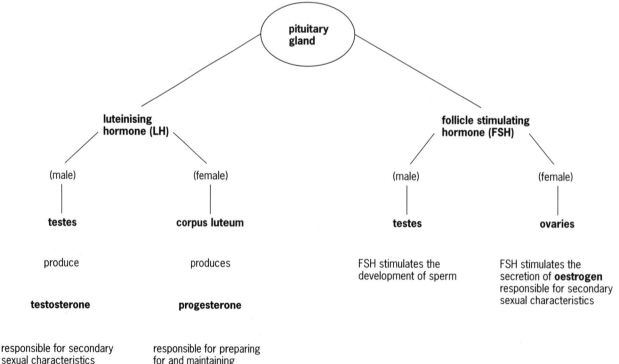

Figure 24.2 Functions of LH and FSH which are produced by the pituitary gland

Sperm production in men is a continuous process, but women produce eggs as a result of the oestrous cycle and are only fertile for short but regular periods of time. The oestrous cycle only occurs in mammals.

The menstrual cycle is a modified oestrous cycle which only occurs in primates (lemurs, tarsiers, monkeys, apes and humans). The mucosa of the uterine wall breaks down approximately 14 days after ovulation, if an ovum has not been fertilised and implanted. This process takes place about every 28 days in humans.

The oestrous cycle is controlled by the secretion of the two pituitary hormones (gonadotrophins) as well as the two hormones (which are steroids) produced by the ovaries.

The control of the oestrous cycle is an example of how one hormone can stimulate or inhibit the secretion of another hormone.

The menstrual cycle is calculated from the first day of bleeding as this point is easy to ascertain. Ovulation takes place on about day 14, although it may vary considerably from one woman to another.

After ovulation, the cells of the follicle in which the ovum develops form a **corpus luteum** which makes progesterone. This, in turn, prevents the release of FSH and LH. If the egg is not fertilised, then on about day 28 the corpus luteum breaks down and little progesterone is produced. This causes the endometrium (uterus wall) to break down and menstruation begins.

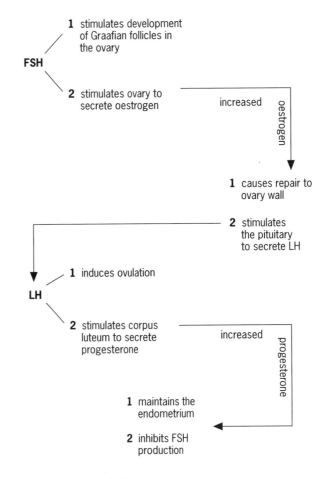

Figure 24.3 Control of the oestrous cycle by hormones

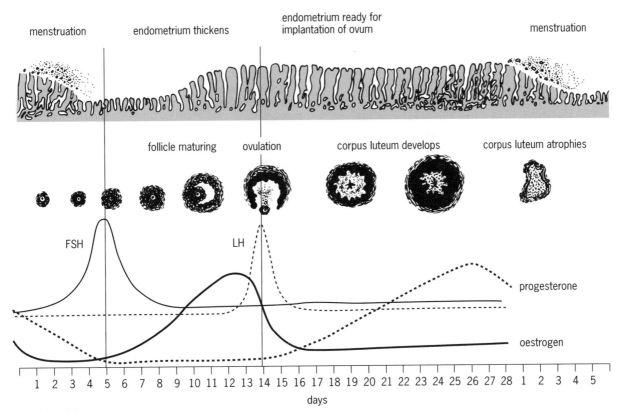

Figure 24.4 The menstrual cycle.

(Adapted from: Mackean D.G., GCSE Biology Second Edition John Murray, 1995)

1 Copy out the table and fill in the spaces to show what happens when blood glucose falls below normal.

	The stimulus of a fall in glucose level of the blood
which endocrine gland is stimulated?	
which target organ is affected?	
how does the body respond to bring about an increase in blood glucose level?	

(3)

2 a Name the hormone which:
   i Stimulates the development of sperm.
   ii Converts excess glucose to glycogen.
   iii Converts glycogen to glucose. (3)
   b Name the cells which produce:
   i glucagon
   ii insulin (2)
   c Name the condition:
   i When blood sugar levels are raised above normal.
   ii When the body produces too little insulin.
   iii When blood sugar levels are lower than normal. (3)

3 The pancreas plays an important role in controlling blood glucose concentration. This is summarised in the diagram below.

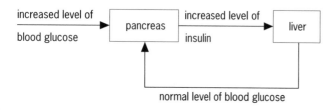

a Use the diagram to name the organs in this control system which represent:
   i The centre which controls the system.
   ii The sensor.
   iii The organ which brings about the response. (3)
   b i Does this system involve negative or positive feedback? (1)
   ii Explain how this feedback system operates. (4)

c People who normally live in Britain and go trekking in the high Himalayas need a period of time to acclimatise. After several weeks they are less breathless because the kidneys produce more hormone **X**. Hormone **X** causes the production of more red blood cells from the bone marrow. Draw a simple diagram to represent this feedback system. (2)

4 Read through the following passage about the human menstrual cycle and the hormones controlling it. List the most appropriate word(s) to fill the spaces.

The development of primary follicles is induced by the anterior pituitary gland releasing _____. After menstruation the lining of the uterus is built up again. This thickening is controlled by the hormone _____ produced by the _____. During ovulation the mature _____ releases a secondary oocyte. The follicle then becomes known as a _____ and this secretes the hormone _____. (6)

5 Copy and complete the table to show the correct function(s) of luteinising hormone, oestrogen and progesterone by marking the appropriate box with a tick. Some hormones may have more than one function. (6)

Function	Luteinising hormone	Oestrogen	Progesterone
immediate cause of ovulation			
immediate cause of repair of the uterine lining after menstruation			
inhibits production of follicle stimulating hormone			
maintains the uterus for implantation			
stimulates formation of a structure which produces progesterone			

Welsh, Paper B3, June 1997

6 Graph **A** below shows the concentration of FSH and LH in plasma during a woman's menstrual cycle. Graph **B** shows the concentration of two hormones **X** and **Y** produced in the ovary during the same menstrual cycle.

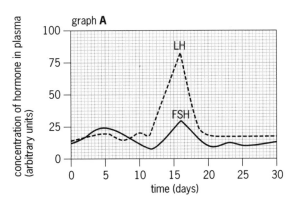

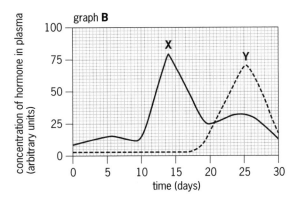

a Where are FSH and LH produced? (1)

b Name the hormones **X** and **Y**. (2)

c **i** On which day did ovulation occur in this woman's cycle? (1)

ii Explain the evidence for your answer. (2)

d Describe an example of negative feedback involving two of these hormones. (2)

NEAB, BY01, June 1997

7 **a** Describe how ovulation occurs. (3)

b Describe how:

i FSH (3)

ii LH (3)

control the oestrous cycle in humans.

c Describe how:

i oestrogen (3)

ii progesterone (3)

play their part in the oestrous cycle in humans.

8 Explain why, in a normal healthy individual, the blood glucose level fluctuates very little. (6)

AQA, Specification A Specimen Paper 6, for 2002 exam

9 Give an account of the principles of chemical co-ordination in mammals with reference to the control of glucose levels in the blood. (10)

Edexcel (London), Paper B3, January 1999

1

	The stimulus of a fall in glucose level of the blood
which endocrine gland is stimulated?	pancreas
which target organ is affected?	liver
how does the body respond to bring about an increase in blood glucose level?	the hormone glucagon is released and converts glycogen stores to glucose, thus raising the glucose level

2 a i follicle stimulating hormone (FSH)
      ii insulin
      iii glucagon
   b i α cells in islets of Langerhans.
      ii β cells in islets of Langerhans.
   c i hyperglycaemia
      ii diabetes mellitus
      iii hypoglycaemia

3 a i pancreas
      ii pancreas
      iii liver
   b i negative feedback
      ii Four of:
      • If the level of blood sugar decreases;
      • the pancreas is not stimulated as much;
      • the pancreas produces less insulin;
      • the level of activity in the liver decreases;
      • leaving more glucose unchanged.
   c *Note: it is necessary to identify the stimulus, receptor, co-ordinator, effector and response from the information given in the question. Then draw a diagram based on the model given at the start of question 3.*

decreased level of blood oxygen → kidney → increase in hormone **X** → bone marrow → increase in red blood cells

normal level of blood oxygen

4 follicle stimulating hormone; oestrogen; thecal cells/ovary; follicle; corpus luteum (yellow body); progesterone

5

Function	Luteinising hormone	Oestrogen	Progesterone
immediate cause of ovulation	✓		
immediate cause of repair of the uterine lining after menstruation		✓	
inhibits production of follicle stimulating hormone		✓	✓
maintains the uterus for implantation			✓
stimulates formation of a structure which produces progesterone	✓		

For each horizontal line one mark will be awarded per tick however incorrect responses will be penalised.

6 a (anterior) pituitary gland
   b **X** = oestrogen; **Y** = progesterone
   c i day 16
      ii Oestrogen reaches a peak (followed by a decline) on day 14.
      Luteinising hormone peaks on day 16. This induces ovulation.
   d *Either*
      as the level of oestrogen builds up; this inhibits the production of FSH by the pituitary.
      or
      as the level of progesterone builds up; this depresses the production of LH by the pituitary.

7 a The Graafian follicle increases in size and moves closer to the surface of the ovary.
      The secondary oocyte is released after the follicle wall breaks down.
      The egg passes into the funnel of the oviduct aided by the beating action of the cilia.
   b i Follicle stimulating hormone, as its name implies, has a stimulatory effect on the development of the follicle.
      Oestrogen is then secreted by the follicle;
      and has an additive effect with LH in stimulating ovulation.
      ii LH completes the development of the follicle.
      LH stimulates ovulation to occur.
      The development of the corpus luteum is stimulated by LH and this then produces progesterone.
   c i Oestrogen stimulates the repair of the endometrium (lining of the uterus).
      The concentration of oestrogen increases to a point where it inhibits the production of FSH;
      and brings about the release of LH from the anterior pituitary gland.

ii Progesterone maintains the lining of the uterus.

It inhibits the release of LH and FSH from the anterior pituitary gland.

Then, as the concentration of progesterone decreases, negative feedback occurs so that FSH is produced again and the cyclic events continue.

8 *Maximum 2 marks for the principles:*
- The glucose level is controlled by hormones.
- Different hormones respond to high and low levels of glucose.

Marks for the detail up to a maximum total of 6:
- A high concentration of glucose leads to an increase in insulin production.
- A low concentration of glucose leads to an increase in glucagon.
- Insulin increases the uptake of glucose by cells.
- Insulin causes excess glucose to be stored as glycogen/glucagon causes the breakdown of glycogen into glucose.
- Both of these processes are activated by enzymes.

9 *First start by describing the general principles of hormonal control and then give specific detail.*

Ten of:
- Homeostasis brings about the regulation of changes in the internal environment.
- Hormones are chemicals secreted by endocrine/ ductless glands.
- They bring about relatively slow effects/their effects are often long lasting.
- They are effective in small amounts (refer to the cascade effect).
- They are transported round the body in the blood/circulatory system;
- affecting target organs/receptor sites on the plasma membranes.
- Explain negative feedback, preferably by referring to insulin.
- Explain positive feedback with reference to oxytoxin/uterine muscle.
- A rise in blood glucose/hyperglycaemia will cause the release of insulin.
- The pancreas detects changes in the blood glucose levels.
- Insulin is secreted by the β cells;
- in the islets of Langerhans.
- Insulin causes increased absorption of glucose by the cells;
- due to the increased permeability of the cell surface membrane to glucose;
- this results in the conversion of glucose to glycogen/glycogenesis.
- Glycogen is stored in the liver/in muscles.
- Refer to diabetes mellitus.
- A fall in glucose level in the blood/hypoglycaemia causes secretion of glucagon;
- by the α cells.
- This results in the conversion of glycogen to glucose/glycogenolysis/gluconeogenesis.
- Refer to the role of adrenaline/cortisol.

25 Plant responses

Although plants may not need to respond as quickly as animals, they rely on responses to stimuli such as light and moisture for their survival. In plants, such responses to directional stimuli are growth movements and are known as **tropisms**. If growth is towards the stimulus, the direction of the response is said to be positive and if away from the stimulus, negative.

Tropic responses in plants

Stimulus	Tropism	Response
light	phototropism	shoots are positively **phototropic**; leaves which are at right angles to the light are **diaphototropic**
water	hydrotropism	roots are positively **hydrotropic**
gravity	geotropism	roots are positively **geotropic**; shoots are negatively geotropic
touch	thigmotropism	pea plants are positively **thigmotropic;** their tendrils twine around objects, clematis climbs by twining around a support
chemicals	chemotropism	pollen grains are positively **chemotropic**; they grow down the stigma, possibly in response to a chemical change

The growth movements of plant roots and stems, resulting in the bending towards, or away from, a particular stimulus, are the result of the unequal elongation of cells on opposite sides of the root or stem.

Growth in plant cells is in two phases:

- cell division of a meristematic (actively dividing by mitosis) cell
- increasing vacuolation of the cell accompanied by elongation of the cellulose cell walls.

Therefore, growth movements occur in those regions where the elongation of cells is taking place. In plants, cell division is apical (the areas are just behind the tips of roots and shoots).

Plant growth substances are 'chemical messengers' and, therefore, hormones. The collective name for these plant growth hormones is **auxin**. There are a number of different substances classed as auxin but the most important is **indole acetic acid (IAA)**.

Experiments were carried out by Charles Darwin in the nineteenth century and by other biologists in the early part of the twentieth century. Classic experiments were carried out on oat coleoptiles which were chosen because of their shape and the ease in which young shoot tips can be decapitated. Experiments have also been carried out on young roots.

All of these experiments illustrate the following facts about the production and properties of auxin.

- The effects of auxin depends on its concentration. It is most concentrated in the tip of shoots and least concentrated in roots.
- High concentrations of auxin stimulate growth in shoots but inhibit it in roots.
- Indole acetic acid is produced at the apices (tips) of shoots and roots and then *transported away from the tips* by diffusion from cell to cell.
- In shoot tips, auxin moves laterally away from the light towards the less illuminated side. Shoots therefore grow towards the light and are positively phototropic. It used to be thought that auxin was destroyed by light but this theory has now been disproved.
- Auxin causes cells to grow by vacuolation and elongation.

All of the experiments that were carried out by Boysen-Jensen in 1913, and others after him, are based on the original work of Charles Darwin in 1880. Darwin showed that oat coleoptiles grow towards unilateral light and that it is the tip of the coleoptile that responds to the stimulus and produces the growth hormone (auxin). This is shown in experiments 1 and 2 in Figure 25.2.

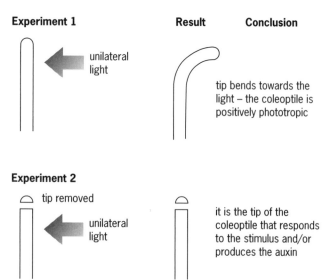

Experiment 1

unilateral light

Result **Conclusion**

tip bends towards the light – the coleoptile is positively phototropic

Experiment 2

tip removed

unilateral light

it is the tip of the coleoptile that responds to the stimulus and/or produces the auxin

Figure 25.2 Experiments 1 and 2

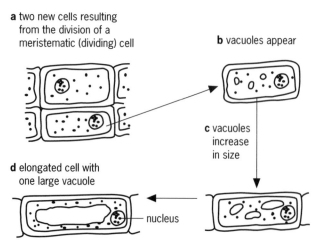

a two new cells resulting from the division of a meristematic (dividing) cell

b vacuoles appear

c vacuoles increase in size

d elongated cell with one large vacuole

nucleus

Figure 25.1 Growth in plant cells

In another of his experiments, Darwin covered the tip of the coleoptile with a cap of lightproof material. The tip did not bend towards the direction of light. Darwin concluded that this was further evidence that the light stimulus must be perceived by the tip of the coleoptile. The following are two further examples of classic experiments that were carried out on oat coleoptiles leading to the discovery of plant growth hormones and how they control plant growth.

In experiment 3 (Figure 25.3) the coleoptile has been decapitated and then a block of agar inserted, separating the tip from the rest of the shoot. Agar is made of gelatin which allows chemicals to diffuse through.

Experiment 3   **Result**

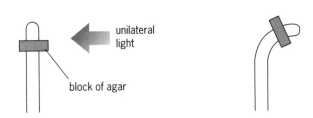

Figure 25.3 Experiment 3

Conclusion: A chemical must have diffused from the tip, through the agar, and into the shoot where a growth curvature has taken place.

If the block of agar is replaced with a thin slice of mica, through which chemicals will not diffuse, no curvature takes place.

Conclusion: As mica conducts electricity, the stimulus must be chemical rather than electrical.

It is believed that unilateral light causes a redistribution of auxin so that more of it accumulates on the shaded side of the coleoptile. The auxin passes down the stem, rather than across it, and this causes greater elongation of cells on the dark side which, in turn, causes the coleoptile to curve towards the light. The amount of curvature is proportional to the concentration of the auxin.

Plants have a system for detecting light involving a light-sensitive pigment called **phytochrome**. There are two forms of phytochrome which are interconvertible, P_R (which absorbs red light) and P_{FR} (which absorbs far red light). When one form absorbs red light it is converted rapidly into the other form and vice versa. Sunlight contains more red light than far red light and so during daylight it is the P_{FR} form that is produced. In the dark it is all converted back to the red form. How phytochromes influence the responses of plants is not yet fully understood, but it is thought that the presence of phytochromes might stimulate the production of plant growth hormones.

In experiment 4 (Figure 25.4) the tip was removed (**a**) and placed on an agar block for several hours (**b**). The agar block was then placed asymmetrically to one side of the decapitated coleoptile (**c**). The coleoptile grew and bent towards the side on which the agar block had *not* been placed.

Experiment 4   **Result**

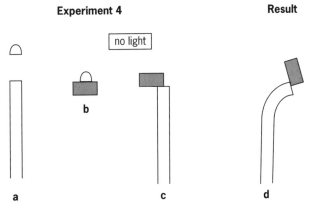

Figure 25.4 Experiment 4

Conclusion: The auxin diffused from the coleoptile tip into the agar block. When the agar block was placed on the decapitated coleoptile the auxin diffused from the agar block and passed down the stem on the same side as the agar block, increasing cell growth and causing bending. Note that there is no light present in this experiment. It is the redistribution of auxin that causes the bending which is due to cell elongation.

Synthetic weedkillers

Auxins control plant growth and are used as selective weedkillers because they have a greater effect on dicotyledonous weeds than on monocotyledonous cereals and grasses. Synthetically produced auxins such as 2,4-D cause the growth of broad-leaved plants to be disrupted so that they die. The upright habitat and narrow leaves of the grasses mean that they receive a lower dosage of the auxin when it is sprayed on the lawn or cereal crop. Even so, the grasses may receive a temporary setback in growth and it is important to carefully control the dosage of weedkiller (auxin) applied.

Gibberellins

Gibberellins are a group of more than 50 growth regulators that stimulate the elongation of stems. Gibberellins were first discovered in Japan in the 1920s. A fungus, named *Gibberella*, attacked the rice crop which then made excessive growth. The rice plants grew tall and weak and then either died or produced a very poor yield of rice. Chemicals extracted from the fungus were found to have the same effect and they were named gibberellins. One of the gibberellins, **gibberellic acid**, is found in plants as well as fungi.

Gibberellins affect plant stems in the same way as auxins, by stimulating elongation of the cells. They do not move in the plant in the same way as auxins but move upwards from the roots via the xylem vessels.

Gibberellins do not affect coleoptiles as auxins do but rather promote the growth of the entire stem, especially when they are applied to the whole plant.

They promote the growth of dwarf peas and beans which are believed to be dwarf because of a genetic mutation which prevents them from producing their own gibberellins and are also involved in the bolting (elongation) of some plants such as carrots.

Gibberellins are abundant in seeds, especially towards the end of their dormant period. Some germinating cereal seeds produce gibberellins which stimulate the synthesis of lipases and carbohydrases. These enzymes are involved in the hydrolysis of food stores and their use by the developing seedling.

Cytokinins

Cytokinins are found in very small quantities in regions of active cell division, especially in fruits and seeds. They promote cell division and differentiation in the presence of auxins.

Abscissic acid

Abscissic acid works antagonistically to gibberellins in dormancy.

Taxes

A taxis (*singular*) is a response by a whole animal or plant to a directional stimulus. The table below gives examples of four different types of taxis.

Taxis	Stimulus	Example
geotaxis	gravity	fruit flies are *negatively geotaxic* and move upwards, away from gravity, in a culture bottle
phototaxis	light	woodlice are *negatively phototaxic* and move away from light
chemotaxis	chemicals	in mosses and ferns the female gametes secrete chemicals into water to attract the male gametes which are *positively chemotaxic*
thermotaxis	temperature	motile algae and bacteria move towards areas of optimum temperature and are *positively thermotaxic*

Nasties

A **nasty** is a growth movement in plants which takes place due to a *non-directional stimulus*.

They may be the result of either changes in the turgor pressure of certain cells or due to growth. The table below gives examples of three different types of nastic movements.

Nasty	Stimulus	Example
photonasty	light	flowers of the wood sorrel open their flowers in the light and close them in the dark
thigmonasty	touch	some insectivorous plants, such as the sundew, rapidly close their leaves when touched
thermonasty	temperature	some flowers, for example the crocus, open their flowers at around 15°C and close them at lower temperatures

The mechanism of nastic movements is not yet known.

1   The growth of oat coleoptiles was investigated in response to unidirectional light. In a similar experiment the oat coleoptiles as shown in the diagram were left for 2 days.

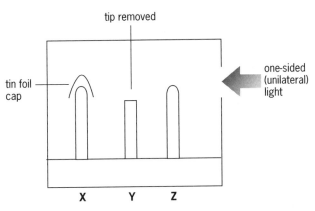

tip removed

tin foil cap

one-sided (unilateral) light

X   Y   Z

a   **i** What would you expect to happen to the height of **X**, **Y** and **Z**? (3)
ii What direction of growth do you expect in **X**, **Y** and **Z**? (3)
b   What is a coleoptile? (1)
c   What is a directional growth response to the stimulus of light called? (1)

2 a   A young broad bean seedling was placed horizontally and given conditions for growth and uniform light intensity.

radicle

plumule

Draw a diagram to show how you would expect it to look 3 days later. (3)
b   What name is given to the response shown by:
i the radicle (seed root)
ii the plumule (seed shoot)? (2)

3 a   Fill in the following table with the name of the growth response expected in each example and state whether this response would be positive, negative or neither of these.

Example	Name of growth response	Positive/negative or neither
a honeysuckle stem is left in contact with an upright stick		
willow roots are growing near to a drain		
the stem of an ivy plant growing in a shed with a tiny gap under the door		

(6)

b   Fill in the following table about the effects of gibberellins and auxins, indicating with a ✓ if the statement is correct and a ✗ if incorrect.

	Gibberellin	Auxin
helps to break dormancy in seeds		
promotes leaf fall		
promotes the formation of adventitious roots in cuttings		
promotes the elongation of cells		

(4)

4   A germinating barley grain was attached to a microscope slide and placed on the stage of a microscope mounted horizontally.

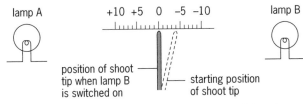

lamp A   +10 +5 0 −5 −10   lamp B

position of shoot tip when lamp B is switched on
starting position of shoot tip

Lamp A was switched on with the tip of the shoot to the right of the zero mark on the scale. The movement of the tip of the shoot towards the lamp was observed through the eyepiece of the microscope. As the tip reached the zero mark on the scale, lamp A was switched off and lamp B switched on. The table shows the position of the tip every three minutes after lamp B was switched on.

Time (minutes)	Position of shoot tip
0	0
3	+3
6	+4.5
9	+5.5
12	+7
15	+8
18	+7.5
21	+6.5
24	+6.5
27	+5
30	+1
33	−3
36	−7

a   Name the type of response being studied in this experiment. (1)
b   **i** Describe the movements of the shoot tip in relation to the direction of light, as shown in the table. (2)
ii Give an explanation for these movements. (3)
c   Describe briefly how you would investigate if the rate of movement of the shoot tip was determined by the light intensity of the lamp. (2)
NEAB, BY04, February 1997

5 The graph shows the effect of different concentrations of auxin on cell elongation in the root and shoot of a plant.

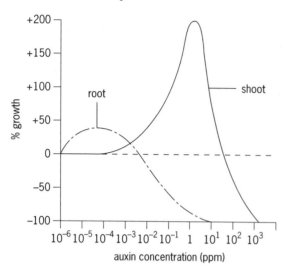

auxin concentration (ppm)

a Use the information in the graph to describe:
  i two similarities
  ii two differences
  between the effect of auxin on the root and the shoot. (4)
b Experiments show that some plants take up auxin sprayed onto them more readily than others. Suggest one commercial use to which this observation could be put. (2)

Welsh, Paper B3, January 1997

6 Describe how auxins can be utilised in horticulture and agriculture. (8)

7 The pigment phytochrome, found in plant tissues in very low concentrations, is involved in the control of flowering in many species. It exists in two forms which can be converted from one state to another as shown below.

$$P_R \ (P_{660}) \xrightarrow[\text{slow conversion in the dark}]{\text{fast conversion in the light}} P_{FR} \ (P_{730})$$

a Seasonal events such as germination and flowering may be regulated by the length of daylight and darkness rather than temperature. Suggest why. (2)
b **i** Suggest how the pigment phytochrome may enable a plant to detect the length of daylight and darkness. (2)
  ii Explain how phytochrome may control flowering in short-day plants. (3)
c Give two similarities between the effect of light on phytochrome in plants and rhodopsin (visual purple) in rods. (2)
d Riboflavin is another pigment found in plants. Illumination of riboflavin and auxin together brings about rapid destruction of auxin. Suggest how this could explain phototropism. (3)

NEAB, BY04, June 1997

1 a i X and **Z** would increase in height. **Y** would remain the same height.
ii X would grow vertically (straight). **Y** would not change. **Z** would bend towards the light.
b A coleoptile is a leaf sheath (typical of grasses).
c phototropism

2 a

The horizontal portion of the seed must be the *same* dimension as the original drawing. The new growth of the radicle must then be shown to curve downwards. The new growth of the plumule must then be shown to curve upwards.
b i positive geotropism   **ii** negative geotropism

3 a

Example	Name of growth response	Positive/negative or neither
a honeysuckle stem is left in contact with an upright stick	thigmotropism	positive
willow roots are growing near to a drain	hydrotropism	positive
the stem of an ivy plant growing in a shed with a tiny gap under the door	phototropism	positive

b

	Gibberellin	Auxin
helps to break dormancy in seeds	✓	✗
promotes leaf fall	✗	✗
promotes the formation of adventitious roots in cuttings	✗	✓
promotes the elongation of cells	✓	✓

4 a phototropism
b i The coleoptile continues to move in the direction of the original light source/away from lamp B for 15 minutes. Between 15 and 18 minutes (or after 18 minutes) it starts to move towards the new light source B.
ii Three of:
• The movement is a growth response;
• due to unequal auxin concentration.
• The delay in response is due to the time taken for a new difference in concentration to be set up;
• due to auxin movement.
c Vary the light intensity keeping other factors constant. Measure the time taken to move a set distance.

5 a i Two of:
• Auxin has an optimum concentration in both.
• Auxin accelerates the growth of root and shoot.
• At higher concentrations auxin inhibits the growth of both.
• Auxin causes growth of root and shoot.

ii Two of:
• Concentrations which inhibit root growth accelerate shoot growth (give values from the graph).
• The root is much more sensitive to auxin than the shoot. (As auxin concentration increases from 10^{-6} to 10^{-4} the elongation of root cells increases directly whereas these concentrations have no effect on the elongation of stem cells.)
• The addition of the specific optimum concentration of auxin has a much greater effect on the elongation of stem cells (+200%) compared with root cells (+40%)/five times greater effect on stem cells.
• The optimum concentration for promoting cell elongation is higher for shoot cells.
b The different rates of uptake could be commercially used to selectively eliminate broad-leaved weeds which have a high rate of uptake of auxin from a crop, e.g. cereals which do not take up the auxin as readily. The weeds would use up their energy supplies in fast growth and die.

6 Synthetic auxins such as 2,4-D are used as weedkillers on lawns and cereal crops since they have a greater effect on dicot weeds than on the monocot grasses. Also the broad-leaved dicots receive a higher dose than the upright narrow grasses. High concentrations of auxin have an inhibitory effect which disrupts the growth of the weeds.
Hormone rooting powders contain auxin which promotes the development of adventitious roots in stem cuttings.
Plants can be produced by tissue culture. Callus tissue formation is aided by the addition of auxin which stimulates mitotic division. Other plant hormones encourage differentiation into stem and root.
Spraying with synthetic auxin encourages fruit to set. It can be used to prevent the 'June apple-drop'. In some cases it can bring about fruit formation from unfertilised flowers (parthenocarpy).

7 a Daylight is a more reliable factor than temperature, because temperature may vary from day to day over a short period of time.
Whereas daylength changes in a consistent way/it is the same every year at the same time.
b i Two of:
• Only the length of the night can be measured;
• as converting P_{FR} to P_R takes time.
• The amount converted to P_R will give a measure of the length of the night.
ii Phytochrome occurs in leaves.
Long nights are required for short-day plants to flower; as during the long night all the P_{FR} will be converted back producing a high P_R.
This high level of P_R will stimulate flowering.
c Two of:
• Both are pigments which absorb light.
• Both can be converted from one form into another.
• Both require a specific wavelength of light.
d Phototropism is a growth response.
The illuminated side would have a lower auxin concentration and grow less;
because auxin promotes elongation of cells.

26 Circulation of the blood

Functions of the blood

Blood is a **connective tissue** consisting of:

1 Three types of cells:
   • **erythrocytes** (red blood corpuscles)
   • **leucocytes** (white blood corpuscles)
   • **thrombocytes** (blood platelets).
2 **Plasma** – a fluid matrix containing:
   • hormones
   • mineral ions
   • fibrinogen – a soluble protein
   • vitamins
   • waste urea plus a small amount of uric acid.

The composition of the blood is maintained within narrow limits by the liver and the kidneys, both of which are **homeostatic** organs. Blood capillaries have walls which are only one cell thick and are 'leaky'. **Tissue fluid** passes through the permeable capillary walls, and leucocytes squeeze between the cells of the capillary wall.

Tissue fluid bathes the cells of the body and thus helps to maintain a constant environment by transporting substances which diffuse, or are actively transported, in and out of cells through the cell membranes. Tissue fluid contains less protein than plasma but is otherwise the same.

The functions of the blood are:

• the maintenance of a constant internal environment (homeostasis)
• defence against harmful bacteria and viruses
• transport:
   – of digested foods (glucose, amino acids and mineral ions) from cells of the small intestine – to all cells bathed in tissue fluid (see **31** Digestion, absorption of food and the role of the liver)
   – of oxygen from the alveoli of the lungs – to all cells bathed in tissue fluid (see also **33** Oxygen and carbon dioxide transport in mammals)
   – of hormones from endocrine glands – to all of the body; specific target organs respond (see **23** The endocrine system: general principles and ADH)
   – of waste urea from liver – to kidneys for excretion (see **34** Structure and function of the kidneys: Ultrafiltration)
   – of carbon dioxide from respiring tissues – to the lungs for excretion (see also **33** Oxygen and carbon dioxide transport in mammals)
   – of heat, from liver and muscles, in particular – to the skin to control body temperature.

Birds and mammals have a **double circulatory system**. The pulmonary and systemic systems are separated but linked by the heart. This helps to maintain blood pressure which would be much less if the blood passed directly from the low pressure in the capillaries of the lungs to the systemic system. Instead, the blood returns to the heart from the lungs and is then re-pumped round the rest of the body.

Arteries, veins and capillaries

The structure of blood vessels is related to their function.

Arteries, and the smaller arterioles, carry blood away from the heart and, with the exception of the pulmonary and umbilical arteries, carry oxygenated blood. The pulmonary artery carries deoxygenated blood to the lungs and the umbilical artery carries deoxygenated blood from the fetus to the placenta. Arteries have thick, elastic walls to withstand the pressure of blood from the heart and do not possess semilunar valves.

Veins, and the smaller venules, carry deoxygenated blood back to the heart, with the exception of the pulmonary vein which carries oxygenated blood to the heart and the umbilical vein which carries oxygenated blood from the placenta to the fetus. The blood pressure in veins is relatively low and they have thin walls. A backflow of blood is prevented by the presence of semilunar valves. Blood flow is maintained when muscles adjacent to veins contract and squeeze the veins, pushing the blood along. Veins can hold a large volume of blood. The **lumen** is relatively large and more than half of the body's total volume of blood is carried in the veins at any one time.

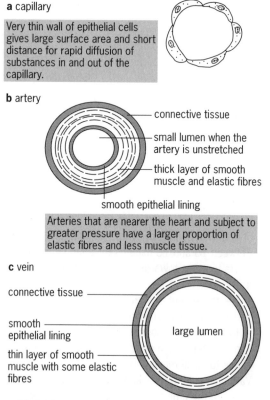

a capillary

Very thin wall of epithelial cells gives large surface area and short distance for rapid diffusion of substances in and out of the capillary.

b artery

connective tissue

small lumen when the artery is unstretched

thick layer of smooth muscle and elastic fibres

smooth epithelial lining

Arteries that are nearer the heart and subject to greater pressure have a larger proportion of elastic fibres and less muscle tissue.

c vein

connective tissue

smooth epithelial lining

large lumen

thin layer of smooth muscle with some elastic fibres

Figure 26.1 The structure of blood vessels

Any backflow of blood will close the semilunar valves present in veins and ensure that the blood flows only towards the heart. Sometimes a vein becomes so dilated that the valves no longer close. Such veins lose their elasticity and varicose veins are formed.

Structure and function of the heart

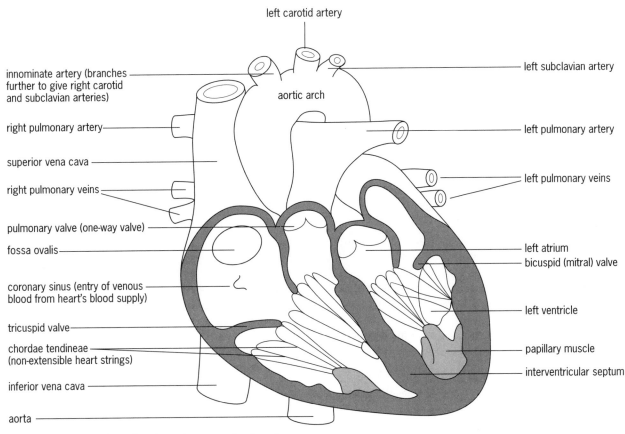

Figure 26.2 Vertical section through mammalian heart

Note the following points:

- The left side of the heart is more muscular than the right side because it has to pump blood to the extremities of the body. The right side of the heart only has to pump blood to the lungs which are situated, together with the heart, in the thoracic cavity.
- The chordae tendineae ('heart strings') enable the bicuspid and tricuspid valves to keep their shape (not be blown 'inside-out') when the ventricles contract. They thus close the entrance between the atria and ventricles.
- The fossa ovalis is the entrance into the right atrium of the venae cavae. (Note the plural form of vena cava.)
- The atria have comparatively thin walls as they only have to pump blood into the ventricles.
- The entrances to both the aorta and the pulmonary artery have semilunar (one-way) valves to prevent a backflow of blood after the ventricles have contracted.
- The heart is made of cardiac muscle which does not become fatigued. The heart has its own cardiac blood supply (see Figure 26.3).

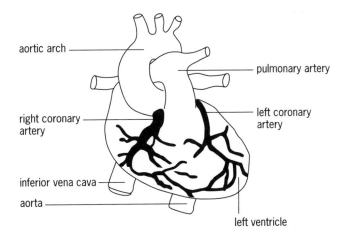

Figure 26.3 Ventral view of mammalian heart

Heart disease

The most common heart disease is **coronary heart disease**.

Atherosclerosis is a condition in which the lumen of the coronary arteries becomes narrowed due to deposits of fat and fibrous tissue on the inside of the vessel wall. This is also known as hardening of the arteries.

Thrombosis is caused by a blood clot in a coronary blood vessel.

These two conditions contribute to heart attacks which, if caused by a blockage to the main coronary arteries, will lead to death. **Smoking** is a major factor leading to both of these heart conditions.

1 Name the following:
   a Blood vessels which carry blood towards the heart.
   b The main veins of the body.
   c The lower chambers of the heart.
   d The valve between the two chambers on the left hand side of the heart.
   e The valve preventing backflow from the pulmonary artery.
   f Another name for red blood cells.
   g The part of the blood which carries urea. (7)

2 a Which chamber of the heart has the thickest wall? (1)
   b Name the blood vessel which carries blood away from this thick-walled chamber (named in a). (1)
   c As blood travels further away from the heart the blood pressure in the arteries decreases. Give an explanation for this. (2)
   d Give two reasons why this decreased arterial pressure helps the capillaries to function efficiently. (2)
   e Explain how the blood in the extremities of the body is returned to the heart. (2)

3 The diagrams below show three types of blood vessel in transverse section. (Different scales have been used.)

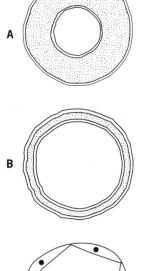

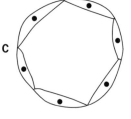

   a Name A, B and C. (3)
   b Describe the role of vessel C. (2)
   c Describe two ways that vessel C is adapted for its function. (2)

4 A small tube called a catheter can be inserted into the blood system through a vein. It can be threaded through the vein and into and through the heart until its tip is in the pulmonary artery. A tiny balloon at the tip can then be used to measure the pressure changes in the pulmonary artery. The diagram shows a section through the heart with the catheter in place. The graph shows the pressure changes recorded in the pulmonary artery.

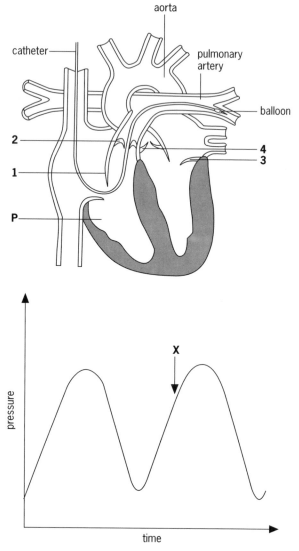

   a Name the chamber of the heart labelled P. (1)
   b Complete the table by placing ticks in the appropriate boxes to show which of valves 1 to 4 will be open and closed at time X on the graph. (2)

Valve	Open	Closed
1		
2		
3		
4		

   c Sketch a curve on the graph to show the pressure changes you would expect if the pressure in the aorta were measured at the same time. (2)

AQA (AEB), Paper 3, Summer 1999

5 a Name the structures **A**, **B**, **C**, **D** and **E** labelled on the diagram of the heart. (5)

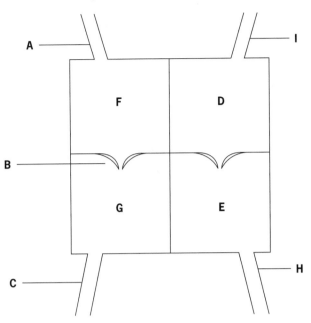

b Refer to the diagram above and give the letter of the structure referred to in the following statements.

i An artery carrying deoxygenated blood.

ii The chamber of the heart where the highest pressure develops.

iii The vessel returning blood from the body to the heart.

iv The artery carrying oxygenated blood to the body.

v A structure preventing backflow of blood. (5)

6 Describe how the following assist in the circulation of the blood.

a The contraction of the smooth muscle layer in the arteriole walls. (2)

b The elastic tissue in the artery walls. (1)

7 The graph shows the changes in pressure which take place in the left side of the heart.

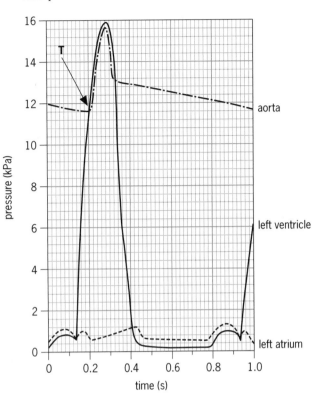

a Use the graph to calculate the heart rate in beats per minute. Show your working. (1)

b i Explain, in terms of pressure, why the valve between the left ventricle and the aorta opens at time T. (1)

ii For how long is the valve between the left atrium and the left ventricle closed? Explain your answer. (2)

c i How would you expect the pressure in the right ventricle to differ from that in the left ventricle? (1)

ii Explain what causes this difference in pressure. (1)

NEAB, BY03, February 1997

8 a Outline the need for transport systems in multicellular animals. (6)

b Explain how the structure of blood vessels enables them to fulfil their functions. (12)

(content 18 marks, quality 2 marks)

Cambridge, Module 4804, November 1996

1 a veins **b** venae cavae **c** ventricles
 d left atrioventricular valve/bicuspid/mitral
 e pulmonary valve **f** erythrocytes **g** plasma

2 a left ventricle **b** aorta
 c As the blood flows there is friction with the arterial wall;
 the lumen of the artery increases.
 d Exchange of materials occurs through leaky capillary
 walls and this is aided by a slower flow.
 Helps to prevent damage to the thin-walled capillaries.
 e One-way watch pocket valves prevent backflow.
 Contraction of muscles during movement squeezes
 blood along the veins towards the heart.

3 a **A** = artery; **B** = vein; **C** = capillary
 b It allows exchange of materials between tissues and blood;
 oxygen/glucose/amino acids/urea/carbon dioxide, etc.
 c Cell walls are only one cell thick.
 There are pores in between the cells of the
 endometrium.

4 a right ventricle
 b

Valve	Open	Closed
1		✓
2	✓	
3		✓
4	✓	

 Valves **1** and **2** correct – 1 mark.
 Valves **3** and **4** correct – 1 mark.
 c The peaks of the two lines must coincide.
 The line drawn must be higher than the peaks and
 troughs for the pulmonary artery.

5 a **A** = vena cava; **B** = atrioventricular valve/tricuspid valve;
 C = pulmonary artery; **D** = left atrium; **E** = left ventricle
 b **i C** **ii E** **iii A** **iv H** **v B**

6 a When the smooth muscle contracts it slows/stops the
 blood flow in the arterioles.
 It alters the flow of blood to the capillary beds/affects
 blood pressure.
 b This allows the elastic recoil of the artery walls and
 evens out the flow of blood.

7 a (*One complete heart beat takes 0.8 seconds/one
 complete pattern. There are 60 seconds in a minute.*)

 $$\frac{60}{0.8} = 75 \text{ beats per minute}$$

 b **i** The pressure in the left ventricle is higher (than that in
 the aorta).
 ii (0.4 − 0.135) = 0.26 to 0.27 seconds;
 since this is when the pressure in the ventricle is higher
 than that in the atrium.
 c **i** The pressure will be lower in the right ventricle.
 ii The wall of the right ventricle is less muscular/thinner.

8 a Six of:
 • In unicellular organisms transport substances need
 only occur over very short distances.

• Multicellular organisms are larger and usually
 metabolically more active.
• Some parts of the body are distant from the
 medium in which it lives/substances need to travel
 distances.
• Distances too great for diffusion/diffusion too slow;
• diffusion is only efficient over short distances.
• The surface area to volume ratio is too small;
• for the uptake of oxygen and nutrients.
• Diffusion alone is often not sufficient to effectively
 remove carbon dioxide/urea/waste products.
• A transport system is required to distribute
 substances (give named examples);
• quickly;
• and in sufficient quantities.
• Give named examples of systems and processes
 which operate efficiently.

b Arteries
• Have thick walls;
• with a large amount of elastic fibres/smooth muscle;
• which allow walls to distend/stretch during systole;
• to accommodate the increased volume of
 blood/withstand high blood pressure.
• Collagen fibres are present in the tunica externa;
• for strengthening/support/prevent rupturing.
• Arteries have a small lumen compared with the
 overall diameter.
• Elastic recoil occurs (during diastole);
• blood pressure is maintained/only small drop in
 pressure along artery/high pressure.
• Blood moved along in a wave/pulse.
• Blood flows at high speed.

Veins
• Have thin walls;
• with not much smooth muscle/elastic tissue.
• Large lumen compared with the overall diameter;
• which reduces the resistance to blood flow;
• the walls can distend to accommodate a lot of blood.
• Blood pressure is low.
• Blood flow is slow.
• Blood flow is non-pulsatile.
• Veins contain valves.
• Unidirectional blood flow is maintained by
 valves/prevent backflow of blood.
• The squeezing action of skeletal muscle aids blood
 flow.

Capillaries
• Are the site of exchange between blood and
 cells/tissues.
• The walls are one cell thick;
• which allows rapid/efficient diffusion;
• of oxygen/carbon dioxide/glucose/ions.
• The walls have pores/gaps;
• to allow the passage of plasma/tissue fluid
 formation.
• Blood travels at low speed.
• Blood is at low pressure.
• Blood flow is non-pulsatile.
• The large surface of the capillary beds;
• increases exchange of materials.

27 Control of the cardiac cycle; the effects of exercise

The cardiac cycle

1 **Systole.** First the two atria contract together, forcing more blood into the two ventricles and this is followed closely by contraction of the ventricles. Blood is pumped out of the heart into the lungs and to the rest of the body.

The sound of the heart beat is described as 'lub-dub'. The 'lub' sound is made when the ventricles *contract* and the blood is forced against the atrioventricular (bicuspid and tricuspid) valves.

2 **Diastole.** Blood enters the two atria, which fill causing the two atrioventricular valves to open, allowing the blood to fill the ventricles as well.

The 'dub' sound is caused by a *backflow* of blood hitting the semilunar valves of the aorta and pulmonary artery when the ventricles relax.

The heart beats at an average of 70 beats per minute, although the rate is much higher in small children.

Control of the cardiac cycle

The heart's rhythm is maintained by a wave of electrical impulses similar to nerve impulses. The stimulus comes from the **sinoatrial node** (pacemaker) which is a small part of the right atrium. Pacemaker waves spread out across the two atria causing the muscle to contract. The pacemaker waves stimulate the **atrioventricular node** which conducts the impulse into the **bundle of His** and then into the **Purkyne tissue** (formerly called Purkinje fibres) which runs through the septum between the two ventricles. The impulse causes the ventricles to contract.

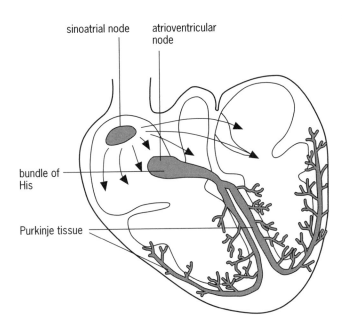

Figure 27.1 Pathway of the cardiac impulse

Cardiac muscle is **myogenic**. This means that impulses are generated spontaneously. However, the heart is connected to nerves of the autonomic nervous system which controls the involuntary activities of the body (see **19** The peripheral and central nervous systems and the brain).

Impulses from the accelerator nerve (sympathetic) increase both the rate of heart beat and the strength of the muscle contractions. More efficient contraction of the ventricles increases the **stroke volume** – the volume of blood pumped at each heartbeat.

cardiac output = stroke volume × heart rate

Impulses from the inhibitory nerve (parasympathetic) decrease the heart rate.

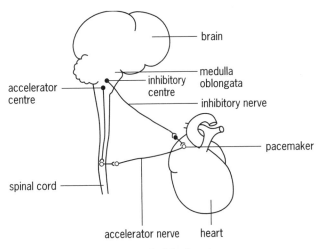

Figure 27.2 Nervous control of the heart

Baroreceptors are small receptors found in the walls of the pulmonary artery, aorta, carotid arteries and venae cavae. A change in blood pressure in any of these areas causes impulses to be sent to the cardiac centres.

A rise in arterial pressure, which distends the artery wall, initiates reflexes that cause the heart to beat more slowly and with less force. The blood vessels also become less constricted.

A fall in blood pressure increases the sympathetic (accelerator) nerve stimulation, causing the heart to beat faster and with more force. The blood vessels become more constricted.

Vasodilation and **vasoconstriction** (the expansion and narrowing of blood vessels) are controlled by the vasomotor centre in the medulla oblongata. Sympathetic nerves stimulate the contraction of the smooth (involuntary) muscle in the walls of arterioles.

Adrenaline is secreted by the adrenal glands in time of stress. Adrenaline excites the pacemaker and the heart rate increases.

High levels of **sodium ions** interfere with the action of calcium ions during muscle contraction and cause the heart rate to slow down.

ECG (electrocardiogram)

Living cells maintain a potential difference across their membranes by creating a difference in the electrical charges between the cell cytoplasm and the extracellular tissue fluids.

There is a change in electrical potential across the membranes of muscle fibres when they contract (see **18** Action potential and the synapse). When electrodes are attached to the surface of the body, the changes in electrical potential of the heart muscle (myocardium) as it contracts can be recorded. Such recordings can be used to detect malfunctioning of the heart.

The pattern produced can be either shown on an oscilloscope screen or traced on paper. The tracing is called an **electrocardiogram** and the apparatus used is an **electrocardiograph**.

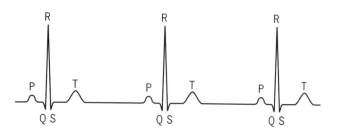

Figure 27.3 P, Q, R, S and T waves in a normal electrocardiogram

A normal tracing, as shown in Figure 27.3, shows five waves which are called P, Q, R, S and T.

- **P** wave = the wave caused by depolarisation as it sweeps over the atria.
- **QRS** wave = the wave caused by ventricular depolarisation as the impulse spreads from the AV node through the bundle of His and Purkinje tissue.
- **T** wave = the wave caused by ventricular repolarisation as the muscles of the ventricles relax.

The physician examines the pattern of waves together with time intervals between cycles and parts of cycles to obtain the information needed to determine whether there is any malfunction.

The pulse rate

Waves of pressure caused by the contraction of the ventricles of the heart can be felt when a firm fingertip pressure is applied to one of the pulse points. The usual pulse points used are the radial artery in the wrist and the carotid artery of the neck.

Pulse rates vary greatly from person to person and how steady the pulse is may be more important than the actual rate. The pulse rate decreases with age. The following figures are average rates for people *at rest*:

- infants   110–140 beats per minute (bpm)
- adolescents   80–90 bpm
- adults   50–85 bpm
- elderly folk   50–70 bpm.

The heart of a normal person at rest pumps about 4 to 6 litres of blood per minute.

The effects of aerobic exercise on the heart

> **Aerobic exercises are those that are designed to increase the efficiency of the body's intake of oxygen.**

Increased levels of physical activity lower the risk of **coronary heart disease** which is characterised by **atherosclerosis**, in which the cavities of the arteries become narrow due to the accumulation of lipid and cholesterol, and blood flow becomes restricted or may even stop. If this occurs in the coronary arteries a heart attack, chest pains or sudden death may occur. One's diet and body weight are important but regular exercise also plays a crucial role in the functioning of the heart.

A programme of training needs to be undertaken for no less than 20 minutes three times a week. Training levels are measured in heart beats per minute. The formula is: 220 minus one's age multiplied by 65%. This means, for example, that the training level for an 18 year old is 131 beats per minute and this level needs to be maintained for the 20 minute session. Such training can include brisk walking, jogging, swimming and aerobic dancing. At the end of 5 minutes after exercising the breathing rate should be approximately:

- less than 120 bpm for those under 50 years
- less than 100 bpm for those over 50 years.

Any fitness programme should also include exercises (such as weight training) that build up muscular strength and increase flexibility. Exercise may also help to relieve high blood pressure (**hypertension**) which damages the arteries in the heart, liver and kidneys, increasing the risk of death from heart failure. Hypertension is caused by atherosclerosis. The heart pumps harder in order to pump the same volume of blood around the body in a given time, and this raises the blood pressure which may even cause arteries to rupture.

1 Name the following:
a The total volume of blood pumped by the heart per minute.
b Pressure receptors found in the walls of the carotid artery.
c The pacemaker of the heart.
d The contraction of arterioles.
e Spontaneous generation of impulses by heart muscle.
f Contraction of the two ventricles of the heart. (6)

2 a What is the effect of the following on the rate of heart beat?
i The sympathetic nerve stimulation of the heart is increased.
ii Adrenaline is secreted into the blood.
iii The blood pressure falls.
iv The concentration of sodium ions in the blood increases. (4)
b The heart has its own pacemaker. It initiates impulses which spread down the heart causing muscular contraction.
i Name the type of muscle of which the heart is made. (1)
ii Name the path of conduction through the septum between the two ventricles. (1)
iii Explain why it is important that after the wave of excitation reaches the atria it is then slowed before reaching the ventricles. (2)

3 The diagram represents some of the forces acting on blood as it passes along a capillary. These result in the formation of tissue fluid.

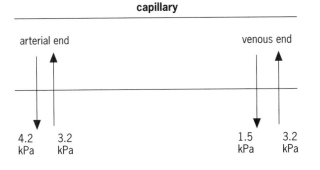

capillary

arterial end   venous end

4.2   3.2   1.5   3.2
kPa   kPa   kPa   kPa

a Name the pressure which forces solutes and water out of the capillary. (1)
b i Name the pressure which tends to draw water back into the capillary. (1)
ii Explain why this is greater at the venous end of the capillary bed. (2)
c Give two reasons why the pressure falls as the blood moves through the capillary bed from the arterial end to the venous end. (2)
d Give two differences between the composition of lymph and the blood in the neighbouring capillaries. (2)
e People suffering from protein deficient diets may have an excessive volume of tissue fluid. Explain this observation. (2)

4 The diagram shows a vertical section through the human heart. The arrows represent the direction of movement of the electrical activity which starts muscle contraction.

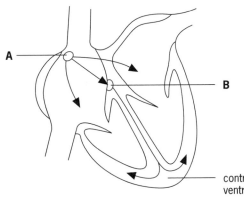

A

B

contraction of ventricles starts here

a Name structure **A**. (1)
b Explain why each of the following is important in the pumping of blood through the heart.
i There is a slight delay in the passage of electrical activity that takes place at point **B**. (1)
ii The contraction of the ventricles starts at the base. (1)
c Describe how stimulation of the cardiovascular centre in the medulla may result in an increase in heart rate. (2)

NEAB, BY03 (Section A), June 1997

5 The diagram shows some nerves associated with the heart and the blood system.

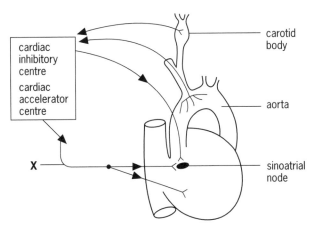

cardiac inhibitory centre

cardiac accelerator centre

carotid body

aorta

X

sinoatrial node

a In which part of the brain are the cardiac inhibitory and cardiac acceleratory centres found? (1)
b Give two stimuli detected in the aorta and the carotid body that may result in a change of heart rate. (2)
c Suggest two ways in which stimulation of the heart by nerve **X** leads to an increase in the amount of blood pumped out by the ventricles. (2)

NEAB, BY01 (Section A), June 1998

6 a The sound of the heart beat is very distinctive and is associated with the movement of the heart valves.
   i What causes the 'lub' sound? (1)
   ii The blood pressure produced by ventricular contraction differs in the aorta and pulmonary artery. Which vessel has the greater pressure? (1)
   b i Explain how the structure of the heart affects the blood pressure in the systemic and pulmonary circulations. (3)
   ii Explain how the mechanisms controlling the heart beat affect the rate of flow of the blood. (6)

7 The rhythmic contraction of the heart muscle during the cardiac cycle is controlled by waves of depolarisation which spread from the sinoatrial node (SAN) across the heart. This electrical activity can be detected using electrodes placed on the surface of the body around the heart and displayed as an electrocardiogram (ECG). The diagram below shows an ECG for a healthy person.

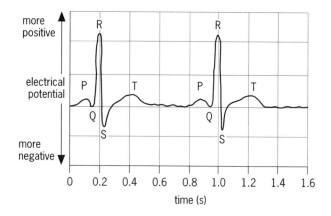

The part of the ECG labelled P represents the wave of depolarisation in the walls of the atria. The parts labelled Q, R and S represent the wave of depolarisation in the walls of the ventricles. The part labelled T represents the recovery of the ventricle walls.
a Explain what is meant by depolarisation. (2)
b i On an ECG the time interval between Q and T is called the *contraction time*. Suggest why it is given this name. (1)
   ii The interval between T of one cardiac cycle and Q of the following cycle is called the *filling time*. Suggest why it is given this name. (2)
c i Explain why there is a time delay between the depolarisation in the walls of the atria P and the start of depolarisation in the walls of the ventricles Q. (3)
   ii Suggest how this time delay helps the heart to carry out its function efficiently. (2)
d i From the ECG, calculate the rate of heart beat in beats per minute. Show your working. (3)
   ii During exercise, the rate of heart beat increases but the contraction time normally remains almost unchanged. What must happen to filling time during exercise? (1)
   Edexcel (London), Paper B3, January 1999

1 a cardiac output
   b baroreceptors
   c sino-atrial node
   d vasoconstriction
   e myogenic
   f (ventricular) systole

2 a i increases
      ii increases
      iii increases
      iv decreases
   b i cardiac
      ii Purkinje/Purkyne fibres
      iii The delay allows the completion of atrial contraction; before ventricular contraction occurs.

3 a hydrostatic/blood pressure
   b i osmotic pressure/solute potential/osmosis
      ii Loss of water from the capillary; but the solute potential due to the large plasma proteins remains fairly constant.
   c Two of:
      • The resistance of the capillary walls.
      • The loss of water from the capillaries reduces the volume of blood.
      • The cross sectional area of the blood vessels increases in total.
   d Two of:
      • More large plasma proteins are present in blood.
      • Less lymphocytes present in blood.
      • More platelets in blood/none in lymph.
      • More red blood cells in blood/none in lymph.
      • Less fat in blood.
   e Such people may have low plasma protein making the solute potential less negative; less return of fluid to the blood capillary.

4 a Structure **A** is the sinoatrial node.
   b i This allows time for the emptying of the atrium before the ventricle contracts.
      ii The aorta/arteries are at the top of the ventricles.
   c Two of:
      • An impulse is sent to stimulate the SAN/pacemaker.
      • The impulse passes along the accelerator/sympathetic nerve/which causes production of noradrenaline at the synapse.
      • This increases the activity of the sinoatrial node.
      • There is a reduction in stimulation/from the vagus.

5 a medulla (oblongata)
   b Two of:
      • Carbon dioxide concentration.
      • Blood pressure.
      • Partial pressure of oxygen/oxygen concentration.
   c Increase in rate; increase in stroke volume.

6 a i Closing of the atrioventricular valve/mitral valve.
      ii the aorta

b i The cardiac muscle is much thicker in the left ventricle than the right ventricle; producing greater pressure in the systemic (body) system than the pulmonary system.
System of one-way valves prevents backflow and allows pressure differences between different chambers of the heart.
   ii Six of:
   • Sinoatrial node initiates the cardiac impulse which triggers heart beat.
   • The medulla oblongata in the brain contains cardio-inhibitory and accelerator centres.
   • Activation by parasympathetic (vagus) decreases the rate of heart beat and reduces cardiac output.
   • Acetylcholine inhibits the pacemaker.
   • Cardio-acceleration is by means of sympathetic stimulation.
   • The pacemaker is stimulated by noradrenaline.
   • Adrenaline is produced in times of fear/excitement/stress.
   • Adrenaline causes an increased heart rate and power of contraction, hence increased pressure.
   • Starling's law – if more blood enters the heart, the cardiac muscle fibres contract with greater force.

7 a It is a reversal as the inside becomes more positive, the outside more negative; of the membrane potential/resting potential.
   b i The contraction time is when the ventricles contract/in systole.
      ii Two of:
      • The filling time is when blood fills the atria.
      • It then fills the ventricles.
      • The heart is in diastole/ventricles relaxing/heart relaxing.
      Note: it is important to make both points. 'When blood fills the heart' is only awarded a single mark.
   c i Three of:
      • There is a layer/septum of non-conducting tissue between the atria and ventricles.
      • Depolarisation begins in the sinoatrial node/pacemaker in the atrium.
      • Depolarisation must pass through the atrioventricular node.
      • The atrioventricular node delays the wave of depolarisation.
      • A wave of excitation/action potential passes down the Purkinje tissue/bundle of His.
      ii Two of:
      • The atria contract before the ventricles;
      • so that blood in the atria passes into the ventricles before ventricular systole/contraction.
      • This maintains a one-way flow through the heart.
   d i Time from T to T/one heart beat = 0.8 seconds.

$$\frac{60}{0.8} = 75 \text{ beats per minute.}$$

Note that two marks would be awarded for working out the beats per minute.
      ii The filling time decreases/reduces/gets shorter.

28 Transport systems in flowering plants 1

Water uptake by roots

Roots form a complex, flexible, branched structure which is well adapted to anchoring the plants in the soil and resisting the strong pulling forces of the wind.

Immediately behind the growing tips of the younger roots are numerous root hairs which are outgrowths of the piliferous layer (a single cell epidermal layer). The root hairs *greatly increase the surface area* of the root through which water and mineral ions are absorbed. Much of the rest of the root consists of vascular tissue (**phloem** and **xylem**).

Surrounding the vascular tissue is a single layer of cells called the endodermis. Each cell has a band of fatty material running around it, called a Casparian strip, which is believed to regulate the movement of water and mineral ions across the root.

The arrangement of tissues in both roots and shoots differs between dicotyledonous and monocotyledonous angiosperms. All of the drawings in **28** and **29** (Transport systems in flowering plants) refer to dicotyledons, which are those flowering plants with two cotyledons (seed leaves) in their seeds.

In Figure 28.1 the cortex consists of parenchyma (thin-walled packing cells) which stores starch. The pericycle also consists of parenchyma cells which may produce lateral roots.

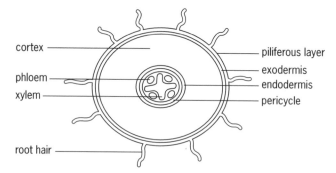

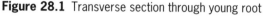

Figure 28.1 Transverse section through young root

How does water move across the root?

Water is continually moved up the xylem and this lowers the water potential of those cells next to the xylem. Water then passes to these cells from neighbouring cells and so on. In this way a water potential gradient is maintained from the root hairs to the pericycle cells (see **4** The transport of substances across cells and through membranes).

However, the passage of water from the soil into the xylem is more complicated than this simple model suggests and water passes by one of three different routes as illustrated in Figure 28.2.

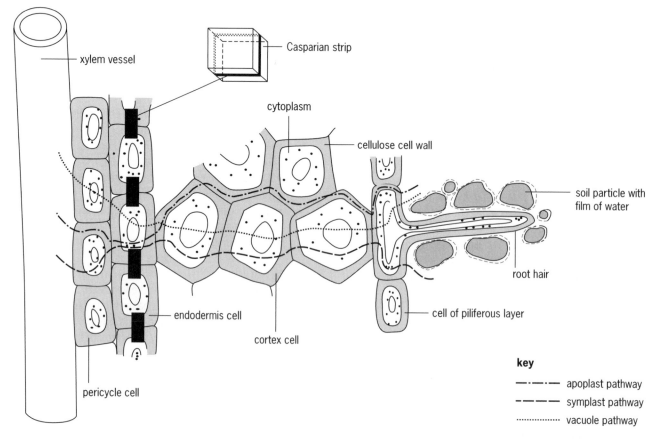

Figure 28.2 Diagrammatic representation of water movement across a root

Apoplast pathway

Cellulose cell walls have a loose structure that can absorb water by capillarity, and as much as 50% of the volume of cell walls can be filled with water.

Mineral ions that are dissolved in the soil water move in by active uptake against the concentration gradient and are transported with the water that enters the cellulose cell walls. Water has strong cohesive forces between its molecules and as water is drawn up the xylem vessels, water in the cell walls is pulled across the root through the cellulose cell walls until it reaches the endodermis. Here, the Casparian strips, which are made of a fatty material called suberin, stop any further movement through the cell walls. At this point the water and dissolved mineral ions must pass into the cytoplasm of the cells, often against a diffusion gradient and therefore active transport must be involved.

Despite this barrier of Casparian strips, the water flows in a continuous stream across the root and finally into the xylem.

It is thought that the Casparian strips are a way by which the amount of water and mineral ions transported into the xylem can be controlled.

Symplast pathway

Cells have connections of cytoplasm called **plasmodesmata** which pass through the cell walls of adjacent cells (see Figure 28.3). Water moves along a water potential gradient from the soil, through the cytoplasm and plasmalemma (**symplast**) of the cells from the root hairs to the xylem.

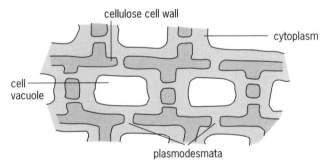

Figure 28.3 Cytoplasmic threads (plasmodesmata) connecting cell cytoplasm of adjacent cells

Vacuole pathway

The water potential of the soil water is higher than the cell sap of the cells in the piliferous layer. This means that the soil water is a more dilute solution. It will therefore move across the cell membranes of the piliferous layer (and in particular the cell membranes of the root hair cells with their large surface area) into the cell sap of these cells by osmosis. The water potential of the root hair cells is thus made higher. The water then moves from vacuole to vacuole, across the root, down a water potential gradient until it reaches the cells of the pericycle. It is not known exactly how the water moves from the cell vacuoles of the pericycle into the xylem but it must be either by the transpiration pull or by active means.

The shoot

The structure of the cells in roots and shoots is basically the same but the arrangement of tissues is different.

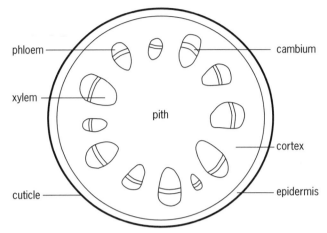

Figure 28.4 TS of young dicotyledonous stem

The cell structure is adapted according to the particular function.

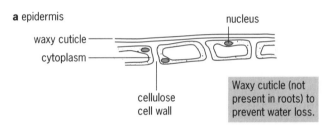

a epidermis

Waxy cuticle (not present in roots) to prevent water loss.

b cambium

Meristematic, unspecialised cells which divide further to form phloem and xylem. No cell vacuoles.

c cortex

Act as packing cells but have walls which are thickened in the corners – **collenchyma**.

d pith (**parenchyma**)

Act as packing cells but have thin walls with intercellular spaces.

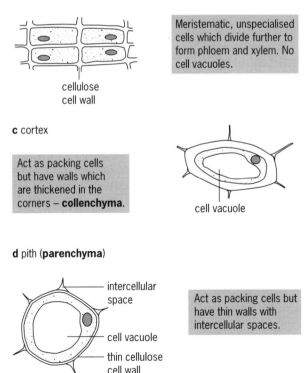

Figure 28.5 Plant cells found in stems

1 Name the following types of plant tissue.
 a The tissue which transports water and mineral salts.
 b Supporting tissue in which the corners of the cells are thickened with extra cellulose.
 c Meristematic cells, full of cytoplasm.
 d The central tissue in young stems. (4)

2 Name the following structures.
 a Cytoplasmic threads passing through pores in the cell walls.
 b Absorptive outgrowths of the piliferous layer of the root.
 c Waterproof area of suberin on the radial walls of the endodermis of the root.
 d A waxy layer found on the epidermis of leaves and stems. (4)

3 The diagrams below show transverse sections of a young stem and root.

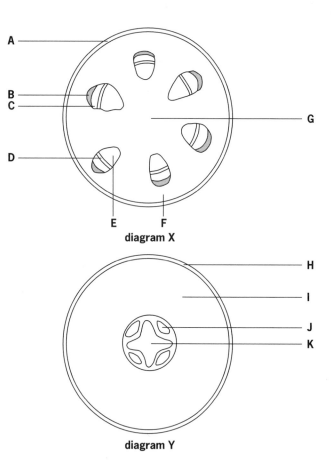

diagram X

diagram Y

 a Which diagram represents the distribution of tissues in the root? (1)
 b Give the two letters which represent each of the following tissues on the diagrams:
  i epidermis
  ii xylem
  iii cortex (6)
 c Give the letter(s) which represent the tissue responsible for:
  i Cell division. (1)
  ii Transport of organic substances. (2)

4 Describe the absorption of water and passage of water across the root, with particular emphasis on the following:
 a The role of root hairs. (2)
 b The apoplast pathway. (3)
 c The symplast pathway. (3)
 d The vacuolar pathway. (3)
 e The role of the Casparian strip. (3)

5 In an experiment to measure the water potential of the water in the soil, it was found to be −40 kPa.
 a i What water potential value would you expect for pure water? (1)
  ii Explain why the value for soil water is different to that for pure water. (1)
 b i Explain how you would expect the water potential to differ in the soil solution of land which has been flooded by sea water. (1)
  ii Suggest why crop plants may die if planted in soil which has been flooded by sea water. (2)

6 The diagrams below show two supporting tissues present in flowering plants.

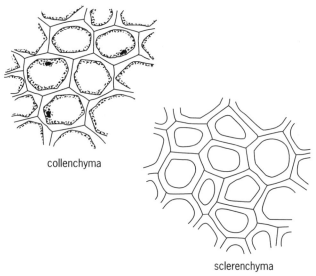

collenchyma

sclerenchyma

 a Give two structural features shown in the diagram which are characteristic of collenchyma. (2)
 b i Give two ways in which sclerenchyma differs from collenchyma. (2)
  ii Collenchyma is often present in the petiole and midrib of leaves. Suggest two reasons why collenchyma is more suitable than sclerenchyma for support in these locations. (2)
 c Both tissues are shown in transverse section. Make a drawing of a sclerenchyma cell as it would appear in longitudinal section. (2)
 London, Paper B6, January 1997

7 a Describe how mineral ions are:
  i Absorbed by the root. (2)
  ii Transported across the root to the xylem. (6)
 b Describe a simple experiment to demonstrate that dissolved minerals are transported in the xylem up the stem. (3)

1 a xylem
b collenchyma
c cambium
d pith (parenchyma)

2 a plasmadesmata
b root hairs
c Casparian strip
d cuticle

3 a Y
b i **A** and **H**
ii **E** and **K**
iii **F** and **I**
c i **D**
ii **C** and **J**

4 a Two of:
- The root hairs increase the surface area of the root.
- Enables greater absorption of water from the soil.
- The cells are thin walled/no cuticle present.

b Three of:
- The apoplast pathway consists of cellulose cell walls and air spaces.
- As water is drawn up the xylem it creates a tension pulling water along this pathway.
- Water molecules are held together by cohesive forces due to hydrogen bonding.
- Capillarity/forces of adhesion occur between the water molecules and the fine spaces between the cellulose fibres.

c Three of:
- The symplast pathway consists of the cytoplasm plus plasmadesmata.
- This movement is by osmosis (diffusion of water across a selectively permeable membrane);
- down a water potential gradient.
- As water is drawn into the xylem the water potential of adjacent cells becomes more negative.

d In the vacuolar pathway water travels from cell to cell via the vacuoles;
down a water potential gradient;
by osmosis across selectively permeable cell membranes.

e Three of:
- The Casparian strip consists of a band of suberin on the radial walls of the endodermis.
- This forces water to pass from the apoplast route into the cytoplasm (symplast) of the endodermis
- allowing development of root pressure
- by *actively* controlling the transport of water into the xylem.

5 a i 0
ii The presence of mineral ions makes the water potential more negative.
b i The water potential in soil solution flooded by sea water would be more negative than normal soil solution.
ii The water potential of the root hair cells may be less negative than the water potential of the soil solution. This would prevent uptake of water by the root hair cells/water might be lost to the soil solution from the root hair cells.

6 a Unevenly thickened walls with the thickening in the corners. Cell with living contents/presence of nucleus/cytoplasm/vacuole.
b i Sclerenchyma has no living cell contents such as nucleus etc.
In sclerenchyma the walls are lignified/with more even thickenings.
ii Collenchyma is more flexible/can stretch as the organs grow (so will bend as the leaves move). Collenchyma develops before sclerenchyma; it can become turgid.

c

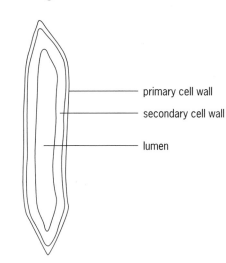

primary cell wall
secondary cell wall
lumen

(*The labels are included for information only. No marks are awarded for labels.*) 1 mark for showing the whole cell with an elongated shape/at least five times as long as wide/with tapered ends.
1 mark for showing the correct wall thickness/correct proportions.

7 a i Mineral salts are present as ions dissolved in the soil water.
They enter the root hair cells of the piliferous region by passive diffusion;
or by active uptake against a concentration gradient.
ii Six of:
- Transported to the xylem along the apoplast route.
- Carried across the root cells passively with the water in the transpiration stream/mass flow.
- Some passive diffusion along apoplast route.
- Some ions are actively pumped across cell membranes of the outer root cells to enter the symplast route.
- Cytoplasmic streaming carries ions through plasmadesmata on the symplast route.
- Active transport of ions from apoplast to symplast occurs at endodermis/Casparian strip.
- Ions are actively pumped from the endodermis to the xylem (via pericycle).

b The cut end of a stem is placed in a solution of a coloured dye such as eosin.
After 1 hour a thin transverse section of the stem is prepared and examined under a microscope.
The stained cells are found to correspond to the known position of the xylem tissue/are seen to consist of xylem vessels.

29 Transport systems in flowering plants 2

Transpiration

The rate of *water uptake* can be measured by using a potometer. If a potometer is placed on a balance, the rate of water uptake can be equated with that of *water loss* if the cells are turgid.

Water evaporates from the intercellular spaces of the leaf via the stomata (see **13** Photosynthesis) into the atmosphere due to the **latent heat of vaporisation**. This is known as transpiration. The energy is supplied by the sun's heat.

Water passes from the ends of the small xylem vessels of the leaf into the leaf spaces of the spongy mesophyll where it evaporates. It then diffuses down a diffusion gradient from the leaf spaces, through open stomata in the lower surface of the leaf, into the external air. Underneath the surface of each leaf there will be a layer of humid air even on a dry or windy day. The thickness of this layer of humid air will depend on the amount of air movement (wind).

Factors affecting the rate of transpiration

- *Temperature*. An increase in temperature increases the amount of water vapour that the air can hold and therefore increases the rate of evaporation from the intercellular leaf spaces and the stomata.
- *Light*. In bright light all the stomata will be open and the rate of transpiration will increase.
- *A shortage of soil water* will cause the stomata to close and the rate of transpiration will fall.
- *Air movement* (wind) causes the rate of transpiration to increase.
- If the *humidity* of the air is high it will lessen the water potential gradient between the leaf and the air, and the rate of transpiration will be reduced. Dry air has the opposite effect.

The opening of stomata

It is not really known why stomata open in the light. The best current theory suggests that light causes potassium ions to be actively transported into the guard cells, lowering the water potential. This, in turn, causes water to move by osmosis into the guard cells from the surrounding cells. The turgid shape of the guard cells causes the stomata to open. It is known that active transport is involved in the movement of the potassium ions because the introduction of a respiratory poison not only stops the opening of the stomata but also the accumulation of potassium ions.

In the dark there is a reversal of the above process.

It used to be thought that because photosynthesis causes an increase in the sugar content of cell sap, water entered by osmosis, making the guard cells turgid. However, some guard cells do not contain chloroplasts but the stomata still open in bright light. The potassium ion theory is the best one to date.

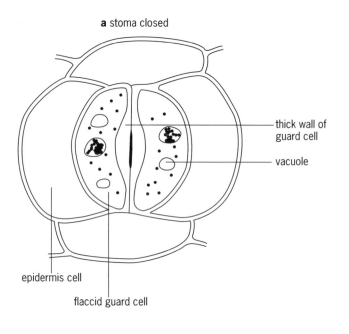

a stoma closed

thick wall of guard cell

vacuole

epidermis cell

flaccid guard cell

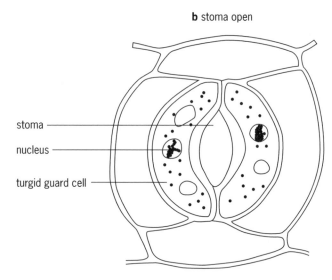

b stoma open

stoma

nucleus

turgid guard cell

Figure 29.1 Open and closed stomata

The guard cells are sausage shaped. The stomatal pore opens when the guard cells are turgid because the walls are unevenly thickened and this causes unequal bending of the cell wall.

Most of the stomata are to be found on the underside of the leaf. There may be some, or no stomata, on the upper surfaces of leaves. Leaves of species that live in shaded conditions are likely to have more stomata on their upper surfaces than those that grow in sunny conditions.

Leaves of aquatic plants have no stomata or waxy cuticles. Oxygen and carbon dioxide in solution diffuse directly through the surfaces of leaves. Aquatic plants do not normally need to conserve water.

The transpiration stream

This refers to the constant stream of water in a living plant due to the loss of water by transpiration.

When molecules of water leave the xylem vessels and either enter a cell by osmosis, or evaporate into the spongy mesophyll of the leaf, a tension is created in the column of water for its whole length down to the roots. Because of the molecular structure of water (see **4** The transport of substances across cells and through membranes) and its dipolar nature, the molecules are held together by hydrogen bonds. This *cohesive force* gives water its high **tensile strength** and the column of water is not likely to break.

When experiments on transpiration are set up it is vital that the leafy stems that are used are cut *under water*. This ensures that the continuous column of water is not broken, causing the 'cut' ends of the column to move rapidly apart because the tension has been removed. If the column is broken, transpiration will effectively cease.

The molecules of water also cling to the walls of the very narrow xylem vessels and the tiny pores of the apoplast pathway. This *adhesive force*, **capillarity**, is sufficient to support the entire column.

Root pressure is due to the pressure caused by water entering the root cells by osmosis. It plays some part in raising water up tall plants like trees but for the most part water is drawn up due to the evaporation of water from the leaves.

Xylem vessels

Xylem consists of living parenchyma cells plus dead, empty, water-conducting tracheids and vessels, and fibres which provide added support.

Xylem vessels are formed from a number of empty tubes with strong lignified walls which are fused longitudinally. There are different patterns of wall thickening. Young xylem vessels (protoxylem) are usually thickened by rings (Figure 29.2) and older xylem vessels (metaxylem) by various patterns as seen in Figure 29.3. The walls have pits which allow a continuity between the cells.

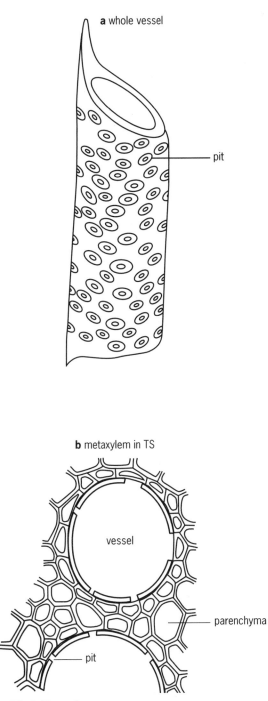

Figure 29.3 Metaxylem

Tracheids are long, slender and tapered. The walls are also heavily lignified and contain pits.

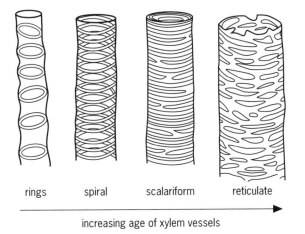

Figure 29.2 Types of wall thickening in xylem vessels

Phloem

Phloem translocates dissolved organic substances such as sucrose from leaves to other cells, where the sucrose may be converted back to glucose for cellular respiration or condensed to starch and stored.

Phloem consists of **sieve tube elements** and **companion cells**. The sieve tube elements fuse together to form long **sieve tubes**. The cross walls between the elements partially break down and form **sieve plates** which are perforated platforms between the elements. The sieve tube elements have walls thickened with cellulose and the cells contain cytoplasm but during their formation the nuclei degenerate and the cytoplasm is dependent for its activities on the nucleus of an adjacent companion cell.

The actual mechanism of transport in the phloem is not fully understood. Munch's **mass-flow hypothesis** (1930) states that leaves have a lower water potential because of the sucrose accumulated, and water enters these cells from the xylem vessels thereby increasing their pressure potential. Cells in the roots have less sugar content and therefore a higher water potential. There is a gradient of pressure potential between the leaves, which are the **source** of sucrose, and the roots which use the sucrose. The phloem links the two and water flows along the sieve tube elements.

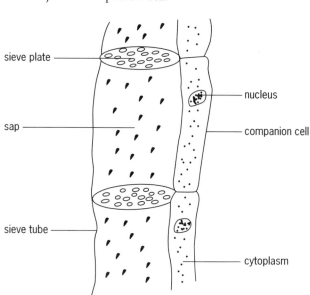

Figure 29.4 Phloem in LS

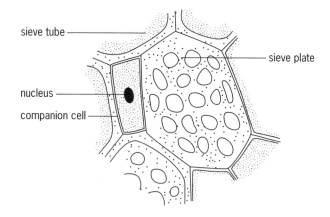

Figure 29.5 Phloem in TS

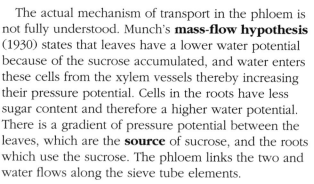
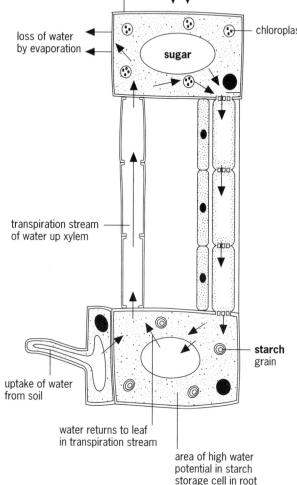

Figure 29.6 Illustration of Munch's mass-flow hypothesis.
(Adapted from: Clegg C. J. and Cox G., Anatomy and Activities of Plants. John Murray 1978.)

1 Complete the table below to show whether the statements about xylem and phloem tissue in plants are true (✓) or false (✗).

Statement	Xylem tissue	Phloem tissue
some of the cells have lignified walls		
can transport substances both up and down the stem		
contains some thin walled living cells		
transports sucrose		
the cross walls are perforated in some of the cells to form sieve cells		

(5)

2 a The diagram shows two xylem vessels thickened with lignin (longitudinal section).

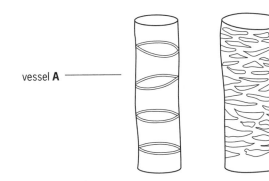

vessel **A**

i Explain why vessels of type **A** are found in the earlier stages of growth. (2)

ii How does lignin enable xylem vessels to function effectively? (2)

b The diagram below shows some cells from plant vascular tissue as seen in an electronmicrograph.

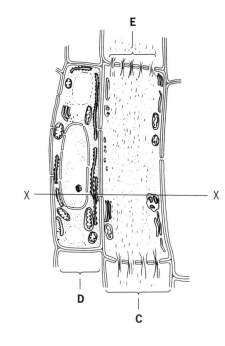

i Name the type of tissue shown above. (1)
ii Name the cells labelled **C** and **D**. (2)
iii What is the name and function of the structure labelled **E**? (2)

c If a cut was made along the line labelled X–X would the view of cells **C** and **D** be of a longitudinal or transverse section? (1)

d It has been observed that during active transpiration the water column inside the xylem vessels is under tension. Explain this observation. (3)

3 The apparatus shown below can be used to compare water losses by transpiration.

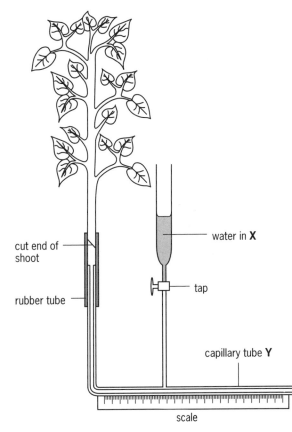

cut end of shoot

water in **X**

rubber tube

tap

capillary tube **Y**

scale

a What is this apparatus called? (1)
b What is the function of **X**? (1)
c Explain why a capillary tube is used at **Y**. (1)
d Explain why the plant shoot should be cut under water, and the apparatus set up under water. (1)
e This apparatus measures water taken up by the plant which is almost the same as the amount of water lost by transpiration. Name two other causes of uptake of water by the leafy shoot. (2)
f Name three environmental factors which should be kept constant when comparing transpiration from two shoots. (3)

4 The diagram represents the mass–flow hypothesis which explains the movement of substances in the phloem.

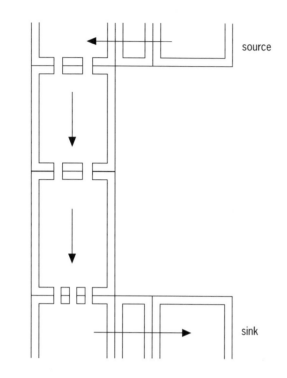

a Suggest a plant organ which is a sink. (1)

b There are companion cells alongside the phloem sieve tubes in a stem. Explain why this does not support the mass flow hypothesis. (2)

c Radioactively labelled molecules applied to illuminated leaves were rapidly transported out of the leaves by the phloem. Radioactively labelled molecules applied to shaded leaves were not transported. Explain how the mass flow hypothesis accounts for these observations. (2)

AEB, Paper 3, June 1997

5 a i Describe how water is lost from the leaves of a plant by transpiration. (3)

ii Explain why the loss of water through the stomata is unavoidable. (2)

There are a number of ways in which the size of stomata can be measured or compared. In the direct observation method, pieces of leaf are placed under a microscope and are often photographed.

The dimensions of the pores can be measured using an eyepiece graticule and a stage micrometer. Using this method a large number of stomata have to be measured.

b i Give one reason why a large number of stomata have to be measured. (1)

ii Suggest two reasons why the leaf surface is often photographed. (2)

The table below shows the approximate size and number of stomata in the leaves of three plants, obtained by using the direct observation method.

Species	Mean number of stomata per mm² on lower epidermis	Mean size of fully open stomata (µm)		Mean area available for water loss through fully open stomata (µm² per mm²)
		Length	Width	
Phaseolus vulgaris (bean)	281	7	3	5901
Hedera helix (ivy)	158	11	4	6952
Triticum sp. (wheat)	14	18	7	

c Using the data in the above table:

i Calculate the mean area per mm² available for water loss through fully open stomata in wheat. Show your working. (2)

ii State why the mean you have calculated in **i** will be inaccurate. (1)

d Suggest two reasons why plants with the same stomatal area might not transpire at the same rate. (2)

Cambridge, Module 4804, June 1997

6 Explain how water travels up the xylem. (6)

7 a Construct a labelled diagram of the surface view of a stoma. (2)

b i Describe the functions of stomata. (2)

ii How is the structure of the guard cell adapted to its functions? (2)

c Discuss the theories to explain the way the stomata open and close. (6)

1

Statement	Xylem tissue	Phloem tissue
some of the cells have lignified walls	✓	✗
can transport substances both up and down the stem	✗	✓
contains some thin walled living cells	✓	✓
transports sucrose	✗	✓
the cross walls are perforated in some of the cells to form sieve cells	✗	✓

2 a i The annular/ring-like thickenings provide flexibility in the young, growing plant.
The rings help to keep the tubes open for transport.
ii Lignin waterproofs the xylem vessels allowing the transport of water.
Lignin strengthens the xylem vessels, providing support.
b i phloem
ii C = sieve tube; **D** = companion cell
iii E = sieve plate
Its function is to contribute to higher pressure by providing some resistance to the movement of contents/to allow the passage of organic materials.
c transverse section
d There is an upward force caused by the evaporation/transpiration of water.
The action of gravity exerts a downward force.
The thin columns of water molecules are bound together by cohesive forces/hydrogen bonds.

3 a potometer
b The reservoir can be used to return the bubble to zero on the scale.
c To improve the accuracy of the readings.
d To ensure that air bubbles do not enter the xylem, so breaking the continuity of the column of water.
e Water is used as a raw material of photosynthesis.
Water may be needed to restore cells to full turgor.
f temperature; humidity; light intensity

4 a One of:
- root
- bud
- apex
- young leaf
- storage organ
- fruit
- shoot
- flower.
b Companion cells produce ATP/energy.
Mass flow is dependent on physical processes/passive diffusion.
c Illuminated leaves make sugar which is loaded into the phloem.
Water follows by osmosis.

5 a i Water travels from cell to cell down a water potential gradient.
Water evaporates from the surface of the mesophyll cells.
Water leaves through the stomata by diffusion.
ii Two of:
- The stomata are open for gaseous exchange/carbon dioxide uptake.
- The atmosphere in the air spaces of the leaf is saturated with water vapour.
- The air outside the leaf is drier.
b i One of:
- To obtain a reliable average since pore size will vary.
- To obtain more accurate results/fair test.
ii Two of:
- Counting takes a long time/measurements can be made at leisure.
- Pore size may change before measurements are complete.
- To avoid the heating/light effects of the microscope lamp.
- To obtain a permanent record.
- A photograph can be enlarged/easier to measure.
c i *(Note: it is important to show your working since the final answer only gains 1 mark.)*

$14 \times 18 \times 7 = 1764$

ii Stomatal shape is not rectangular/not regular.
d Two of:
- Not all the stomata might be open.
- Plants could be under different conditions.
- Leaves in some plants have stomata on the upper surface/hairy leaves.
- Sunken stomata/thickness of cuticle might vary.
- There may be smaller pores in a larger number of stomata/stomatal distribution near edge of leaf may differ.

6 The main cause is the transpiration stream pulling water from above.
Capillarity, adhesive forces between the sides of the xylem vessel and water molecules (increases in inverse proportion to diameter of the vessel).
However capillarity alone is not sufficient to explain the rise of water to the top of tall trees.
Strong cohesive forces between (polar) water molecules mean that the thin column of water under tension in the xylem acts like a rope being pulled from above.
Root pressure due to active pumping of ions into xylem; uses ATP from respiration to make the water potential of xylem more negative.

7 a (Labels ½ mark each)

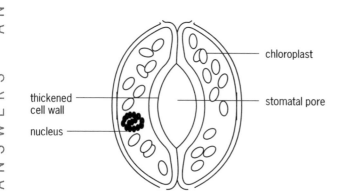

chloroplast

thickened cell wall

stomatal pore

nucleus

b i Two of:
- The function is to open to allow exchange of gases for photosynthesis (in light).
- Carbon dioxide enters, oxygen leaves, by diffusion.
- To close in the dark, or when water loss exceeds water gain.

ii The walls of the guard cell are thicker near the pore. When the guard cells are turgid this results in more stretching of the thinner wall, so guard cells bend away from each other and the aperture opens.

The guard cells are the only epidermal cells containing chloroplasts. This allows formation of ATP in the light to produce a lower water potential.

c When the water potential of the guard cells becomes more negative, water will enter by osmosis.

Theory 1 – It was thought that chloroplasts in guard cells photosynthesised in the light, producing sugar which made the water potential more negative. However guard cells do not have Calvin cycle enzymes.

Theory 2 – Starch may be hydrolysed to sugar when conditions become more alkaline (due to removal of acidic carbon dioxide during photosynthesis).

Theory 3 – ATP from photosynthesis (photosystem 1) may be used to actively transport potassium ions into guard cells in the light (and out in the dark).

This would make the water potential more negative but probably involves exchange of other ions to maintain electroneutrality.

30 Diet and the ingestion of food

Autotrophic nutrition

Auto means 'self' and trophic means 'feeding'.

Autotrophic feeders (*autotrophs*) use an inorganic form of carbon, such as carbon dioxide, to synthesise (build up) complex organic compounds such as carbohydrates. This process requires energy:

- green plants use light energy in *photosynthesis*
- bacteria use chemical energy in *chemosynthesis*.

Heterotrophic nutrition

Heterotrophs cannot use light or chemical energy to synthesise their own foods. They must, therefore, feed on materials previously synthesised by other organisms.

Parasitic nutrition

Parasites include members of all the five main kingdoms of organisms (Prokaryotae, Protoctista, Fungi, Plantae and Animalia).

> **A parasite is a heterotrophic organism which is dependent on a host organism from which it obtains its organic nutrition.**

The organic food, which may consist of the host's body tissues and fluids or the contents of the host's gut, is usually in solution and easily absorbed into the parasite's body. Examples of parasites include the fungus which causes 'athletes foot', threadworms which live in the human gut and *Plasmodium* the malarial parasite.

The pork tapeworm – an example of an endoparasite

Taenia is one genus of a group of Platyhelminthes belonging to the class Cestoda (the tapeworms), all of which are endoparasites (i.e. living within the host). They have a *large surface area/volume ratio* and absorb digested food directly from the host's gut. They possess *suckers and hooks* with which they attach themselves to the host, a *tough outer coat which resists the digestive enzymes of the host* and *no gut* of their own. They may be longer than 3 metres and they *produce vast numbers of eggs*. Each segment has both male and female reproductive organs which degenerate after fertilisation leaving a uterus packed with eggs. Ripe proglottids (segments) detach from the rest of the tapeworm diet leave the body in the faeces. The eggs then need to be ingested by a secondary host (such as a pig) before they can find their way back into the human gut when insufficiently-cooked and infected meat is eaten.

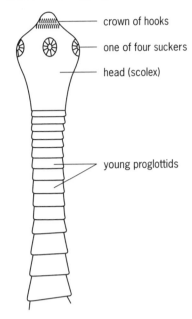

a head and young proglottids

- crown of hooks
- one of four suckers
- head (scolex)
- young proglottids

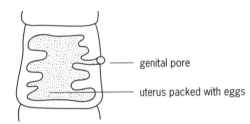

b mature proglottid

- genital pore
- uterus packed with eggs

Figure 30.1 *Taenia solium* – the pork tapeworm

The tapeworms are some of the best-adapted parasites inasmuch as they inflict minimum harm to their hosts which continue to live and reproduce normally, even though they may be very debilitated. The death of the host usually means the death of the parasite!

Saprobiontic nutrition

This type of nutrition occurs mainly among the bacteria, protoctists and fungi. Digestion is either extra-cellular or the food may be already in a soluble form.

> **Saprobionts are heterotrophic organisms which obtain their organic material from dead or dying organisms or organic wastes.**

The great importance of saprobionts is that they are *decomposers* and assist in the breakdown of dead or dying biological material which is then returned to the soil and atmosphere. In this way they play an important part in the nitrogen cycle. *Rhizopus*, the common bread mould, is a good example of a saprobiontic fungus. It secretes **carbohydrases**, **lipase** and **proteases** and so can feed on a wide variety of substrates (Figures 30.2 and 30.3).

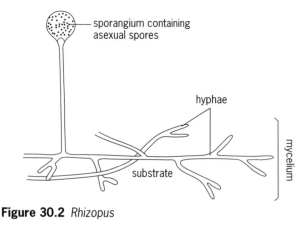

Figure 30.2 *Rhizopus*

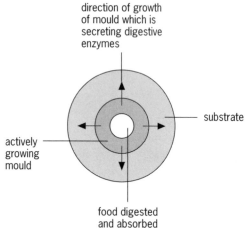

Figure 30.3 Growth pattern of a mould

Holozoic nutrition

> **Holozoic nutrition is feeding off solid organic matter derived from the bodies of other organisms.**

Holozoic nutrition is dealt with in **31** Digestion, absorption of food and the role of the liver.

A balanced diet for humans

The seven necessary components of a balanced diet are carbohydrates, fats, proteins, water, fibre, minerals and vitamins. Students should already be aware of which foods are good sources of each of the above.

Nutritional requirements

The required daily intake of food varies with *age, sex* and *occupation*. Pregnant women, and those breast-feeding babies, need extra nourishment and a healthy diet that will provide all the nutrients required by the developing fetus or baby.

The amount of food needed is usually expressed in terms of joules. Apart from water, plus some bulk, almost all of our food is carbohydrate, fat or protein. Each of these can be respired to release energy and are therefore measured in terms of joules available per gram of intake. Carbohydrates and proteins, in excess of the body's requirements, are converted to fat and

stored under the skin and around internal body organs. This process will eventually cause **obesity** which results in a shorter life span.

Nutrient	Required for
carbohydrates	*energy* – 1 mole of glucose releases 2880 kJ
fats	*energy* – a concentrated source – 1 gram of fat releases a little more than twice as much energy as carbohydrates
proteins	*body building* – of the 20 amino acids needed, 11 can be formed by transamination but the rest (essential amino acids) cannot and must be included in the diet. Proteins in excess of the body's requirements can provide energy
water	essential for *body fluids* and *life processes* – about 70% of body weight
fibre	fibre (roughage) cannot be digested but is needed to provide *bulk* for peristalsis
mineral ions	*growth* and *body maintenance* – the only inorganic components of a healthy diet, apart from water. Minerals required include calcium, magnesium, phosphorus, potassium, sodium and iron. In addition, trace elements (e.g. iodine and copper) are needed in very small amounts
vitamins	essential for *good health*

The amount of energy available from a given mass of food is measured in units of 1000 joules which equals 1 kJ (kilojoule). The following table gives an idea of the number of kJ needed per day by moderately active children, men and women. The governments of most western countries have produced their own Recommended Dietary Allowance tables which differ slightly from each other in the actual figures quoted. For example, those living in hot climates will need fewer joules to maintain body temperature. At best, these figures are an approximation but, nevertheless, a useful guide.

Age (years)	Sex	Body weight (kg)	Protein required (grams)	kJ required
11–14	female	45	46	9 500
11–14	male	45	45	10 500
19–50	female	60	50	9 250
19–50	male	79	62	12 200
Over 75	female	65	50	8 000
Over 75	male	75	62	9 600

Figures for protein have been quoted separately because although the body can convert carbohydrates into fats, and vice versa, proteins for the body's growth and maintenance must be ingested as amino acids. (*Remember* – proteins contain the element nitrogen but carbohydrates and fats do not – see **2** Biological molecules.)

Too much fat in the diet can be a health hazard. Unsaturated fats (of plant origin and liquid at room temperature) are better in the diet than saturated fats (of animal and tropical plant origin – solid at cool room temperature). See **2** (Biological molecules) for the chemical structure of fats. Examples of the two types of fats/oils are:

- Saturated fats: fatty meat, butter and cream, coconut and palm oils.
- Unsaturated fats: corn oil, soya bean, sunflower.

Obesity is generally a condition in which the body weight is 20% or more over the optimum. Apart from being aesthetically undesirable, people suffering from obesity are more likely to die prematurely of diseases of the kidneys, heart and arteries as well as diabetes.

Excess cholesterol in the blood is excreted by the liver into the bile which passes into the gall bladder. If the amount of cholesterol in the blood is excessive, gall stones may form in the gall bladder. These can cause much pain as the stones may block the bile duct. Conversely, if the amount of cholesterol is insufficient for the body's needs it can be synthesised by the liver. Levels of cholesterol in the blood plasma above 4.9 millimoles per litre (equivalent to 190 mg of cholesterol per 100 ml) are progressively more likely to cause a condition known as **atherosclerosis**.

In atherosclerosis, lipoproteins (which carry cholesterol which is otherwise insoluble) are deposited over a period of years onto the inner lining of arteries thus restricting the flow of blood. *Calcification* and *scar tissue* may make the vessels less elastic and this condition is known as **arteriosclerosis** or hardening of the arteries.

An inadequate supply of oxygen to the heart due to fatty deposits on the walls of the coronary arteries causes a condition known as **coronary heart disease**.

There are two major conditions caused by atherosclerosis:

- **Angina** – in which the coronary arteries of the heart become blocked, causing pain and the death of part of the heart muscle. Treatment includes:
  - a **coronary by-pass** (fresh veins are taken from another part of the body, usually a leg, and used to provide a supply of blood to the affected part of the heart)
  - **angioplasty** in which a tube (catheter) with a balloon end is inserted into the affected arteries to dilate them.
- **Strokes** – in which cerebral blood vessels become restricted (cerebral arteriosclerosis), often resulting in a degree of paralysis or even death.

Essential fatty acids

There is a small group of fatty acids which are essential in the diet. Linoleic acid is the most important of these because the body can convert it into any of the other essential fatty acids. It also tends to lower the amount of cholesterol in the plasma.

Some consequences of malnutrition

Kwashiorkor (protein malnutrition)

This is a condition found particularly in developing tropical and subtropical countries when young children are weaned on starchy foods with insufficient protein in the diet. Symptoms include dry skin and skin rashes, swollen abdomen and general weakness. Kwashiorkor may also cause stunted mental development.

Anorexia nervosa

This results from an emotional aversion to food. As is the case with bulimia, the disease occurs mostly in young women. Unlike famine victims, those with anorexia are often able to maintain their normal strength and activities and do not feel hungry. However, a small percentage of cases result in death. The symptoms are a reduction to as little as half normal body weight and a failure to menstruate. Vomiting may be self-induced.

Bulimia

Individuals show a great concern for their body shape and weight but are, in fact, close to their normal expected weight. It is a psychological disorder in which there is a pattern of overeating followed by induced vomiting and purging. This cycle of events can cause stomach rupture or dehydration and may be fatal.

How can one ensure a balanced diet?

We have considered the consequences of malnutrition as well as the number of kilojoules needed per day (from fats and carbohydrates) and the daily protein requirement. In addition we need sufficient minerals, vitamins and roughage. But how do we know what to include in our diet, in what proportions and how much?

The last point is the easiest to answer. You should eat regular meals and not eat more than leaves you feeling that you could just enjoy a little more – not always easy! If we overeat, the stomach stretches so that in the future we eat more before our stomach feels full. This can be progressive.

What to eat, and in what proportions, is best illustrated by the 'Food Guide Pyramid' which was developed by the US Department of Agriculture (see Figure 30.4).

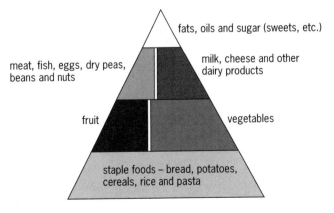

Figure 30.4 The Food Guide Pyramid

DIET AND THE INGESTION OF FOOD

1 Read through the following passage about the different types of nutrition and then write a list of the most appropriate word(s) to fill the gaps.

Organisms which are able to manufacture complex organic compounds from simple inorganic molecules are described as having _____ nutrition. This group includes two types of such organisms, those which are able to photosynthesise and the others which _____.

The nutrition of all other types of organism is collectively known as _____. This includes organisms such as humans who obtain their food by _____ nutrition. The dead bodies of organisms are recycled by a group of organisms known as _____. This group digests the food externally by the secretion of extra-cellular _____ onto the substrate and then absorbs the soluble products. (6)

2 The terms autotrophic, saprophytic and parasitic are terms used to describe the ways in which different organisms obtain their nutrients.

The table shows a list of organisms. Copy out and complete the table by ticking the box or boxes which apply to each organism.

It may be necessary to tick more than one box to fully describe the way in which the nutrients are obtained.

Organism	Autotrophic	Heterotrophic	Saprophytic/ saprobiontic	Parasitic
apple tree				
human immuno-deficiency virus				
dog				
nitrogen-fixing bacteria				
Penicillium sp.				
Phytophthera infestans (potato blight fungus)				

Welsh, Paper B3, January 1997

3 The diagram below shows a part of a beef tapeworm, *Taenia saginata*.

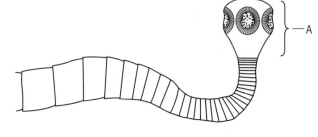

a Explain the importance of the part labelled **A** in the life of the tapeworm (2)
b **i** Describe how the tapeworm obtains its nutrition. (2)
ii How does the nutrition in *Rhizopus* differ from that of the tapeworm. (2)

London, Paper B2, June 1997

4 *Rhizopus*, the bread mould, is a ubiquitous saprobiont, being able to grow on many types of substrate such as the starch and gluten (protein) in bread.

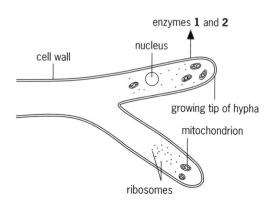

a Suggest names for enzymes **1** and **2**. (2)
b After digestion by enzymes **1** and **2**, what monomers would be absorbed by the hypha? (2)
c Explain why the presence of many ribosomes and mitochondria is helpful in the nutrition of *Rhizopus*. (2)

5 A cow obtains most of its nutritional requirements from mutualistic organisms living in its rumen. The diagram summarises the biochemical processes carried out by these organisms.

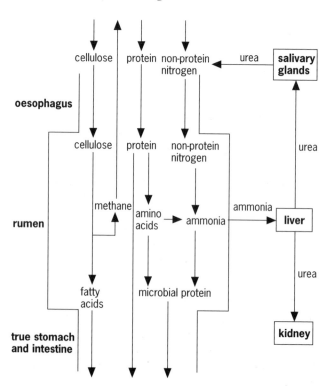

a Use the information in the diagram to help explain why:

i The relation between the cow and the micro-organisms which live in its rumen may be described as mutualistic. (2)

ii It is possible for a cow to live on a diet which is low in protein. (2)

iii Ruminant animals such as cows are less efficient than non-ruminant animals in converting energy in food into energy in their tissues. (2)

b Aphids are small insects which feed on plant sap. The table shows the relationship between the amount of soluble nitrogen in plant sap and the body mass and reproductive rate of one species of aphid.

Soluble nitrogen in plant sap as percentage of dry mass	Mean adult body mass (mg)	Reproductive rate (number of young produced per day per aphid)
2.0	1.8	1.3
2.5	1.6	1.0
3.0	1.3	0.7
3.5	1.0	0.4
4.0	0.7	0.1

Explain how the amount of soluble nitrogen in plant sap affects the body mass and reproductive rate of aphids. (2)

AQA, Specification A Specimen Paper 6, for 2002 exam

6 Eating disorders are becoming increasingly more common in western society. Give **four** health risks or symptoms for **anorexia nervosa** and **four** health risks for **obesity**. (8)

OCR, Human Health and Disease, March 1999

7 The chart shows the risk of developing coronary heart disease during the next 10 years in relation to a number of risk factors.

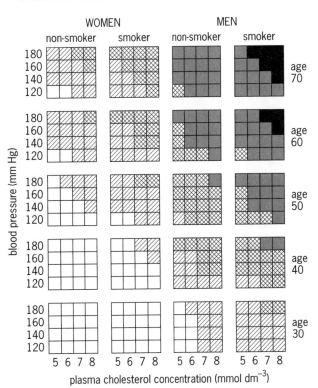

(Adapted from: Odyssey, the Glaxo Wellcome Journal of Innovation in Healthcare, and is reproduced with permission.)

a According to the chart, what are the characteristics of the people who have the highest risk of developing coronary heart disease? (2)

b Explain the relationship between the risk of developing coronary heart disease and:

i plasma cholesterol (2)

ii the sex of the person concerned. (1)

c Explain the limitations of using the chart as a method of prediction of the risk of developing coronary heart disease. (3)

AQA (NEAB), BY08, March 1999

8 Write an essay on 'The causes and effects of overnutrition'. (20)

Edexcel (London), Paper B6, January 1999

1   autotrophic (holophytic); chemosynthesise; heterotrophic; holozoic; decomposers/saprophytes; enzymes

2

Organism	Autotrophic	Heterotrophic	Saprophytic /saprobiontic	Parasitic
apple tree	✓			
human immuno-deficiency virus		✓		✓
dog		✓		
nitrogen fixing bacteria	✓			
Penicillium sp.		✓	✓	
Phytophthera infestans (potato blight fungus)		✓		✓

3   a   **A** has four suckers for secure attachment to intestine wall;
        and to maintain its position in the body of the host.
     b   i Food already in simple soluble digested form in the intestine of the host.
            Simple diffusion of monomers across the cuticle into segments of tapeworm.
            ii *Rhizopus* produces digestive enzymes, none produced in tapeworm.
            Rhizopus is a decomposer/saprophyte/saprobiont/lives on dead organic products, tapeworm is a parasite/lives on a living host.

4   a   Carbohydrase/amylase (to digest the starch).
        Protease (to digest the protein).
     b   monosaccharides/glucose; amino acids
     c   Ribosomes are needed for the production of proteins for enzymes/raw materials of growth.
        Mitochondria are needed to provide energy for the growth of hyphae into new areas of substrate/to provide ATP to drive chemical reactions, etc.

5   a   i Two of:
          • Both organisms have a nutritional advantage.
          • The cow gains fatty acids/proteins.
          • The micro-organism gains cellulose/protein/urea.
          ii Two of:
          • The cows obtain non-protein nitrogen.
          • Micro-organisms convert this into microbial protein;
          • which the cow can then digest.
          iii Some food energy is used by the micro-organism.
          Some energy is lost in methane.
     b   Two of:
          • Growth and reproduction require the production of new tissues;
          • which has a high protein requirement.
          • Nitrogen is an essential part of protein.

6   **Anorexia nervosa**
    Extreme loss of body mass.
    Failure to menstruate/amenorrhoea.
    Neurotic reaction to weight reduction/schizophrenic delusions resulting in abhorrence of food.
    Digestive disturbances with vomiting.
    Obesity – four of:
    • High blood pressure.
    • Strain on joints/arthritis.
    • Increased risks in surgery/childbirth.
    • Increased risk of coronary heart disease.
    • Increased risk of mature onset diabetes mellitus.

7   a   Male smokers.
        Aged 60 with cholesterol above 7/aged 60 with blood pressure above 160/aged 70 with cholesterol above 6/aged 70 with blood pressure above 140.
     b   i Two of:
          • Because of the formation of atheroma/deposition of fatty material in artery walls the heart is weakened leading to aneurysm.
          • Leads to narrowing of arteries.
          • Increases the chance of a clot obstructing the artery.
          ii The presence of oestrogen protects women against coronary heart disease (CHD).
          Risk factors will change over 10 years.
          The smoking in the chart is not quantified.
          Other risk factors are involved – stress/heredity/high salt diet/exercise.

8   *Note: in this type of essay, marks are awarded for its scientific content, the cover of the topic and for its coherence. The essay needs to start with a general introduction and should then discuss most of the areas listed below with suitable examples.*

    • A definition of overnutrition, deviation from a balanced diet, components of the diet which are taken in excess, resulting in overweight, obesity.
    • Discussion of the nature of foods with respect to their calorific content.
    • Definition of obesity, reference to BMI, how measured, how used to assess overweight, calculation.
    • Effects to include reference to links with coronary heart disease, high intake of fats, consequences, possible relationship of high intake of fats, sodium, cholesterol.
    • Risks of developing mature onset diabetes mellitus, strain on the pancreas, consequences of this condition.
    • Other risks associated with overnutrition, in surgery, in strain on joints and arthritis, difficulty in conception, childbirth.
    • Discussion of remedies, reference to the basis of a slimming diet, relationship between restricted energy intake, physical activity and weight loss.

    (Scientific content – 17 marks [balance (good coverage, no irrelevance) – 3 marks]; clarity and coherence (well written, good grammar and spelling – 3 marks; maximum total – 20 marks)

31 Digestion, absorption of food and the role of the liver

Holozoic nutrition is feeding on solid organic material derived from the bodies of other organisms. It is the main method of feeding used by animals.

> **Digestion is the breaking down and hydrolysis, of complex food molecules into small, soluble and diffusible molecules. Enzymes are involved in chemical digestion.**

Digestion

Physical digestion

Large particles of food are broken down into smaller particles. In mammals this is achieved with the aid of teeth plus tongue and hard palate (roof of the mouth).

The mastication of food into smaller particles produces a larger surface area on which enzymes can act, and is therefore a very important prerequisite to chemical digestion.

Chemical digestion

The hydrolysis of food compounds (see **2** Biological molecules) involves the action of digestive enzymes. The particles to be digested need to be small, giving larger surface areas for a given mass of food, so that there is sufficient time for both digestion, and the absorption of the products of digestion. Otherwise large particles would not be completely digested before elimination. In mammals this is achieved in several ways.

1   The food is chewed (masticated) by teeth, lubricated by mucin in the saliva and formed into a bolus (sphere) before being swallowed. Waves of muscular contraction (**peristalsis**) pass the food along the oesophagus to the stomach.
2   Food is retained in the stomach for several hours. It is churned and mixed with enzymes before being released in small 'manageable' amounts. Herbivores have specially adapted stomachs which enable the digestion of cellulose with the help of bacteria which are present. Without these bacteria the cellulose cannot be digested. In carnivorous and omnivorous animals any cellulose in the diet cannot be digested but provides fibre (bulk).
3   The length of the gut in adult humans is about 9 metres. This increases the time that food takes to pass along its length and also the surface area available for absorption. Most of the digestion and all of the absorption takes place in the small intestine which is 5 to 6 metres in length.
4   The surface area of the small intestine is vastly increased by the presence of numerous folds called villi. The villi are themselves covered with a **brush border** of microvilli which number as many as 1700 per cell. Altogether the surface area of the small intestine is about 200 m^2.

Figure 31.1 shows the structure of the gut wall as seen in the oesophagus, which is not highly specialised.

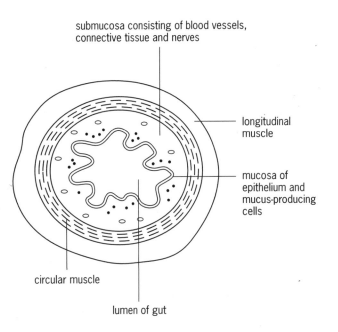

Figure 31.1 TS of oesophagus of mammal (low power map)

Other parts of the gut are modified to perform particular functions.

- The lining of the stomach is covered with gastric pits, which are themselves lined with mucus-secreting cells, oxyntic cells which produce hydrochloric acid (giving the gastric juice a pH of about 2) and peptic cells which produce the peptidase, pepsin.
- The small intestine is lined with villi, which give a very large surface, well-supplied with capillaries. Digested products pass by diffusion and active transport through the epithelial cells of the villi into the capillaries. Fats are digested into fatty acids and glycerol which can pass directly into the capillaries. There is an alternative path, particularly after a fatty meal. The glycerol and fatty acids recombine to form minute droplets of fat which pass into the lymph via a lacteal which is present in each villus. Each fat droplet becomes coated with protein to form a lipoprotein. Eventually these lipoproteins return to the bloodstream and are reconverted back to fatty acids and glycerol before being absorbed by cells.
- The large intestine (colon and rectum) is much shorter than the small intestine but the lumen is much larger. The colon's contents include indigestible food, dead epithelial cells that have been sloughed off, mucus, bile and a lot of water. The function of the colon is to reabsorb as much of the water as possible. The longer the waste remains in the colon, the more water is absorbed.

The action of digestive enzymes

Carbohydrases (general term for all enzymes that catalyse the digestion of carbohydrates)

Salivary amylase is produced by the salivary glands, situated in the cheek and lower jaw, as a result of the stimulation of the parasympathetic nervous system by the smell or taste of food. Salivary amylase hydrolyses starch to maltose (see **2** Biological molecules).

Pancreatic amylase is released into the duodenum (a short loop – the first part of the small intestine) from the pancreas as one of the constituents of pancreatic juice (pH 7).

Amylase is also produced in the ileum (the main part of the small intestine) by the epithelial cells, but is not secreted into the lumen of the gut. Both the amylase and maltase (an enzyme which catalyses maltose to glucose) work in the membranes of the villi.

Proteases (general term for all enzymes that catalyse the digestion of proteins into amino acids)

Proteases include the peptidases, of which there are two types: **endopeptidases** and **exopeptidases**. Both are secreted as pancreatic juice and work on the long chains of amino acids that make up proteins. Enzymes are usually specific for particular substrates (see **9** How enzymes work) and exopeptidases only work on the bonds in the final link at the end of the amino acid chains. **Carboxylase** hydrolyses the link at the carboxyl end and **aminopeptidase** hydrolyses the amino acid end of the molecule.

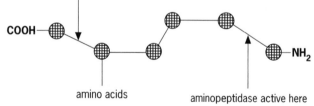

Figure 31.2 Action of exopeptidases

Endopeptidases break down the long chains of amino acids into shorter ones which greatly speeds up the digestive process. There are many endopeptidases as they only act on the bonds between specific amino acids.

Peptidases are secreted in an inactive form to avoid self-digestion. For example, **trypsin** is secreted by the pancreas as an inactive precursor, **trypsinogen**, which is later converted to trypsin in the small intestine by the action of another enzyme called **enterokinase**.

Lipase and bile

Lipase, which digests fats, is secreted by the pancreas and released into the duodenum. Bile is produced by the liver, stored in the gall bladder and released into the duodenum where it neutralises stomach acid and emulsifies fat droplets, thus increasing their surface area for digestion.

Co-ordination

It is important that the correct enzymes are present at the right time and in the required concentration. In mammals the secretion of digestive juices is under the control of both nerves and hormones. The following are examples of this dual control.

- The salivary glands are stimulated by the sight and smell, or even the thought, of food to produce saliva. Such a response is a conditioned reflex (see **17** Nerves and reflexes).
- Actual contact with the food in the mouth causes the taste buds on the tongue to send impulses to the brain, which is stimulated to send impulses to the gastric glands in the stomach wall to produce gastric juice. This is another example of an unconditioned reflex.
- Acid chyme (food in a semi-fluid state plus hydrochloric acid) enters the duodenum from the stomach, and stimulates cells lining the wall of the small intestine to produce enzymes plus the hormones **secretin** and **cholecystokinin** which are carried in the blood, and eventually stimulate the gall bladder, liver and pancreas.
  Secretin stimulates:
  – the liver to secrete bile
  – the pancreas to secrete the alkaline non-enzyme parts of the pancreatic juice.
  Cholecystokinin stimulates:
  – the gall bladder to contract and release bile into the duodenum via the bile duct
  – the pancreas to secrete the enzymes of the pancreatic juice.
- When there is food present in the stomach, the hormone **gastrin** is released into the blood. It travels back to the stomach and stimulates the **oxyntic** cells in the stomach wall to produce HCl. The hydrochloric acid kills bacteria and small organisms swallowed whole, and also stimulates the secretion of some enzymes.

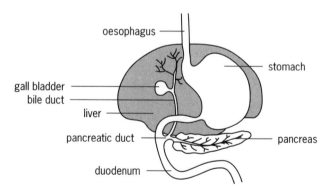

Figure 31.3 Liver, gall bladder and pancreas in the mammal

The control of the digestive system by nerves as well as hormones ensures that the digestive processes work only as and when required.

Note that different groups of cells in the pancreas work as either exocrine or endocrine glands.

The role of the liver in the control of blood sugar

Digested food is transported in the blood from the small intestine via the **hepatic portal vein** to the liver. After the liver has adjusted and controlled the composition of digested food in the blood, the blood returns to the heart via the **hepatic vein**.

The liver is the largest and most complex organ of the body. It performs a number of vital tasks and requires a good supply of oxygenated blood. Thirty per cent of the blood is supplied by the hepatic artery and the remaining 70% arrives via the hepatic portal vein.

A portal vein is any vein which carries blood from one set of capillaries to another set of capillaries, rather than transporting the blood directly back to the heart. This is known as a **portal system**.

The liver has an important role to play in glucose homeostasis; the maintenance of a constant blood sugar concentration. Glucose supplies from the gut arrive when carbohydrates are being digested. The body needs a system by which excess glucose can be quickly and efficiently converted to insoluble carbohydrate (glycogen or animal starch) which can be stored. This process is known as **glycogenesis**. The liver can store about 75 g of glycogen, and some glycogen is stored in muscle cells as an immediate reserve.

The blood needs to maintain a constant blood sugar level of about 80 to 100 mg of glucose per 100 ml of blood.

Some hours after a meal and/or during periods of exercise the blood sugar level falls. Glucose can be added to the blood in one of two ways:

- by **glycogenolysis** – the hydrolysis (see **2** Biological molecules) of glycogen stored in the liver
- by **gluconeogenesis** – the breakdown of proteins into glucose.

For details of the role of the hormones **insulin** and **glucagon** in the control of blood sugar, see **24** (The endocrine system: Insulin and glucagon; sex hormones).

Protein metabolism

Proteins cannot be stored in the body. Excess proteins can be converted to carbohydrate or fat, as necessary, and the nitrogen atoms excreted. But carbohydrates and fats contain no nitrogen and therefore they cannot be converted into amino acids or proteins.

The breakdown of proteins and the removal of the amino group (NH_2) is known as **deamination**. The amino group is actually removed as ammonia (NH_3) which is then converted to urea or uric acid.

Deamination of excess amino acids

Ammonia is a very soluble gas which diffuses rapidly and is also very toxic. Some aquatic animals (especially unicellular ones with a relatively large surface area) excrete ammonia, because it can be rapidly diluted in the large volume of surrounding water.

Terrestial animals convert their waste nitrogen into a safer compound – either insoluble uric acid as with insects, reptiles and birds, or soluble, but much less toxic, urea. Humans excrete urea in solution as urine.

(Note the importance of insoluble, non-toxic uric acid being excreted within the shell, during the embryonic development within the egg of a bird, for example. The excretion of a soluble, toxic substance, such as urea, would quickly kill the embryo.)

The chemistry of deamination can be simplified (Figure 31.4).

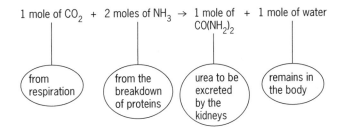

1 mole of CO_2 + 2 moles of NH_3 → 1 mole of $CO(NH_2)_2$ + 1 mole of water

| from respiration | from the breakdown of proteins | urea to be excreted by the kidneys | remains in the body |

Figure 31.4 Deamination in mammals

In fact, there are a series of enzyme-controlled reactions known as the **ornithine cycle** (Figure 31.5).

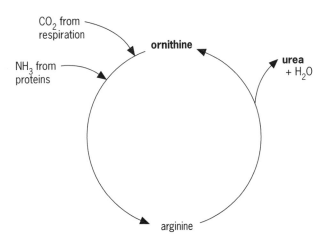

CO_2 from respiration

NH_3 from proteins

ornithine

urea + H_2O

arginine

Figure 31.5 The ornithine cycle

Transamination

The hepatocytes (liver cells) are also concerned with transamination in which non-essential amino acids, which may not be available in the diet, can be synthesised from other amino acids.

(Non-essential amino acids are those that *can* be synthesised. Of the 20 amino acids necessary for protein synthesis, half are non-essential.)

Transamination involves the combination of one amino acid with a keto acid. With the aid of an enzyme, called transaminase, a *new* keto acid is formed together with the *new* amino acid. A keto acid is an organic acid with a ketone group present, for example:

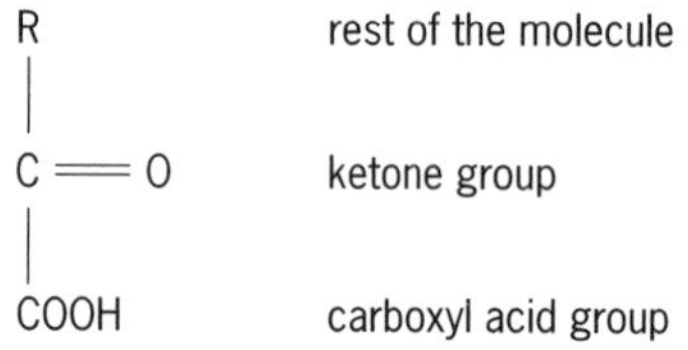

R   rest of the molecule
|

C $=$ O   ketone group
|

COOH   carboxyl acid group

Figure 31.6 A keto acid

Plasma proteins – albumin and globulin

Plasma proteins are made in the liver. They circulate in the blood plasma for a limited period of time and are then broken down by the liver.

The most abundant plasma protein is albumin. Its main function is to retain water in the blood by its osmotic effect.

The remaining plasma proteins have a wide range of both structure and function and are collectively called globulins. These include the antibodies, which are also known as immunoglobulins. They are produced as a response to a specific antigen. Other plasma proteins, such as fibrinogen, are concerned with the coagulation (clotting) of the blood.

Lipid (fat) metabolism

Cholesterol is an important steroid needed by the body for making cell membranes and for producing hormones. The liver makes cholesterol but the hepatocytes remove it if there is an excess. The excess is excreted in the bile and may precipitate to form gall stones in the gall bladder.

The homeostatic function of the liver in removing excess cholesterol is an important health factor because a raised cholesterol level may result in it being deposited on the walls of the blood vessels. This obstructs the vessels and may cause the blood to clot. If the clot is in the region of the heart it may cause a 'heart attack' or coronary thrombosis (see **26** Circulation of the blood).

The body does not 'waste' food which is ingested in excess of its needs. Excess proteins will be deaminated and converted to carbohydrate or fat as appropriate. Excess carbohydrate will be converted to fat and, together with any excess fat, will be stored. For humans such a diet in excess of normal requirements is unhealthy. For hibernating animals, however, deposits of fat are needed to insulate the body (even though the core temperature of the animal is considerably lowered). During hibernation the fat stores will be used up to produce body heat.

1 a **i** Name the process of chewing food.
ii Name the waves of muscular contraction which move food through the gut.
iii Name the part of the gut responsible for reabsorption of water.
iv Give an alternative name for the gut.
v Name the acidic, partly-digested contents of the stomach. (5)

b Name the following secretions:
i The hormone released into the blood when food is present in the stomach.
ii The alkaline secretion which emulsifies lipids.
iii The enzyme which converts trypsinogen to its active form.
iv The hormone which stimulates bile production.
v The hormone which stimulates the release of bile from the bile duct. (5)

2 a Fill in the following table, giving details of the general names of enzymes and their digestive reactions.

General name for class of enzymes	Example of polymer	End products (monomers)
carbohydrase	starch	
	lipid	
		amino acid

(5)

b Complete the following table to give details of the digestive system in humans.

Enzyme	Area of production	Products of this reaction
maltase	wall of small intestine	
	salivary glands	
	stomach	

(5)

3 The diagram shows the structure of a part of a liver lobule.

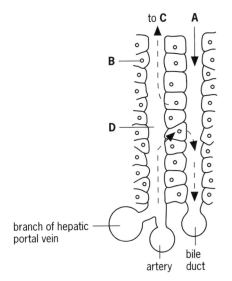

a Name **A**, **B**, **C** and **D**. (4)
b Explain why hepatocytes possess:
i many microvilli
ii mitochondria. (2)
c Explain the difference between an essential and a non-essential amino acid. (1)

4 What term is used to describe the following processes which occur in the liver?
a The hydrolysis of glycogen into its monomers.
b The chemical conversion of protein into glucose.
c The conversion of glucose into glycogen. (3)

5 The diagram of an electronmicrograph shows some of the cells which form the lining of the mammalian small intestine.

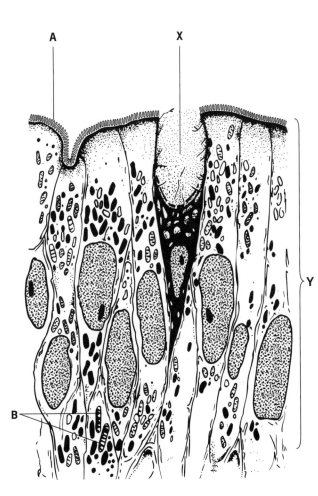

a What is the general name given to a tissue such as that labelled **Y**? (1)
b **i** Name the features labelled **A** and **B** in the diagram. (2)
ii Explain fully how **A** and **B** function in this tissue. (4)
c **i** Name the secretion labelled **X**. (1)
ii State two functions of this secretion. (2)

Welsh, Paper B3, June 1997

6 a **i** Name the region of the digestive system in which the digestion of fat begins. (1)
ii Name the enzyme which is responsible for the digestion of fat. (1)
iii State the site of secretion of this enzyme. (1)
iv Describe fully the part played by the liver in the digestion of fat. (3)

b 'Olestra is a calorie-free fat substitute which, it is argued, can be used to fight obesity and excessive fat intake. It mimics fat but is not absorbed by the body. In Olestra sucrose replaces glycerol, and attaches six, seven or eight fatty acids.'

(Adapted from: New Scientist, 25 November 1995)

The diagram shows a molecule of fat.

$$CH_2 — OOC — R$$
$$CH — OOC — R$$
$$CH_2 — OOC — R$$

i Circle the part of the molecule which is replaced by sucrose in Olestra. (1)
ii Suggest why Olestra is not absorbed by the body. (2)
iii During digestion, fat soluble vitamins attach themselves to fat molecules. Some people are concerned that Olestra will attract vitamins from other foods and pass through the body, increasing the risk of vitamin deficiency.

Suggest one way in which this problem could be overcome. (1)

Welsh, Paper B3, January 1997

7 The drawing shows a section through the human stomach wall.

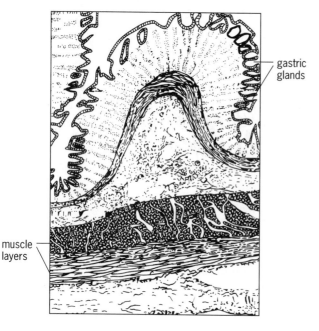

(Adapted from: Freeman W.H. and Bracegirdle B., Atlas of Histology. Heinemann Educational.)

a Describe one function of the muscle layers of the stomach wall. (1)
b **i** Name the type of protein digesting enzyme secreted by the gastric glands (1)
ii Describe the function of this type of enzyme. (2)
iii Name the hormone which stimulates the release of this enzyme. (1)
c Explain briefly why a different enzyme is required for each type of food substance. (2)

NEAB, BY03, March 1998

8 The diagram shows the formation of urea from an amino acid.

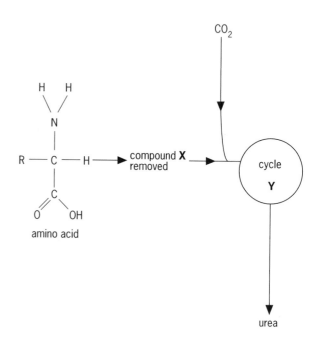

a **i** Name the compound **X** which is removed from the amino acid. (1)
ii Draw ring(s) around the atoms removed from the amino acid to form compound **X**. (1)
iii Name the chemical process which results in formation of **X**. (1)
iv Name the cycle of chemical reactions **Y** which results in the formation of urea. (1)
b Name the organ which is responsible for the formation of urea. (1)
c Name the organ which is responsible for the excretion of urea. (1)
d **i** Name the nitrogenous excretory compound produced by insects. (1)
ii Explain why this is a safer excretory compound for terrestrial animals than the ammonia produced by many marine invertebrates. (1)

9 The diagram shows a section of a human cell membrane which contains receptors **X** for insulin.

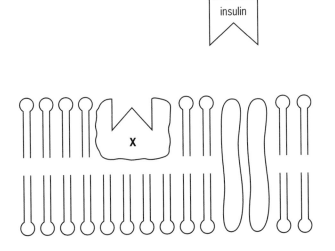

a Explain how glucose enters the cell. (2)
b Explain how insulin affects the rate of uptake of glucose by the cell. (2)
c What general name is given to mechanisms such as this which involve molecules with complementary shapes? (1)
d Name the two main components of cell membranes. (2)

10 Explain the following statements with reference to the functions of the liver:
a The effects of alcohol wear off several hours after drinking.
b Eating a diet containing liver helps to prevent anaemia.
c A human being can survive and grow on a diet deficient in some amino acids.
d Most of the erythrocytes in the body are less than 120 days old. (4)

11 The diagram shows the main blood vessels going to and coming from the liver.

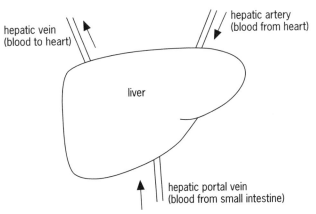

a In a healthy person the blood glucose level in the hepatic vein fluctuates much less than in the hepatic portal vein. Explain why this is so. (3)
b Blood sugar level is more or less constant even if a person has not eaten for several days. How does gluconeogenesis help to maintain this constant blood sugar level? (1)
c Suggest why people suffering from diabetes are advised to eat their carbohydrates in the form of starch rather than as sugars. (2)
NEAB, BY03, June 1997

12 a Describe the functions of plasma proteins produced by the liver. (5)
b Explain the role of the liver in the maintenance of nutrient levels in the body. (13)
Cambridge, Module 4804, November 1997

1 a i mastication
    ii peristalsis
    iii colon
    iv alimentary canal
    v chyme
  b i gastrin
    ii bile
    iii enterokinase
    iv secretin
    v cholecystokinin

2 a

General name for class of enzymes	Example of polymer	End products (monomers)
carbohydrase	starch	monosaccharides (glucose)
lipase	lipid	fatty acids glycerol
protease (peptidase)	protein (or named)	amino acid

  b

Enzyme	Area of production	Products of this reaction
maltase	wall of small intestine	glucose
amylase	salivary glands	maltose
pepsin	stomach	peptides

3 a A = canaliculus
    B = hepatocyte
    C = central vein/branch of hepatic vein
    D = sinusoid
  b i The microvilli increase the surface area for absorption of metabolites.
    ii The mitochondria provide ATP for active transport of bile/driving other metabolic processes.
  c An essential amino acid must be obtained from proteins taken in as food whereas non-essential amino acids can be made by transamination of other amino acids.

4 a glycogenolysis
  b gluconeogenesis
  c glycogenesis

5 a epithelium (not endothelium/epidermis)
  b i A – microvilli/brush border
    B – mitochondria
    ii A – increase the surface area; for absorption.
    B – provide energy/ATP; for absorption against a concentration gradient/active transport.
  c i mucus/mucin (not intestinal juice)
    ii Two of:
    • Coats the small intestine.
    • Protects the small intestine against acid pH of stomach contents.
    • Acts as a lubricant.
    • Protects against protease activity.

6 a i duodenum
    ii lipase
    iii pancreas
    iv The liver produces a mixture of bile salts/sodium glycolate and taurocholate.
    Two of:
    • The bile salts reduce the surface tension of the fat globules.
    • The bile salts emulsify the fat globules.
    • A larger surface area is exposed to enzyme activity.
    • It raises the pH to the optimum for lipase.
  b i

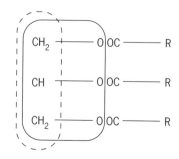

    ii Olestra cannot be digested because there is no enzyme to break down the bonds between fatty acids and sucrose /fatty acids protect the sucrose from enzyme activity. The Olestra molecule is too large to be absorbed/pass through the cell membrane.
    iii The problem could be overcome by saturating Olestra with vitamins so that vitamins are not absorbed from the food/take supplementary fat-soluble vitamins.

7 a Churning/mechanical digestion/mixing/ peristalsis/moves food.
  b i endopeptidase pepsin
    ii Two of:
    • Breakdown of protein into polypeptides.
    • By hydrolysing/breaking of peptide bonds.
    • Hydrolysing/breaking bonds deep inside the protein.
    iii gastrin
  c Molecules in food have different shapes/active site and substrate have complementary shapes.
    Enzymes are therefore needed with different shaped active sites.

8 a i ammonia
    ii

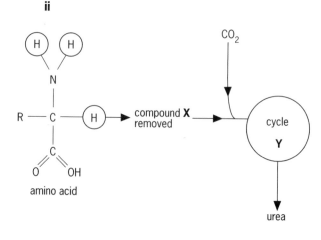

    iii deamination
    iv the ornithine cycle

b the liver

c the kidney

d i uric acid

ii Uric acid is insoluble and requires very little water to eliminate it from the body, whereas ammonia is soluble and very toxic. It must be diluted many times but this is not a problem in a watery habitat.

9 a Glucose passes through protein-lined channels within the cell membrane.
This occurs by diffusion.

b Insulin will increase the rate of uptake of glucose by the cell.
The formation of glycogen inside the cell decreases the glucose concentration so that glucose continues to diffuse in.

c lock and key

d protein; phospholipid

10 a The liver cells remove toxins from the blood and break them down.

b The liver cells store iron.

c The hepatocytes are able to synthesise (non-essential) amino acids by transamination.

d Old red blood cells are engulfed by the liver cells and broken down.

11 a The blood glucose in the hepatic portal vein increases after a meal/varying amounts are absorbed from the gut.
The excess glucose;
is converted to glycogen in the liver;
(*or* if there is insufficient blood glucose;
then stores of glycogen in the liver are broken down).

b Glucose is produced from molecules other than glycogen such as lipids/proteins/amino acids.

c The starch has to be digested; therefore there is a slower uptake of sugar/no rapid rise in the blood sugar.

12 a Five of:
- The plasma proteins fibrinogen and prothrombin are important in blood clotting.
- The soluble fibrinogen is converted into insoluble thrombin.
- This converts inactive prothrombin to thrombin.
- The globulins/globular proteins have a role in transporting thyroxine/insulin/phospholipids/lipids/cholesterol/iron/vitamins ($A/E/B/_{12}/D/K$).
- The plasma protein albumin has a role in transporting hormones/thyroxine/drugs/aspirin/penicillin/vitamins (A/C)/acetylcholine/bilirubin/calcium/copper/zinc.
- Plasma proteins exert a solute potential, maintaining the osmotic pressure of the blood.
- They contribute to the viscosity/density of the blood which is important in determining the pattern of blood flow in vessels.
- The plasma proteins have a buffering effect.

b *This part can be organised by considering the following four groups of nutrients.*

Control of blood glucose levels
Five of:
- When the blood sugar rises, insulin is secreted.
- This causes glycogenesis – the conversion of glucose to glycogen.
- Respiration increases/the liver cells take up glucose.
- When the blood glucose level falls, glucagon is secreted.
- This causes glucogenesis – the conversion of glycogen to glucose.
- Gluconeogenesis – the production of glucose from non-carbohydrate sources may occur.
- The secretion of adrenaline causes the conversion of glycogen to glucose.

Control of blood amino acid levels
- Excess amino acids are deaminated.
- The amino group is converted to ammonia.
- This is then converted to urea; via the ornithine cycle.
- The remainder of the amino acid molecule is used in respiration.
- Amino acids can be transaminated/converted into different amino acids.

Control of blood lipid levels
- Excess carbohydrate/protein can be converted into lipid; for storage.
- Lipid can be used as a respiratory substrate to produce energy.
- Lipid is used in the synthesis of cholesterol; for use in cell membranes/steroid hormones/waterproofing the skin.
- The liver removes excess cholesterol/phospholipids from the blood and breaks them down.
- Lipoprotein formation occurs in the liver; for the transport of lipids.

Vitamins/minerals
- The liver is a storage site for vitamins; e.g. $A/B_{12}/D/C/E/K$.
- It is a storage site for minerals; e.g. copper/zinc/iron/cobalt/molybdenum/potassium.
- Iron is stored as ferritin.
- In times of shortage of vitamins/minerals the stored materials can be released.

(A maximum of 6 marks can be awarded for any group of nutrients; max. 13 marks; quality of language 2 marks)

32 Respiratory systems

Students need to be able to distinguish clearly between *breathing*, *gaseous exchange* and *cellular respiration.*

Breathing is the process by which air is taken into the lungs (inspiration) and then expelled from the lungs (expiration) after gaseous exchange has taken place.

Structures which make such gaseous exchange possible are referred to as *respiratory systems.* All living organisms have a respiratory system of some kind and the term respiratory system is *not* confined to vertebrates with lungs.

Gaseous exchange is the *physical* exchange of gases (oxygen and carbon dioxide) by diffusion across respiratory surfaces.

Cellular respiration is a catabolic process (the breaking down of complex substances into simpler ones) which takes place in the cytoplasm and mitochondria of cells and by which energy is released and used for other metabolic processes (see **14** and **15** Respiration).

Respiratory systems

Unicellular organisms have a large surface area/volume ratio. This allows sufficient oxygen to be obtained, and carbon dioxide to be excreted in solution across the cell membrane by diffusion down a concentration gradient. Unicellular organisms are either aquatic, or they live in damp moist places.

As the ratio of the surface area to the body volume in larger animals decreases, so the efficiency of diffusion also decreases. Simple diffusion across the body surface is no longer sufficient to supply the cells with oxygen or remove carbon dioxide quickly enough. The requirements for efficient diffusion are dealt with in **4** The transport of substances across cells and through membranes. When simple diffusion across the body surface is insufficient to supply the body's needs, a respiratory surface is required. They have certain features in common:

- The surface area needs to be large. The larger the animal, and the more active it is, the larger the surface area needs to be.
- The surface must be moist so that gases can diffuse across in solution.
- Diffusion is only rapid over very short distances and so the respiratory surface must be thin – one cell thick.
- Where the gases are carried in the blood there needs to be a rich supply of blood in thin-walled capillaries which are very close to the respiratory surface. In this way the distance that the gases have to diffuse in solution is kept to an efficient minimum.

Fick's law states that the rate of diffusion is proportional to:

$$\frac{\text{surface area} \times \text{difference in concentration}}{\text{thickness of membrane}}$$

Respiratory systems in dicotyledonous plants

As plants become larger their surface area/volume ratio also decreases but there are no specialised systems in plants for the following reasons:

- The gas exchange surface for respiration is the moist cell walls of all the living cells in a green plant. (The walls of the mesophyll cells of the leaves are the most important for gaseous exchange in photosynthesis.)
- Diffusion of oxygen and carbon dioxide in solution occurs through a system of interconnecting air spaces plus some unused xylem vessels.
- Plants are not as metabolically active as animals and therefore comparatively less gas exchange is needed.
- Green plants produce carbon dioxide during cellular respiration but they also use it in daylight hours for photosynthesis. There is therefore less overall gas exchange than for respiration alone.

Note: efficient respiratory surfaces also create problems of limiting water loss from the gaseous exchange surfaces (see also **29** Transport systems in flowering plants 2).

In land plants, undue water loss is controlled by the thick cuticle on the upper surface of leaves and the presence of stomata which are wholly, or largely, confined to the lower surface.

Respiratory systems in terrestrial insects

Insects are covered with an exoskeleton and a waterproof cuticle. Diffusion of gases across the body surface is not possible. Nevertheless, insects are active and require a good system for gaseous exchange.

Air enters along the side of the body through a series of holes, or spiracles, of which there are ten pairs. These may be as much as 1 mm across and are clearly visible in large adults and larvae.

During periods of inactivity all of the spiracles may be closed. This minimises loss of water. When activity increases, the carbon dioxide produced is detected by chemoreceptors, and more of the spiracles open. The spiracles are not simply holes. They are surrounded by hairs which help to retain water vapour and they are closed by a system of valves which are operated by small muscles.

Each spiracle is connected to a chitinous tube (**trachea**) within the body of the insect. The chitin, which is the same substance as forms the cuticle, stops the tracheae collapsing when the pressure is low. It is also impermeable to gases and so very little gaseous exchange takes place in the tracheae. The tracheae are connected in turn to a complex series of branching tubes called **tracheoles** which pass right into the tissues and even penetrate the cells. The tracheoles give a very large surface area for gas exchange.

The ends of the tracheoles are fluid-filled. When the insect is active the respiring cells produce lactic acid which raises their osmotic potential and water is drawn from the tracheoles into the cells by osmosis. More space is then made available for oxygen. When activity slows down, the lactic acid is oxidised, and the watery fluid once more enters the ends of the tracheoles.

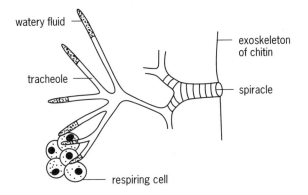

Figure 32.1 Tracheal system of an insect

Ventilation of the tracheoles may be improved by contraction of the insect's abdominal muscles which flatten the body and reduce the volume of the tracheoles. When the muscles relax, fresh air is drawn in. The insect's tracheal system is efficient but since it relies purely on diffusion the size that can be attained by an insect is limited.

Respiratory systems in fishes

Fishes have a respiratory system which is efficient for animals living constantly in water. Gases cannot diffuse across a body covering of scales, and a system has evolved which:

- maintains a constant flow of water over the respiratory surfaces (gills) which could otherwise be a problem when the fish stops swimming
- provides a large surface area with thin walls and a good blood supply for gaseous exchange.

The bony fish (this excludes the sharks and flat fish such as skate and rays) have a covering over the delicate gill slits called an **operculum**. It not only protects them but makes for a more efficient respiratory system. When the floor of the **buccal cavity** (mouth) is lowered, the pressure drops and water is drawn in through the mouth. At the same time the opercula are pressed tightly against the sides of the body.

When the floor of the buccal cavity is raised, the pressure is increased and the water is forced over the gills and expelled via the opercula. The pressure causes flaps of skin to close the mouth and prevent loss of water. The gills overlap each other at the ends and this slows down the passage of water, thus increasing the time for gases to diffuse across the gill membranes.

The gills consist of a number of gill lamellae stacked on top of each other. They need water to keep them apart and prevent them from collapsing. The gill filaments, which are the main site for gaseous exchange, run at right angles to the lamellae.

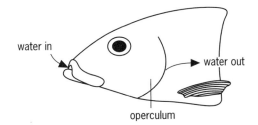

Figure 32.2 The passage of water in a bony fish

The blood in the vessels which supply the gill lamellae flows in the opposite direction to the water passing over the lamellae. This ensures that the concentration gradient is kept as steep as possible because products are transported away before they have time to accumulate. A high rate of diffusion is maintained. This is known as a **countercurrent exchange system**.

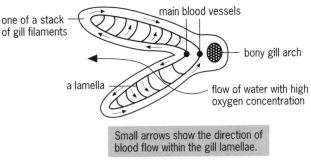

Small arrows show the direction of blood flow within the gill lamellae.

Figure 32.3 Flow of water over the gill lamellae in a fish

Respiratory systems in mammals

In mammals air is forced into the lungs by air pressure. Intercostal muscles move the ribs in an upward and outward direction at the same time as the muscular diaphragm flattens from the domed shape adopted when relaxed. This increases the volume of the chest cavity and air is forced in by the greater air pressure outside the lung cavity. The structure of the lungs provides all the necessary requirements for efficient diffusion that have already been discussed in this topic. The surface area of the alveoli in the lungs increases tenfold between birth and being adult. In an adult human the lungs have a surface area of approximately $90\,m^2$. For each gram of body weight there is about $7\,cm^2$ of lung surface provided by 300 million alveoli which are, in turn, supplied by 275 million capillaries. Students are not expected to learn such figures but they convey an impressive picture of the very efficient respiratory surfaces possessed by mammals.

Animals, as well as plants, have problems of limiting water loss. Invertebrates, such as insects, have a waterproof cuticle and gaseous exchange takes place via small openings (spiracles) along the side of the abdomen into which the tracheae open (**33** Oxygen and carbon dioxide transport in mammals). The spiracles possess valves which can close and thus prevent undue water loss. Other invertebrates, with soft bodies, live in shells, in burrows or under stones and other moist places. Reptiles, birds and mammals have lungs which are situated deep within the body.

Breathing in humans

The rate and depth of breathing is controlled by a part of the brain called the **breathing centre** which is situated in the **medulla oblongata** (see **19** The peripheral and central nervous systems and the brain). The breathing centre consists of two areas: the **inspiration centre** and the **expiration centre**. Control is by both nerves and hormones.

Nervous control of breathing

The muscles involved in breathing are voluntary and therefore can be controlled by the will but, because control is brought about by a series of reflex actions, respiration movements are largely automatic.

1   Spontaneous nerve impulses from the inspiration centre cause the intercostal muscles of the ribs and the muscular diaphragm to contract – volume of the thoracic cavity increases.
2   **Stretch receptors** in the walls of the lungs become excited and send impulses (via the vagus nerve) to the inspiration centre which causes the intercostal muscles and diaphragm to relax – volume of thoracic cavity decreases.
3   Stretch receptors are no longer stimulated and so the cycle begins again.

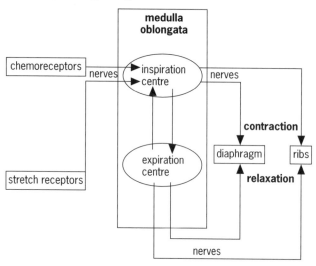

Figure 32.4 Nervous pathways involved in breathing control

Chemical control of breathing

The brain is sensitive to minute changes in the concentration of carbon dioxide (and hence pH) in the blood. These changes are detected by **chemoreceptors** (areas of nerve tissue) situated in the walls of the carotid artery and aorta. Excess carbon dioxide causes a lowering of the blood's pH which stimulates the chemoreceptors to send an impulse to the inspiration centre; this stimulates deeper breathing to reduce the level of carbon dioxide in the blood. This is another example of a **negative feedback mechanism,** see also **23** and **24** (Endocrine system) for further examples.

Lung capacity

Although the total lung capacity of an adult averages about $6\,dm^3$ (6 litres) only 10% is changed when at rest. This is known as the **tidal volume**. If one expels as much air as possible there is still a **residual volume** of approximately $2\,dm^3$ of air left in the lungs which helps to keep the volume of air in the alveoli fairly constant. This residual air mixes with incoming air and will be gradually expired.

Figure 32.5 illustrates the various capacities of the lungs and the terms commonly used.

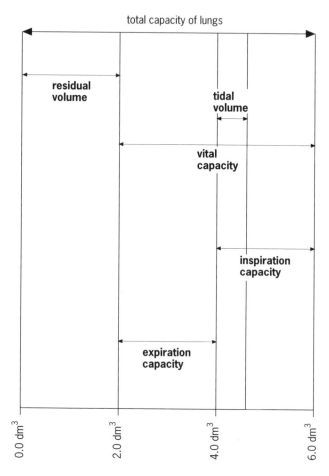

Figure 32.5 Adult human lung capacities

During exercise we breathe faster and deeper in order to reduce the concentration of carbon dioxide in the blood and *not* (as far as the breathing centre is concerned) to obtain more oxygen. This is why it is possible to stay under the water for a longer period of time (when diving etc) by not completely holding one's breath but by expelling small quantities every now and again. This is despite the fact that small quantities of oxygen are 'lost' when letting out small breaths. The important thing is that carbon dioxide is also being expelled and this helps to keep the concentration of carbon dioxide in the blood as low as possible in the circumstances.

Note: the requirements vary for different exam boards, so it is important to check your syllabus. In many cases the syllabuses stress the importance of realising that gaseous exchange also results in water loss and that this needs to be minimised, e.g. by stomata or spiracles closing or having structural adaptations.

1 a Name the processes described below:

i The net movement of molecules of a liquid or gas from a region of higher concentration to a region of lower concentration.

ii The chemical oxidation of a substrate within a cell resulting in release of energy.

iii The movement of gases by diffusion across cell surfaces to supply raw materials for respiration and photosynthesis and remove the waste materials of these processes.

iv Air, containing oxygen, is drawn into the lungs, and air with less oxygen and more carbon dioxide is expelled from the lungs. (4)

b **i** Name the part of the human lung where oxygen uptake occurs.

ii Name the region in an insect where oxygen diffuses into the tissues.

iii Which cells in a plant will require oxygen for respiration? (3)

2 The diagram below shows part of the route taken by air to the cells of an insect.

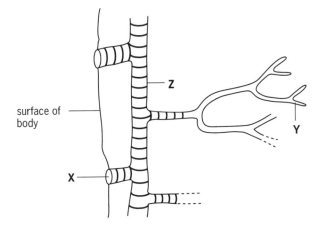

a Name the parts **X**, **Y** and **Z**. (3)

b Name the process by which gases normally move to and from the cells in a small insect. (1)

c In larger insects a process of ventilation may occur during strenuous exercise.

i Describe how this can be brought about. (2)

ii Explain why ventilation may be necessary in this example. (2)

d The part labelled **X** in the diagram may be surrounded by hairs in some insects. Suggest a reason why hairs are present in some insects. (1)

e Part **Z** has rings of chitin along its length.

i What is the function of these rings of chitin? (1)

ii Structures with a similar function are found in airways leading down to the lungs in a mammal. Name the material of which these mammalian structures are made. (1)

3 a **i** A thin surface and a diffusion gradient are both features of gas exchange surfaces. Describe how these are achieved at the gas exchange surfaces of:

i a mammal (3)

ii a leaf (3)

b Explain how excess water loss from the gas exchange surface is prevented in:

i a mammal (1)

ii a leaf (1)

AQA (NEAB), BY03, March 1999

4 There are many species of annelid worm. Some are very small, only a few millimetres in length. Others, such as lugworms, are much larger. The drawing shows a lugworm and part of one of its gills.

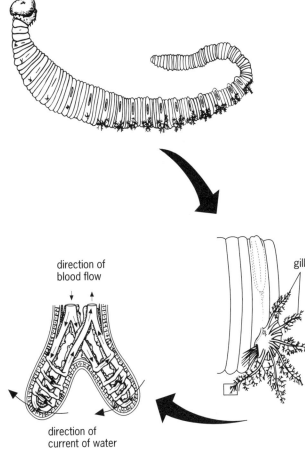

direction of blood flow

gill

direction of current of water

a Smaller species of annelid do not have gills. Explain why these small worms do not need gills to obtain sufficient oxygen. (2)

b In many of the lugworm gills, the blood flows in the opposite direction to the current of water passing over them. Explain the advantage of this arrangement. (2)

c Explain two ways, other than that described in part **b**, in which the structure of a lugworm gill is adapted for efficient gas exchange. (2)

d Explain why water is always lost from the gas exchange surfaces of terrestrial organisms. (2)

The table shows the ratio of the amount of water lost to the amount of oxygen gained for two terrestrial animals, an annelid worm and an insect.

Organism	Ratio	$\dfrac{\text{mass of water lost/mg}\,g^{-1}\,\text{minute}^{-1}}{\text{volume of oxygen taken up/cm}^3\,g^{-1}\,\text{minute}^{-1}}$
annelid worm	2.61	
insect	0.11	

e Both the annelid and the insect take up oxygen at a rate of $2.5\,cm^3\,g^{-1}\,minute^{-1}$. Calculate the rate at which oxygen would be lost in meeting these requirements in:

i The annelid (answer ____ $mg\,g^{-1}\,minute^{-1}$).
ii The insect (answer ____ $mg\,g^{-1}\,minute^{-1}$). (2)

f Give two explanations of why the rate of water loss during gas exchange is very low in most insects. (4)

AQA (AEB), Module 3, June 1999

5 a Explain fully what is meant by diffusion. (2)
b The table below shows various dimensions of cubes of animal tissue.

Length of side	Volume (V)	Ratio of surface area:volume
1 cm	1 cm^3	6:1
2 cm	8 cm^3	3:1
3 cm	27 cm^3	2:1

i State, in words, the quantitative relationship between length and the surface area/volume ratio. (1)
ii Explain fully how this relationship indicates that increasing size in animals makes respiratory exchange more difficult. (3)
iii Suggest three ways in which animals overcome this problem. (3)

Welsh, Paper B4, June 1997

6 a The graph below shows the change in lung volume during inspiration.

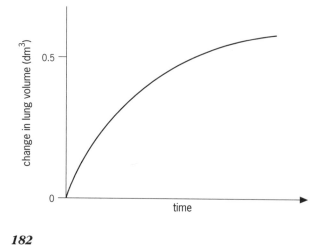

Sketch a curve on an axis copied from the figure below to show the expected change in alveolar pressure during inspiration. (1)

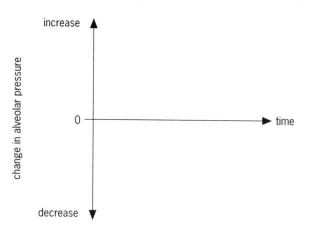

b Breathing rate changes as the concentration of carbon dioxide and of oxygen in inspired air, and the pH of arterial blood, change. This is shown below.

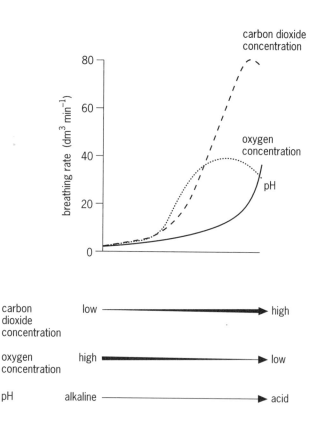

i Which factor has the greatest effect on breathing rate? Give evidence from the graph to support your answer. (1)
ii Divers who do not use air tanks may inhale and exhale deeply several times in quick succession before diving. As a result, they may lose consciousness and drown. Using information in the graph, explain why they may lose consciousness. (3)

c The diagram shows how the ventilation cycle is controlled.

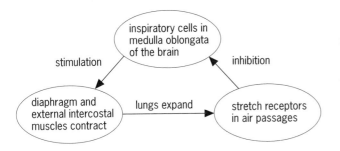

Explain how the stretch receptors are involved in controlling the ventilation system through negative feedback. (3)

AQA (NEAB), Paper BY10, March 1999

7 a Fick's law can be written as:
The rate of diffusion is proportional to

$$\frac{\text{surface area} \times \text{difference in concentration}}{\text{thickness of membrane}}$$

Explain how the structure and ventilation of fish gills provide the conditions necessary for a high rate of diffusion. (4)

b Some fish use *ram ventilation* when swimming fast. In ram ventilation, the normal method of using muscles to pump water over the gills stops operating and the fish swims with its mouth open.
i Explain why, when swimming at the same speed, the fish needs less energy using ram ventilation than when using the gill pumping mechanism. (2)
ii In an experiment, the oxygen concentration of the water surrounding a fish swimming at high speed was gradually reduced. The diameter of the open mouth of the fish increased as the oxygen concentration fell. Suggest an explanation for this observation. (1)

AEB, Paper 3, June 1997

8 Compare the way gas exchange occurs in flowering plants and mammals. (10)

9 The graph shows the effects of changes in the concentration of respiratory gases in the air on the volume of air inhaled in one minute by a person.

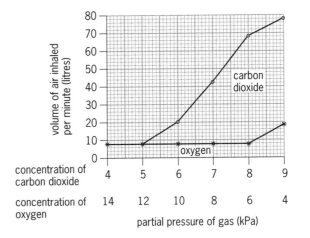

a Calculate the percentage increase in the volume of air inhaled per minute when the partial pressure of carbon dioxide rises from 5 kPa to 8 kPa. (1)

b Describe the effect of changing the oxygen concentration on the volume of air inhaled per minute. (2)

c The volume of air inhaled per minute is regulated by the respiratory centre. In which part of the brain is the regulatory centre located? (1)

AQA (NEAB), BY03, March 1998

1 a i diffusion
ii respiration
iii gaseous exchange
iv breathing
b i alveoli
ii end of tracheoles
iii All the cells require oxygen although some is supplied as a waste gas in photosynthesising cells in the light.

2 a X = spiracle; **Y** = tracheole; **Z** = trachea
b diffusion
c i Ventilation is brought about by alternate contraction (to draw in air) and relaxation of some of the abdominal muscles; coupled with opening and shutting of spiracles.
ii Two of:
- Larger insects will have a higher respiratory requirement for oxygen/need to expel more carbon dioxide because of the larger mass of tissues.
- Rapid exercise will require more energy/respiration/increase the oxygen required and therefore more rapid gaseous exchange will be needed.
- Diffusion alone will not be sufficient to supply the oxygen required/ventilation will increase the volume of oxygen supplied.

d Hairs will reduce water losses from the insect's body.
e i To keep the airways open.
ii cartilage

3 a i Three of:
- The wall of the alveoli/capillaries has a single epithelial layer/layer of cells/the alveoli and capillaries are close together.
- The epithelium is flattened/pavement epithelium.
- Ventilation maintains a high oxygen/low carbon dioxide concentration in the alveoli.
- Blood flow/circulation maintains a high carbon dioxide/low oxygen concentration in the blood.

ii Three of:
- The leaf is very thin/only a few cells thick.
- Intercellular spaces expose the cell surface membrane/wall directly to gases.
- Production of oxygen in photosynthesis maintains a high oxygen concentration.
- Use of carbon dioxide during photosynthesis maintains a low carbon dioxide concentration.

b i Enclosed within the body cavity/limited contact with the outside air.
ii Wax layer/guard cells close the stomata.

4 a Large surface area : volume ratio in the smaller species. This meets their requirements by diffusion (over the body surface).
b This maintains a concentration gradient. Allows diffusion of oxygen across the width of the gill.
c Two of:
- The lugworm gill has a large surface area for diffusion/oxygen uptake/gaseous exchange.
- There is a short diffusion path.
- The gill is highly vascularised/has many blood capillaries for oxygen transport.

d Two of:
- Gas exchange surfaces are permeable to small molecules.
- There is a higher concentration of water inside the cells than out/ψ gradient.
- Water will diffuse outwards/evaporation.

e i $2.61 \times 2.5 = 6.53$
accept 6.5
ii $0.11 \times 2.5 = 0.275$
accept 0.28 or 0.3

If both answers are correct, 2 marks, but this answer can gain 1 mark if you have clearly shown your working out but made an error in calculation.

f *Either*
- *Refer to the spiracles; which limit the exposure of the respiratory surface/spiracles can be closed.*
or
- *Sunken spiracles/hair round spiracles; trapping moist air.*
or
- *The trachea is lined with cuticle. The insect only loses water through the tracheoles.*
or
- *The trachea/tracheoles are inside; limiting the exposure of the respiratory surface.*

5 a Two of:
- Diffusion is the movement of molecules (ions or particles, not substances);
- usually in solution or gas;
- from a higher concentration to a lower concentration;
- which occurs passively (without input of energy).

b i The magnitude of the ratio is inversely proportional to the length; (e.g. doubling of the length results in a halving of the ratio).
ii Three of:
- Uptake of oxygen is proportional to the area of the absorbing surface.
- Consumption of oxygen is proportional to the volume of tissue.
- Therefore a low ratio means that demand tends to exceed supply.
- An increase in volume (size) means that oxygen must diffuse over a greater distance.

iii Three of:
- By possessing a flat/thin shape rather than a cuboid shape.
- By remaining small.
- By development of circulatory systems.
- By the development of special, large gas exchange surfaces such as gills or lungs.

6 a A decreasing line/initial drop followed by an increasing line (U-shaped line).
b i Carbon dioxide because this factor covers the biggest range/greatest increase/the most marked response.
ii Exhaling deeply results in decreased carbon dioxide concentration in their blood.
The breathing centre is no longer stimulated.
The resulting lack of oxygen causes unconsciousness.

c Three of:
- Stimulus – lung expansion causes activation of the stretch receptors.
- Action – which prevent the inspiratory/medulla/brain cells from generating impulses.
- Response – so the diaphragm and intercostal muscles are unable to contract;
- causing stretch receptors to be no longer stimulated;
- so nerve impulses are generated again by the medulla;
- leading to lung expansion.

7 a *(Do not be put off if you have never heard of Fick's law. This is a question about gaseous exchange at the respiratory surface of fish. Answer by recalling the characteristics of a respiratory surface and giving details of how this is achieved in fish.)*
Four of:
- A large surface area is achieved by gill plates/filaments/lamellae.
- A large surface area is achieved by the extensive capillary network.
- The difference in concentration is maintained by the countercurrent mechanism, and by blood removing oxygen from the exchange surface.
- The ventilation mechanism brings more oxygenated water.
- The thin membrane is achieved by the thin squamous/pavement epithelium.

b i The muscles used in normal ventilation are not required in ram ventilation.
ATP/energy is required for muscle contraction.
ii Increasing the diameter of the mouth will result in an increased volume of water flowing over the gills.

8 *(Note: the question asks you to **compare**. It is therefore important that you do not describe gas exchange in mammals followed by a description of gas exchange in plants. Key concepts to emphasise in your answer are the factors which increase diffusion and the characteristics of a respiratory surface.)*

- Gas exchange in both flowering plants and mammals takes place over respiratory surfaces.

- Refer to the lungs in mammals.
- Refer to the surface of the mesophyll cells in leaves of plants.
- Access to the respiratory surface is via trachea/bronchi etc, in mammals.
- Access to the respiratory surface is via stomata/pores between the guard cells in flowering plants.
- Refer to lenticels in cork tissue of woody stems.
- In mammals there is an active system of ventilation to draw in the air.
- Give brief details of the ventilation mechanism.
- In plants there is passive movement/diffusion of gases.
- The respiratory surface for gaseous exchange must be large; this is provided by the alveoli in mammals and by the cell surfaces of the spongy mesophyll in flowering plants.
- Thin surfaces aid rapid diffusion by providing a short diffusion path.
- The walls of the alveoli are one cell thick in mammals.
- Mesophyll cells have thin cell walls.
- Moist surfaces aid diffusion of gases.
- A film of moisture lines the alveolar surface.
- Plant cell walls are moist.
- Respiratory surfaces must be permeable to gases.
- In mammals there is a capillary network around the alveoli (none in plants);
- which maintains a high concentration gradient;
- enabling more rapid diffusion.

9 a 750% increase
(Increase = 68 − 8 = 60
divide by original volume = 60/8
× 100 to obtain the percentage increase)
b • A decrease in the oxygen concentration in the range 14–6 kPa has no effect.
- Reduction to below 6 kPa partial pressure of oxygen causes an increase in volume inhaled.
or
- An increase in oxygen concentration from 4 to 6 kPa results in a decrease.
- Further increase in the partial pressure of oxygen has no effect.
c medulla

33 Oxygen and carbon dioxide transport in mammals

The interchange of gases

Diffusion takes place down a concentration gradient and therefore:

1 Oxygen diffuses across the alveoli walls of the lungs into the blood capillaries where it is transported as oxyhaemoglobin together with a small amount in solution in the plasma.
2 Carbon dioxide, which has been transported in the blood in solution and also as bicarbonate ions, diffuses from the blood capillaries, across the alveoli walls into the alveolar spaces.

Atmospheric pressure at sea level is approximately 100 kilopascals (kPa). One pascal has a pressure of one Newton per metre squared ($1\,N/m^2$). The pressure exerted on the walls of the alveoli by the mixture of gases in the air is the same as atmospheric pressure. Each gas in the mixture exerts a part of the pressure proportional to its relative atomic mass and concentration. It is said to exert a partial pressure.

Oxygen uptake and transport

The main part of the blood is plasma which makes up about 55% of the total volume. The majority of blood cells are the red blood cells, **erythrocytes**, which constitute approximately 45% of the volume and contain a respiratory pigment called haemoglobin. It is the haemoglobin which enables sufficient oxygen to be transported to meet the needs of an active, warm-blooded animal such as a mammal or bird. Haemoglobin is a complex molecule made of a molecule of the protein **globin**, attached to a **haem** group which contains an iron atom.

Figure 33.1 Haemoglobin molecule

The oxygen combines loosely with the haemoglobin to form **oxyhaemoglobin**. The iron of the haem group is not oxidised but some chemicals, such as carbon monoxide, which are poisonous, oxidise the iron, producing substances which can no longer carry oxygen. As much as 15% of carboxyhaemoglobin is found in the blood of smokers, hence one of the dangers of pregnant women smoking.

In the blood, the oxygen concentration is measured as partial pressure (also referred to as **oxygen tension**). Since the air is 21% oxygen, and atmospheric pressure at sea level is 100 kPa, the partial pressure of the air is 21 kPa.

The haemoglobin in the blood capillaries which surround the alveoli is saturated with oxygen. There is a relationship between haemoglobin saturation and oxygen partial pressure which can be shown in an **oxygen dissociation curve**.

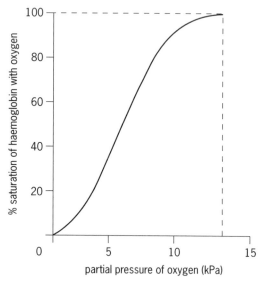

Figure 33.2 Oxygen dissociation curve for haemoglobin in an adult human

Note in Figure 33.2 that the partial pressure of oxygen in the lungs is only about 13 kPa (saturation occurs when the partial pressure is 13 kPa). This is because the air entering the alveoli is diluted with air that has not been removed during expiration – the alveoli do not collapse! Nevertheless the haemoglobin in the lung capillaries normally becomes saturated with oxygen.

Four oxygen molecules combine with one haemoglobin molecule to form oxyhaemoglobin.

$$Hb \quad + \quad 4O_2 \quad \rightarrow \quad HbO_8$$
(haemoglobin)

The first oxygen molecule that combines with the haemoglobin alters the shape of the molecule in such a way that it is easier for the next oxygen molecule to join, and so on. The same thing happens in reverse when the oxygen is unloaded. It becomes progressively harder for the oxyhaemoglobin to give up its oxygen. These facts account for the shape of the curve in Figure 33.2. A small increase in the partial pressure of the oxygen in the alveoli causes the blood to become rapidly saturated with oxygen.

Also, a small drop in the oxygen level of respiring tissues will result in oxygen readily being unloaded by the oxyhaemoglobin.

Adaptations to low oxygen concentration

There are a number of haemoglobins, which differ in their affinity for oxygen. Animals that live at high altitude, such as the llama, have haemoglobin which loads more readily with oxygen. Such animals show a dissociation curve which is displaced to the left of 'normal'.

The more a curve is displaced to the left, the more readily it loads with oxygen but the less readily it releases it.

Figure 33.3 compares the oxygen dissociation curve for the llama with that for humans.

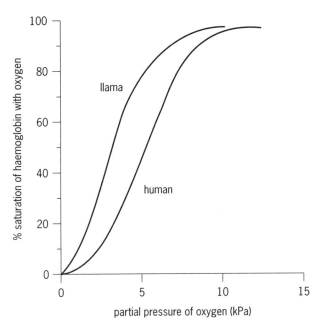

Figure 33.3 Oxygen dissociation curves for llama and human

Similar displacements to the left will be found for any animal living in an environment where the oxygen concentration is low. One example is marine worms living in muddy waterlogged burrows.

Animals which have high metabolic rates such as flying birds and small mammals with a large surface area/volume ratio, have haemoglobin which releases oxygen readily to the tissues. In these examples the oxygen dissociation curve is displaced to the right.

The more a curve is displaced to the right, the less readily it loads with oxygen but the more readily it releases it.

Humans that move to a high altitude show signs of high altitude sickness. This is caused by the body breathing at a faster rate to obtain more oxygen. This removes more carbon dioxide from the body, which in turn removes the stimulus to the brain for breathing. The results are headaches, dizziness and even unconsciousness. Acclimatisation, which is only fully achieved by those who are born at high altitude, includes a sustained increase in the rate of breathing together with a higher cardiac output plus an increase in the number of red blood corpuscles.

The Bohr effect

Once blood has travelled to the body tissues, oxygen is released because of a drop in the partial pressure of the oxygen and a rise in the partial pressure of carbon dioxide in respiring cells. A rise in the partial pressure of the carbon dioxide lowers the affinity of haemoglobin for oxygen which is therefore released. This is called the Bohr effect or the Bohr shift.

An increase in the partial pressure of carbon dioxide shifts the oxygen dissociation curve to the right.

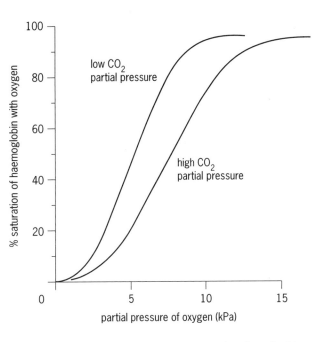

Figure 33.4 Effect of the partial pressure of carbon dioxide on the oxygen dissociation curve

An increase in temperature, such as that which occurs in muscle during exercise, also lowers haemoglobin's affinity for oxygen and extra oxygen is therefore unloaded from the blood. The oxygen dissociation curve shifts to the right.

Fetal haemoglobin

During pregnancy the developing fetus pushes upwards against the mother's diaphragm. This would lower her ability to obtain oxygen but there is a widening of the thorax, plus a 40% increase in breathing rate, which together achieve an actual increase of about 20% in oxygen consumption.

The fetus has a very high oxygen demand. The fetal haemoglobin is of a type which has a higher affinity for oxygen than the mother's haemoglobin. Oxygen is therefore readily unloaded from the mother's blood to the fetal blood. A graph for fetal haemoglobin shows a shift to the left from the curve for adult haemoglobin.

At birth, the fetal haemoglobin is replaced by adult haemoglobin.

1 a i Name the cells in a mammal responsible for the transport of oxygen. (1)
ii Name the chemical in these cells which is responsible for carrying oxygen. (1)
iii Name the chemical formed when the human respiratory pigment combines with oxygen. (1)
iv Name the chemical formed when the human respiratory pigment combines with carbon monoxide. (1)
v Name the unit which is used to express the pressure of a force in Newtons acting on an area measured in square metres. (1)
vi Name the effect of carbon dioxide in shifting the oxygen dissociation curve of haemoglobin to the right. (1)

b Describe the effect on the normal oxygen dissociation curve in the following:
i An increase in carbon dioxide tension.
ii Fetal haemoglobin.
iii Haemoglobin of animals accustomed to living at high altitude.
iv Haemoglobin of animals living in burrows where oxygen is scarce. (4)

2 State three properties which make haemoglobin an effective respiratory pigment. (3)

3 a Graph A below shows the dissociation curves for human haemoglobin in the absence of carbon dioxide and in the presence of 5% carbon dioxide.
Graph B shows similar curves for crocodile haemoglobin.

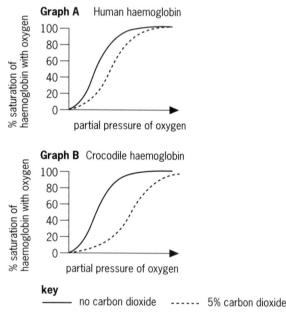

i Use Graph A to describe the effect of carbon dioxide on human haemoglobin. (1)
ii Explain how this effect enables respiring tissues to obtain oxygen. (2)
iii Crocodiles are able to stay under water longer than humans. Explain how the different effect of carbon dioxide on the dissociation of their haemoglobin helps them to do this. (2)

b The diagram below shows the structure of crocodile haemoglobin.

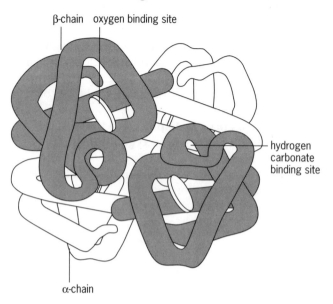

What is the evidence that this protein has:
i a tertiary structure (1)
ii a quaternary structure? (1)
(For revision of structure of proteins, see **2** Biological molecules.)

c In terms of molecular shape, suggest how the presence of hydrogen carbonate binding sites might account for the amount of oxygen released by the crocodile haemoglobin. (2)
AEB, Paper 3, June 1997

4 The diagram below shows exchange taking place between a red blood corpuscle (RBC) and the fluid surrounding active muscle tissue.

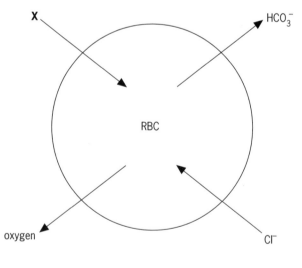

a Name **X**. (1)
b Name an organ of the body where this pattern of exchange would not occur. (1)
c Explain fully how the release of oxygen is brought about. (4)
d Suggest a reason for the movement of chloride ions shown in the diagram. (1)
Welsh, Paper B4, June 1997

5 The graph below shows the oxygen dissociation curve for mouse haemoglobin and elephant haemoglobin. The experiment was performed under conditions of constant pH and temperature and carbon dioxide partial pressure.

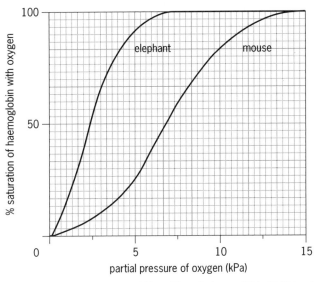

(*Adapted from: Lorimer and Schmidt Nielson.*)

a Explain why it is important to control the pH in this experiment. (1)

b Describe the effect of decreasing the partial pressure of oxygen on the percentage saturation with oxygen in mouse haemoglobin. (3)

c **i** Which animal has haemoglobin with the lowest affinity for oxygen? (1)
ii Give evidence which supports your answer to question **c i.** (1)

d A mouse is a small but metabolically active mammal compared to an elephant. How are the differences in the two haemoglobin dissociation curves related to this fact? (2)

e Draw on the graph a line to represent the haemoglobin dissociation curve you would expect for a cat. (1)

6 A small amount of oxygen diffuses from the blood into the small intestine of a mammal. Some parasite platyhelminths living in the small intestine can make use of this oxygen. A graph shows oxygen dissociation curves for human haemoglobin and for the haemoglobin of a parasitic platyhelminth which lives in the human intestine.

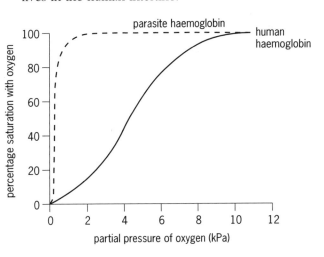

a Use the graph to help you to explain how human haemoglobin releases oxygen when it reaches the cells of the wall of the small intestine. (3)

b Explain one advantage to the parasite of having haemoglobin with an oxygen dissociation curve like that shown on the graph. (2)

AEB, Paper 3, June 1996

1 a i erythrocytes/red blood cells
  ii haemoglobin
  iii oxyhaemoglobin
  iv carboxyhaemoglobin
  v Pascal
  vi Bohr effect/shift
 b i Shifts to the right.
  ii Shifts to the left.
  iii Shifts to the left.
  iv Shifts to the left.

2 • Its four haem (Fe) groups have an affinity for oxygen/can carry oxygen/it can exist in two forms, haemoglobin and oxyhaemoglobin.
 • The shape of the molecule is changed by the uptake of one oxygen molecule, making it progressively easier for the next oxygen to combine, resulting in the steep sigmoid dissociation curve/haemoglobin becomes rapidly saturated with oxygen at the normal partial pressures found in the lungs.
 • The oxygen dissociation curve is shifted to the right by an increase in carbon dioxide concentration/decrease in pH/Bohr shift/releasing oxygen more readily to actively respiring cells/as oxygen partial pressure drops.

3 a i Carbon dioxide displaces the curve to the right/decreases the affinity for oxygen/unloads at a higher partial pressure/Bohr effect.
  ii Respiring tissues produce carbon dioxide;
  so haemoglobin can remain less saturated with oxygen/more oxygen is released to the tissues.
  iii In the crocodile there is a larger displacement of the curve/Bohr effect;
  so it gives up more of the oxygen carried by the haemoglobin.
 b i The twisted shape of the haemoglobin chains/it is folded into a globular shape.
  ii The molecule is made of more than one chain.
 c Hydrogen carbonate binds to the binding site;
  and
  altering the shape of haemoglobin;
  or
  which can no longer bind to oxygen.

4 a X = carbon dioxide
 b the lungs
 c Oxygen is carried in the RBC as oxyhaemoglobin.
  Carbon dioxide combines with water to form carbonic acid.
  This ionises to hydrogen carbonate ions and hydrogen ions.
  The raised hydrogen ion concentration (increased acidity) causes decomposition of oxyhaemoglobin to release oxygen.
 d The movement of the chloride ions preserves the electrical balance on either side of the membrane/replaces the negative hydrogen carbonate ions diffusing out.

5 a A change in pH will shift the curve (decrease in pH – shift to the right)/pH will alter the saturation with oxygen.
 b Three of:
  • Mouse haemoglobin is saturated with oxygen at 14 kPa partial pressure of oxygen.
  • As the partial pressure of oxygen falls the percentage saturation decreases slowly at first (give figures from the graph).
  • Percentage saturation decreases more quickly between 10 and 5 kPa.
  • The rate of decrease slows as percentage saturation falls below 5 kPa/between 5 and 0 kPa.
 c i mouse
  ii At the lower partial pressures of oxygen mouse haemoglobin has a lower saturation than elephant haemoglobin.
 d Mouse haemoglobin will release oxygen at higher partial pressures of oxygen than the elephant haemoglobin. Mouse cells will have a higher oxygen requirement to supply energy to maintain body temperature and drive chemical reactions (mouse has higher surface area to volume ratio).
 e Line of similar shape drawn between the mouse curve and the elephant curve.

6 a Three of:
  • There is a low partial pressure of oxygen in the cells of the small intestine.
  • Human haemoglobin is only partially saturated at this partial pressure (at the intestine)/haemoglobin releases oxygen at low partial pressures of oxygen.
  • Oxygen diffuses out of the blood, into the cells.
  • The effect of increased carbon dioxide results in the Bohr shift.
 b The parasite has very little oxygen in its environment. The parasite can gain more oxygen because its haemoglobin becomes saturated at a low partial pressure of oxygen.
  or
  The parasite haemoglobin holds on to oxygen even when there is very little present since it has a high affinity with oxygen.
  This functions as an emergency store for use when needed.
  or
  The parasite haemoglobin can get oxygen from the human haemoglobin;
  since it has a higher affinity for oxygen than human haemoglobin.

34 Structure and function of the kidneys: Ultrafiltration

The kidneys are organs of both excretion and homeostasis.

> **Excretion is the getting rid of the waste products of metabolism. Homeostasis is the maintenance of a constant internal environment.**

In the liver, waste nitrogen from the breakdown of amino acids is incorporated into urea. The kidneys excrete urea which passes in the urine along the ureters to the bladder where it is temporarily stored.

The excretory function of the kidneys cannot be separated from their role as homeostatic organs. The kidneys regulate the water content of the blood and tissue fluids, together with the concentrations of salts and acids. Control of the water (osmotic) potential in organisms is known as **osmoregulation**.

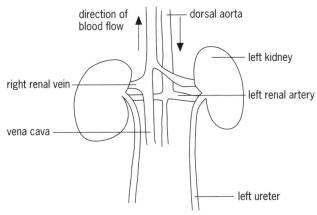

Figure 34.1 Kidneys and associated blood vessels in mammals

Ultrafiltration in the mammalian kidney

Urea is filtered under pressure from the blood as it passes through the kidneys – a process known as **ultrafiltration**.

The kidneys are protected by a layer of hard fat. They are about 8–10 cm in length and 3–4 cm wide. Each kidney is packed with blood vessels and about 1 million **nephrons**, giving a total of about 25 miles of tubules for each kidney. Each nephron consists of a **Malpighian body** (or Malpighian corpuscle) together with its associated kidney tubule.

The cortex, which is a darker pinkish-brown colour, contains the Malpighian corpuscles and the convoluted parts of the kidney tubules. The medulla, which is paler, contains the loops of Henle and the collecting tubules.

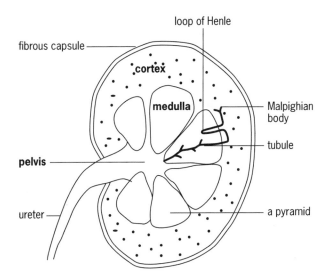

Figure 34.2 Internal structure of a mammalian kidney

To understand how ultrafiltration works we need to look in detail at the structure of a Malpighian body. Each Malpighian body consists of a cluster of capillaries within a cup-shaped Bowman's capsule.

Note that both the **efferent** and **afferent** blood vessels are arterioles. After leaving the glomerulus, the efferent vessels once more form capillaries which surround the tubules before rejoining to form venules in the usual manner.

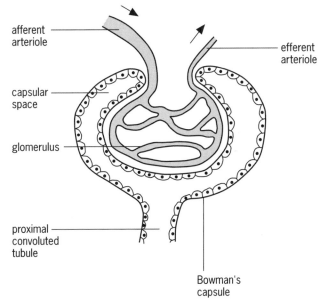

Figure 34.3 A Malpighian body

The kidneys are situated close to the dorsal aorta and are supplied with blood by short renal arteries. This causes the kidneys to have a higher blood pressure than other organs, and this pressure is maintained because the afferent arterioles, carrying blood into the glomeruli, are larger in diameter than the efferent arterioles. Therefore the blood in the glomeruli is filtered under pressure.

The capillaries of the glomeruli consist of a single layer of thin endothelial cells perforated by minute pores about 0.1 μm in diameter which allow all the contents of the blood to pass through except blood cells.

The endothelial cells are very close to a basement membrane which is the real site of ultrafiltration as plasma proteins cannot pass through.

Specialised cells in the wall of the Bowman's capsule, called **podocyte cells**, also let all of the contents of the blood, apart from blood cells, pass through. Since filtration has already been effected by the basement membrane, the precise function of the podocyte cells is not understood.

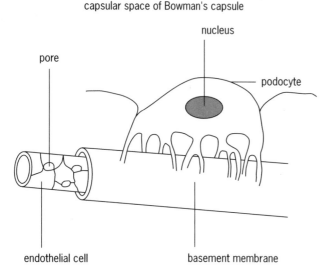

Figure 34.4 Podocyte cell and blood capillary from the inner layer of the Bowman's capsule

The fluid which is forced out of the capillaries of the glomerulus into the capsular space of the Bowman's capsule contains urea and uric acid, amino acids, glucose, vitamins, hormones and water.

Ultrafiltration is a passive process and which substances are filtered depends entirely upon their molecular size. The system is efficient but some of the substances which have been filtered are required by the body; as the filtrate passes along the kidney tubules both passive and active uptake takes place in order to:

- reabsorb useful molecules such as glucose and amino acids which are needed by body cells
- reabsorb most of the water plus sodium and chloride ions.

Without this reabsorption we would produce nearly 200 dm³ (litres) of urine a day instead of about 2 dm³ which is the approximate volume produced.

Reabsorption in the kidney tubule

The kidney tubule can be divided into three distinct areas:

- proximal convoluted tubule
- loop of Henle
- distal convoluted tubule.

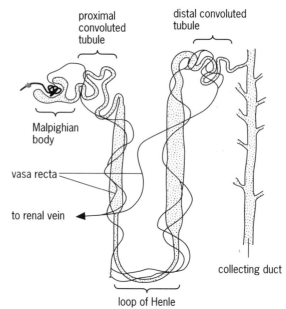

Figure 34.5 A kidney nephron

To enable efficient reabsorption, the kidney tubule is adapted as follows.

- The proximal and distal tubules are long and winding. Therefore both the surface area, and the time available for reabsorption, are increased.
- The tubules have walls only one cell thick. (Diffusion is efficient only over very small distances.)
- The cells of the proximal and distal convoluted tubules have a brush border of microvilli (Figure 34.6). These greatly increase the surface area for reabsorption. They also contain numerous mitochondria which are the sites of aerobic respiration. The energy is required for active uptake when substances are moved across cells against the diffusion gradient.
- The bases of the proximal tubule cells are irregular in shape where they are adjacent to a blood capillary and there are numerous intercellular spaces. The cells are ideally adapted for the diffusion of absorbed substances into the capillary network that surrounds the tubules.

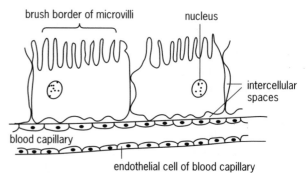

Figure 34.6 TS of cells from the proximal convoluted tubule

1 Describe the following processes with reference to the kidney:
   a homeostasis   (2)
   b ultrafiltration   (2)
   c excretion   (2)
   d osmoregulation   (2)

2 Write a list of the most suitable word(s) to fill the gaps in the following account.

   The kidneys are supplied with blood by the _____ artery, which splits into arterioles supplying the knot of capillaries known as the _____ inside the Bowman's capsule of each _____. The blood pressure is high because the diameter of the _____ arteriole entering the capillary knot is _____ in diameter than the arteriole which leaves the capillary knot. This causes the process known as _____ to occur.   (6)

3 **a** Name the parts labelled **A**, **B**, **C** and **D** on the diagram of the kidney and associated parts. (4)

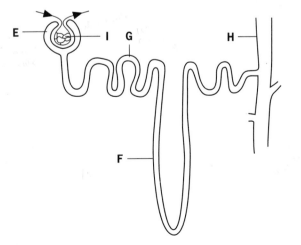

   b Name the parts labelled **E**, **F**, **G**, **H** and **I** on the diagram of a kidney tubule.   (5)

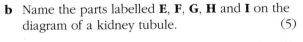

   c In which region of the kidney would you expect to find **E**?   (1)

4 Explain why eating extra protein may have a greater effect on the urea content of the urine of an adult than in a child.   (3)

5 The graph below shows how the rates of filtration and reabsorption of glucose in human nephrons vary with concentration of glucose in the blood plasma.

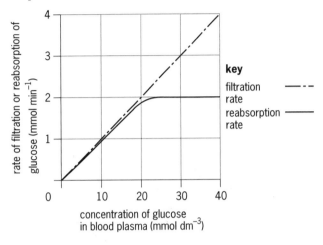

   a **i** From the graph determine the maximum rate of glucose reabsorption.   (1)
       ii Comment on the relationship between filtration rate and reabsorption rate as shown by these two curves.   (3)
   b If the concentration of glucose in the blood plasma exceeds $20\,mmol\,dm^{-3}$ glucose appears in the urine. Suggest an explanation for this observation.   (2)

       London, Paper B3, January 1997

6 The graph shows the rate of glucose reabsorption in, and excretion from, a mammalian kidney in relation to the glucose concentration in the plasma.

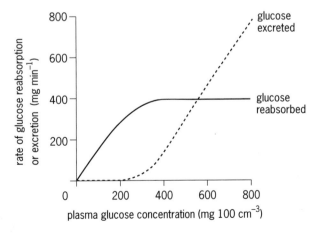

   a Draw a line on the graph to show the rate of filtration of glucose in the renal capsule   (1)
   b In which part of the nephron is glucose reabsorbed?   (1)
   c Explain the shape of the glucose reabsorption curve.   (3)

   AQA, Specification A Specimen Paper 6, for 2002 exam

7 a Name the kidney region where the proximal convoluted tubule is found. (1)

The graph shows the glucose concentration at various points from the beginning of the proximal convoluted tubule.

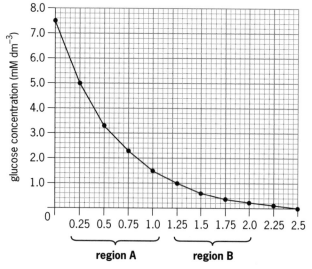

distance from beginning of tubule (mm)

b Assume that the fluid moves along at a speed of $0.5\,\text{mm}\,\text{s}^{-1}$.
Showing your working:
i Calculate the average rate of change in $\text{mM}\,\text{dm}^{-3}$ for region A (0 to 1 mm). (2)
ii Calculate, as a percentage of the concentration at 0 mm, the fall in the glucose concentration between 0 and 0.25 mm. (2)
iii Over region B (1.25 to 5 mm) the values corresponding to those you have calculated are $0.53\,\text{mM}\,\text{dm}^{-3}\,\text{s}^{-1}$ and 10.67%.

Compare these values with your calculated values and suggest the main factor influencing the uptake of glucose. (1)

c Where else in the body would glucose concentration be very similar to that at point 0 mm? (1)

d i Suggest a medical condition which would result in higher values at point 0 mm. (1)
ii Explain the simplest way in which a doctor could detect this condition. (1)

Welsh, Paper B1, June 1997

1 a Homeostasis is keeping conditions inside the body (the internal environment) constant.
The kidney is concerned with the control of water/solute/osmotic potential.

b Ultrafiltration is the process by which small molecules in the blood are forced under pressure across the basal membrane of the endothelial cells of the Bowman's capsule into the nephron.
It is a passive process.

c Excretion is a process which eliminates the waste products formed by chemical reactions occurring in the cells.
The kidney excretes the toxin urea and excess mineral salts.

d Osmoregulation means controlling the water and solute balance.
The kidney controls the osmotic potential of the blood, lymph and tissue fluids under the influence of ADH (produced by the hypothalamus and released by the posterior pituitary).

2 renal; glomerulus; nephron; afferent; wider; ultrafiltration

3 a **A** = vena cava; **B** = ureter; **C** = renal vein; **D** = left kidney

b **E** = Bowman's capsule; **F** = loop of Henle; **G** = first (proximal) convoluted tubule; **H** = collecting duct; **I** = glomerulus

c cortex

4 Protein is first digested to amino acids.
Excess amino acids are deaminated and the amine group converted to urea.
Children will use a relatively high proportion of amino acids for growth compared with adults.

5 a i $2\,mmol\,min^{-1}$.

ii Three of:
  - Filtration rate proportional to concentration of glucose in blood plasma.

  - Filtration rate and reabsorption rate are the same up to $10/18$–$20\,mmol\,dm^{-3}$ glucose concentration.
  - Filtration rate and/or reabsorption rate proportional to concentration of glucose in blood up to 18–$22\,mmol\,dm^{-3}/2\,mmol\,min^{-1}$;
  - above this level reabsorption rate remains constant/reabsorption rate stays level/levels off.

b Two of:
  - Filtration rate exceeds/greater than/more than reabsorption rate;
  - therefore more glucose in glomerular filtrate/in first tubule/proximal convoluted tubule, than can be reabsorbed;
  - because reabsorption rate at maximum level/rate does not increase above $2\,mmol\,min^{-1}$/reason for limitation of reabsorption such as reference to ATP or insufficient number of carrier molecules available.

6 a Line continues from glucose reabsorption, directly proportional to glucose concentration.

b Proximal/first convoluted tubule.

c Glucose is taken up by active transport.
The more glucose taken up, the more the sites on the transporter protein are in use;
these are eventually all saturated.

7 a the cortex

b i $7.5 - 1.5 = 6$ in $1\,mm$
At a speed of $0.5\,mm/s$, $1\,mm$ will take $2\,s$.
Rate of change $= 6 \div 2 = 3\,mM\,dm^{-3}\,s^{-1}\,s$

ii $7.5 - 5.0 = 2.5$
$$\frac{2.5}{7.5} = \frac{33.3}{100} = 33\%$$

iii The main factor is the glucose concentration (or the gradient decreasing).

c In the blood.

d i Diabetes/hyperglycaemia (this word means high sugar).

ii A doctor could detect this by the presence of sugar in the urine or by elevated levels of sugar in the blood.

35 Structure and function of the kidneys: Homeostasis

The proximal convoluted tubule

This is the longest part of the nephron. More than 80% of the glomerular filtrate is reabsorbed back into the blood.

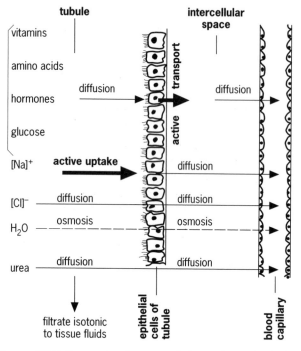

Figure 35.1 How molecules move during reabsorption in the proximal convoluted tubule

Note the following points:

- The vitamins, amino acids, hormones and glucose diffuse into the epithelial cells of the proximal convoluted tubule. They are then actively transported into the inter-cellular spaces between the epithelial tubule cells and the blood capillaries. Finally they diffuse into the surrounding capillaries. Because these substances are actively transported into the intercellular spaces, and are being constantly removed by the bloodstream, a diffusion gradient is maintained and further molecules diffuse into the tubule cells from the lumen of the kidney tubule.
- The sodium ions (Na$^+$) are actively transported into the tubule cells. This raises the osmotic potential in the cells and water enters by osmosis. The water molecules, together with the chloride ions, then move along diffusion gradients into the capillary network.
- About 50% of the urea which was filtered out now passes back into the blood capillaries by diffusion. Not all of the urea can be excreted first time round!
- The net result is a tubular filtrate which is isotonic with (i.e. the same concentration as) the blood in the surrounding capillaries. Fine adjustments take place in the following parts of the tubule.

The loop of Henle

Mammals are constantly losing water through sweating (or panting), breathing, lactation, urination and defecation. Some animals may take in more water than is needed. The control of osmotic potential in organisms is called **osmoregulation**. Osmotic potential changes when water levels change.

In a dry environment, or in the sea, organisms (both plants and animals) tend to lose water and may have adaptations to conserve water (for example, in a wet environment, or in fresh water, organisms may have adaptations to remove excess water).

The loop of Henle in mammals reabsorbs water so that **hypertonic** urine (i.e. urine more concentrated than tissue fluids) can be produced and water conserved.

The function of the loop of Henle is to create a water potential gradient between the renal filtrate and the peritubular (medullary) fluid which bathes the loop.

The length of the kidney nephrons varies depending on an animal's natural habitat. Animals such as the otter, which live in an aquatic environment, tend to have much shorter loops of Henle than those, such as the rabbit, living in an environment in which the conservation of water may be a very important factor. The longer the loop, the more water can be reabsorbed.

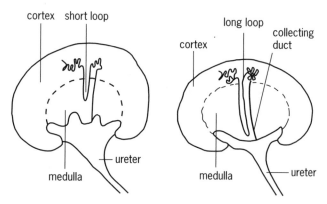

Figure 35.2 Loop of Henle in **a** an otter's kidney, and **b** a rabbit's kidney

The loop of Henle creates a water potential gradient between the renal filtrate and the peritubular fluid (tissue fluids surrounding the tubule) by establishing a **countercurrent exchange**. This mechanism depends on two fluids flowing in opposite directions and therefore the descending and ascending limbs need to be considered together. Note the following points.

- The filtrate which enters the loop of Henle is isotonic with the peritubular fluid.
- Only the thin part of the descending limb is freely permeable to water.

- The sodium and chloride ions which are actively pumped out of the ascending limb into the peritubular fluid increase the concentration of the peritubular fluid. This causes water in the descending limb to be drawn out by osmosis. The water moves straight into the capillaries (**vasa recta**) surrounding the loop of Henle and the peritubular fluid is not diluted.
- The effect is cumulative and therefore the filtrate at the bottom of the descending loop is more concentrated than at either end.
- The system only works because the fluid in the two limbs flows in opposite directions. This, and the fact that the effect is cumulative, accounts for the term that is used – **countercurrent multiplier**.
- The filtrate entering the distal tubule is hypotonic to (less concentrated than) the peritubular fluid. **Aldosterone**, a hormone produced by the cortex of the adrenal glands, causes active uptake of sodium ions from the filtrate in the ascending limb of the loop of Henle and the distal tubule into the plasma. Water passes out by osmosis from the distal tubule, thus further concentrating the filtrate.

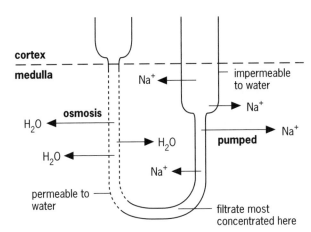

Figure 35.3 Loop of Henle – a countercurrent multiplier.

The permeability to water of the wall of the distal tubule and collecting duct is controlled by the hormone **ADH** (antidiuretic hormone). Accurate control of the water and salt balance of the body is maintained. If the walls are permeable, water leaves by osmosis and more concentrated urine is produced. If the walls are impermeable, the urine will be less concentrated.

If no ADH is produced a disease known as *diabetes insipidus* occurs. Because the tubes are permanently impermeable to water, large quantities of very dilute urine are produced and the patient is constantly thirsty.

In addition, the distal tubule controls the pH of the blood by excreting hydrogen ions and retaining hydrogen carbonate ions if the blood is acid, and the reverse if the pH rises.

$$CO_2 + H_2O \rightleftharpoons H_2CO_3 \rightleftharpoons [H]^+ + [HCO_3]^-$$

The carbon dioxide comes from respiration and the carbonic acid is present in the blood.

Excretion by terrestrial insects

Insects are covered with a waxy cuticle on the exoskeleton which prevents the loss of water by evaporation. Water loss from excretion is minimised by the production of uric acid instead of urea. The uric acid passes into the Malpighian tubules of the gut which are bathed in the animal's blood in a blood cavity, or **haemocoel**. Here the uric acid is further concentrated into uric acid crystals. These crystals eventually pass into the rectum and are mixed with the waste materials of the digestive system to form a very concentrated product for excretion.

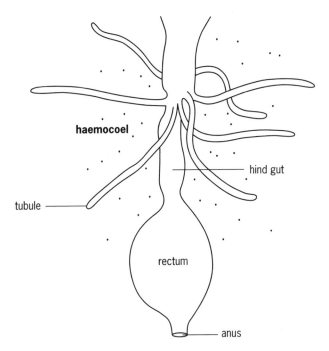

Figure 35.4 Malpighian tubules of an insect

Excretion by fresh water animals

Animals which live in fresh water have the problem that their aquatic environment is hypotonic to their body fluids and they are therefore subject to the osmotic uptake of water.

Fishes have a covering of scales which is impermeable to water but the gills are highly permeable. Apart from taking up water, the gills of fresh water fishes are the site of salt loss and gaseous exchange and are the main excretory organs of the body.

The kidneys have a large number of glomeruli. Salts are selectively reabsorbed into the blood but there is still some salt loss, and this deficiency must be made up from the diet and by active uptake from the surrounding water. The net result is the production of copious amounts of very dilute urine.

Ammonia is the nitrogenous product excreted. Although it is very toxic, it is also very soluble and has a rapid rate of diffusion. For these reasons most of the ammonia is excreted by the gills although some is excreted by the kidneys.

1 a Name five ways in which water is lost by mammals. (5)

b Name the nitrogenous excretory product of:
  i mammals
  ii terrestrial insects
  iii fish (3)

c Explain the term:
  i isotonic
  ii hypotonic (2)

d Name the site of uptake of much water by osmosis in fresh water fish. (1)

2 The graph shows the concentration of the solutes in the fluid in the different regions of a nephron from a human kidney. Curve A shows the concentration in the presence of the hormone ADH. Curve B shows the concentration when no ADH is present.

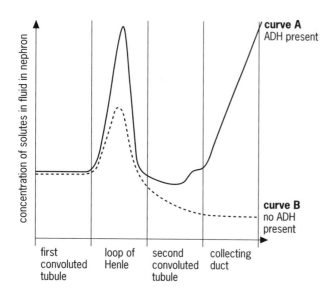

a Although solutes such as glucose are reabsorbed in the first convoluted tubule the concentration of these solutes does not change as the fluid flows along this part. Explain why. (1)

b What causes the concentration of the solutes to increase, then decrease as the fluid flows through the loop of Henle? (4)

c Explain the difference in concentration of the fluid in the collecting duct in curves A and B. (2)
NEAB, BY03, February 1997

3 a State where in the nephron glucose is reabsorbed. (1)

b Describe how glucose is reabsorbed. (2)

c Name two other substances which are reabsorbed in the same part of the nephron as glucose. (2)
London, Paper B3, January 1997

4 A diagram of a cell from the wall of the proximal convoluted tubule is given. In this region of the nephron the concentration of glucose inside the tubule falls to zero.

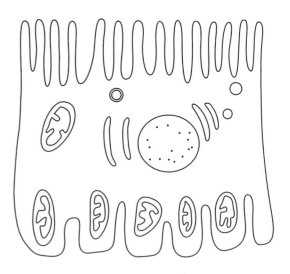

a Explain how the glucose concentration is decreased. (3)

b Explain how the structure of the cell is adapted to bring about this process. (4)

c What other substances are removed from the nephron filtrate in the proximal tubule? (4)

5 The diagram below shows the simplified structure of a kidney tubule (nephron).

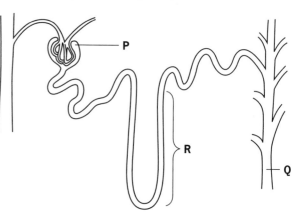

a In the table below, columns 1 and 2 show the quantities of water, glucose and urea passing through **P** and **Q** in a 24 hour period. Columns 3 and 4 show the quantities and percentages reabsorbed during the same period. Complete the table by writing the correct figures in the boxes labelled **i** to **iv**. (4)

Substance	Quantity passing through P	Quantity passing through Q	Quantity reabsorbed	Percentage reabsorbed
water	$180\,dm^3$	$1.5\,dm^3$	$178.5\,dm^3$	**i**
glucose	180 g	**ii**	180 g	100
urea	53 g	25 g	**iii**	**iv**

b Describe how **R** is involved in adjusting the concentration of the filtrate as it passes through the medulla of the kidney. (3)
London, Paper B3, June 1996

1 a Sweating/panting, breathing, lactation, urination and defecation.
   b i urea
      ii uric acid
      iii ammonia
   c i Isotonic means of the same osmotic (solute) potential.
      ii Hypotonic means of a less negative osmotic (solute) potential.
   d the gills

2 a Water is also removed.
   b Four of:
      • In the descending limb water moves out (it is permeable to water);
      • by osmosis/down water potential gradient.
      • The descending limb allows the entry of sodium and chloride by passive diffusion (slow process).
      • Chloride and sodium ions are removed from ascending limb;
      • by active transport into the tissue surrounding it.
      • The walls of the ascending limb are impermeable to water.
   c With ADH present (curve A) the collecting duct becomes more permeable to water. Water will be removed by osmosis/down water potential gradient.

3 a First/proximal convoluted tubule.
   b Two of:
      • Diffusion/moves along concentration gradient.
      • Active transport/uptake.
      • Detail of mechanism, e.g. Na^+ co-transport/reference to carrier molecule/use of ATP.
      • Microvilli/brush border in epithelial cells lining the proximal convoluted tubule/many mitochondria/folded cell membrane.
   c Any two of:
      • Ions/named e.g. K^+/Na^+/Cl^-/Ca^{2+}/Mg^{2+}/SO_4^{2-}/PO_4^{3-}/HCO_3^-/NH_3/NH_4^+/organic anions.
      • Amino acids.
      • Vitamins/named example.
      • Water.
      • Urea.
      • (Small) proteins/albumin.
      • Peptide hormones.
      • Uric acid.

4 a The glucose diffuses into the wall (epithelial) cell.
      Active transport is used to pump the glucose out of the cell into the intercellular spaces.
      The glucose then diffuses into the blood and is carried away.
   b Microvilli provide a greater surface area;
      for reabsorption of glucose from the tubule.
      Mitochondria provide energy;
      for active transport.
   c Four of:
      • vitamins
      • amino acids
      • some water
      • some urea
      • some Na^+
      • some Cl^- ions.

5 a

Substance	Quantity passing through P	Quantity passing through Q	Quantity reabsorbed	Percentage reabsorbed
water	180 dm^3	1.5 dm^3	178.5 dm^3	(178.5/180 x 100) **99.17**
glucose	180 g	(180–180) **0.0 g**	180 g	100
urea	53 g	25 g	(53–25) **28 g**	(28/53 x 100) **52.8**

   b Three of:
      • R is impermeable to water but permeable to sodium chloride ions.
      • Chloride/sodium ions are moved out by active transport from R.
      • As the sodium/chloride ions only ever enter the descending limb slowly;
      • the surrounding tissue/medulla/interstitial spaces will become more concentrated than the filtrate.
      • Water will pass out of Q/descending limb/distal convoluted tubule/collecting duct;
      • by osmosis;
      • the filtrate becomes more concentrated.

36 Sexual reproduction in plants and animals

> **Sexual reproduction is a method of reproduction in which male gametes fertilise (fuse with) female gametes to form zygotes. The zygote develops into an individual organism which will eventually produce gametes during meiosis.**

The advantage of sexual reproduction is that variation occurs due to **meiosis** (**37** The continuity of life: Meiosis). The *variation* is due to crossing over and the independent segregation of **gametes**. There is also random fusion of gametes. Compare this with mitosis in which the chromosome number is not halved and there is no variation.

The names of gametes in animals and plants

	Plants	Animals
male gamete	generative nucleus in the pollen grain	sperm
female gamete	egg cell (found in the embryo sac)	ovum (egg)

Gametes are formed during a process known as **gametogenesis**. The following descriptions apply to *humans and other mammals*. (In plants the processes are similar but more complex.)

1 **Spermatogenesis** is the formation of spermatozoa in the male.
2 **Oogenesis** is the formation of ova (eggs) in the female.

Note the following points:

- Meiosis is a reduction division, in which the number of chromosomes is halved from the diploid state ($2n$) to the haploid state (n), followed by a mitotic division with the production of *four* haploid gametes.
- Ova (eggs) are produced in relatively small numbers compared to spermatozoa (sperm) and the only function of the polar bodies produced during oogenesis is to receive chromosomes during cell divisions. As the ovum develops, the polar bodies degenerate.
- The secondary oocyte does not divide to form the ovum until fertilisation takes place. There is no pause between the two divisions of meiosis in the testes. The spermatids which are formed pass into the seminiferous tubules (sperm ducts) of the testes where they mature into motile spermatozoa.

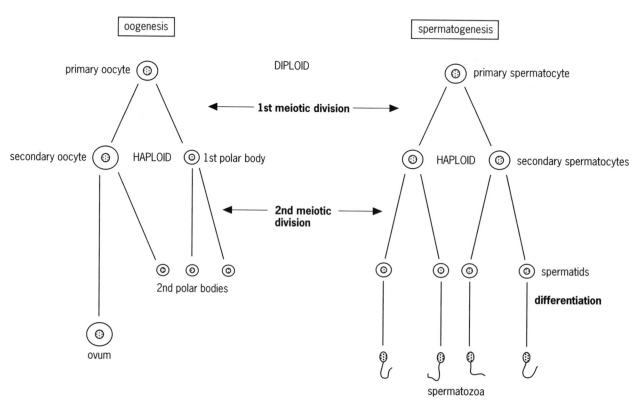

Figure 36.1 Gametogenesis

Sexual reproduction in flowering plants

Structure of a typical insect-pollinated flower

The shape, perfume and colours of the flower, the presence of nectaries (which produce nectar which feeds many insects) and markings, which guide insects (especially bees) to the nectaries, all play a part in attracting insects to flowers. The visiting insects enable pollination to take place.

> **Pollination is the transfer of pollen from the anthers to the stigma.**

Most flowering plants have bisexual flowers – they are **hermaphrodite**. Some, however, have separate male and female flowers on the same plant and are said to be **monoecious** (e.g. hazel). Other flowering plants are unisexual (**dioecious**) and bear flowers of only one sex (e.g. willow).

The **gynoecium** constitutes all the female parts of a flower:

- **Stigma** – sticky surface on which pollen is deposited. The sugary surface aids germination of the pollen grains.
- **Style** – a tube connecting stigma to ovary.
- **Ovary** – produces the ovules which contain the gametes (egg cells). After fertilisation the wall of the ovary develops into the fruit (which may be dry or succulent) and the ovules ripen into seeds.

Collectively, the stigma, style and ovary are known as the pistil.

The **androecium** constitutes all the male parts of the flower:

- **Anthers** – produce the pollen. Anthers typically have four pollen sacs within which the pollen grains (which contain the male gametes) are produced during meiosis. When ripe, the anthers dehisce (split) to release the pollen. Typically, the anthers ripen some time before the stigmas of the same flower. This mechanism avoids self fertilisation and is known as **protandry**. The adjective is **protandrous**. Plants whose female gametes ripen before the male gametes are said to be protogynous.
- **Filaments** – thin flexible stalks which bear the anthers.

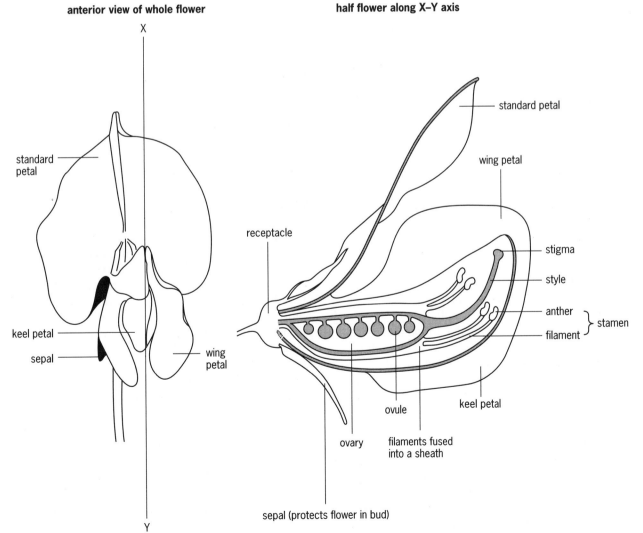

anterior view of whole flower

half flower along X–Y axis

Figure 36.2 Parts of the sweet pea flower (pollinated mainly by bees)

Structure of a typical wind-pollinated flower

Wind-pollinated flowers appear early in the season before the leaves (e.g. willow catkins) or else they are borne on long stalks which raise them so they catch the wind (e.g. grasses). In addition:

- they possess no petals (which would interfere with the pollen being blown away), scent or nectaries
- the stigmas are feathery – which gives them a greater surface area for catching the wind-blown pollen
- the filaments of the stamens are often long and thin so that the anthers are well exposed to the wind.

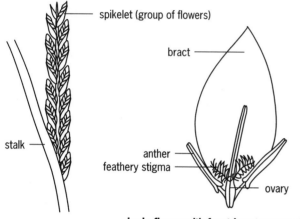

single flower with front bract removed

Figure 36.3 Structure of perennial rye-grass (*lolium perenne*); wind-pollinated, 10–90 cm high

Fertilisation

> **Fertilisation is the fusion of male and female gametes to form a zygote.**

In animals

Fertilisation may be *external* (in an aquatic environment – e.g. fishes and amphibians) or *internal* (e.g. reptiles, birds and mammals). In internal fertilisation the sperm, which have to swim to the egg(s), are ejaculated in a fluid (seminal fluid). Sperm + seminal fluid = semen. The seminal fluid is produced by the seminal vesicles (which also store sperm) plus the prostate gland.

Fertilisation in humans normally takes place in the upper part of an oviduct (fallopian tube) if an egg has been released during ovulation during the previous 8 to 24 hours. (For details of ovulation see **24** The endocrine system: Insulin and glucagon; sex hormones.)

The head of a sperm contains the **acrosome** which is derived from the **Golgi apparatus** (see **1** Cells and tissues).

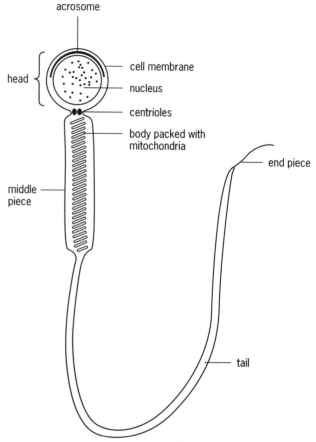

(length approximately 120 μm)

Figure 36.4 Diagram of human sperm

Enzymes from the acrosome are released when its membrane ruptures on contact with an egg. These enzymes aid penetration of the egg membrane which then allows the sperm nucleus to transfer and fuse with the egg nucleus. When the sperm enters the egg cell, the nucleus of the egg undergoes its second meiotic division to become the ovum proper.

Only one sperm enters the egg during fertilisation.

Immediately after fertilisation, the **zygote** starts to divide and a cluster of 16 to 32 cells (now an embryo) passes into the uterus where the embryo becomes embedded in the endometrium (uterine wall). The embryo produces peptidase (a protein-digesting enzyme) which digests part of the endometrium, thus making a space for it to become implanted. This is known as **implantation**.

In mammals the developing embryo is nourished by the placenta with its large surface area of villi and microvilli for the absorption of food. The mother's and fetus's blood supplies are kept separate and a steep diffusion gradient is maintained by the counter-flow of the two blood systems. This is known as a countercurrent exchange system. It is dependent on the two blood systems flowing in opposite directions and having a concentration gradient between them (see also **35** The structure and function of kidneys: Homeostasis).

In flowering plants

After the pollen grains have been formed during meiosis they secrete a thick wall and then a further mitotic division of the haploid nucleus results in the formation of two cells: one with a **generative** nucleus which will eventually fertilise the egg cell of an ovule and a second cell with a tube nucleus (see Figure 36.5).

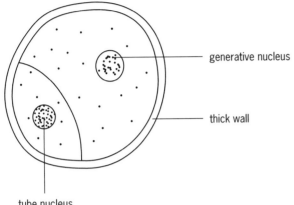

Figure 36.5 Mature pollen grain

After pollination, the pollen grains germinate and grow **pollen tubes** down the style and into the ovary where they enter the ovules via small apertures known as **micropyles**. The haploid pollen tube nucleus is situated at the tip of the pollen tube and is followed by the generative nucleus which divides by mitosis in the pollen tube to form two haploid male nuclei.

In the ovule, a **megaspore mother cell** divides by meiosis to form four haploid **megaspores**. *One* of the megaspores then grows in size and undergoes three mitotic divisions to form eight haploid cells which are arranged in a characteristic way to form the **embryo sac** (see Figure 36.6). Meanwhile, the other three megaspore cells degenerate.

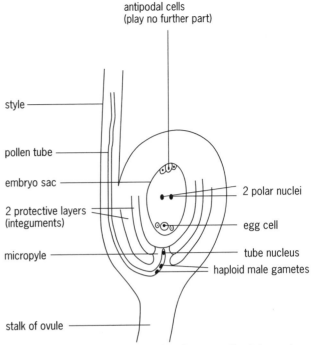

Figure 36.6 Male gametes passing down a pollen tube and entering an ovule of a flowering plant

On reaching the ovule the pollen tube disintegrates and one of the male nuclei (gametes) fuses with the nucleus of the egg cell to form a diploid zygote. The second male nucleus fuses with the diploid nucleus of the **embryo sac**. This diploid nucleus was formed by the fusion of the two haploid **polar nuclei**. This **double fertilisation**, which is unique to flowering plants, results in a **triploid nucleus** ($3n$), which then develops into the endosperm which surrounds the developing embryo and serves as a food source. **Endosperm** contains insoluble food reserves which are made available to the developing embryo by the action of enzymes which hydrolyse food reserves into soluble products. Enzyme production is initiated by the hormone **gibberellin**.

1 a Explain the importance of the part played in gametogenesis by:

  i mitosis (2)

  ii meiosis (2)

  b Fill in the following table to compare the features of insect-pollinated and wind-pollinated flowers. (4)

Feature	Insect-pollinated flower	Wind-pollinated flower
stigma		
pollen		
presence of nectaries		
anthers		

2 The diagram below shows a vertical section through a flower of the family Papilionaceae.

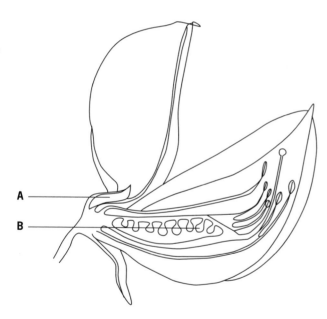

  a State one function for each of the parts labelled **A** and **B**. (2)

  b This flower is insect-pollinated. Describe three features, visible on the diagram, which are characteristic of insect-pollinated flowers. (3)

  c **i** State one similarity between a human sperm and a male gamete of a flowering plant. (1)

  ii State one difference between a human sperm and a male gamete of a flowering plant. (1)

  Edexcel (London), Paper B6, January 1999

3 The diagram shows the sequence of events which take place when the nucleus of a sperm enters the cytoplasm of an egg.

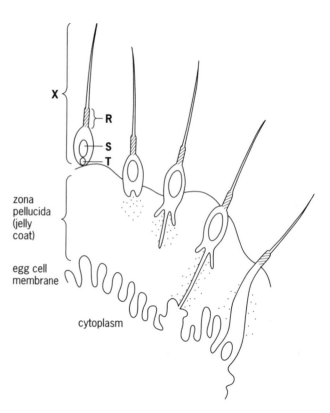

  a Name the part of the reproductive tract in which these events take place. (1)

  b **i** Complete the table to show the **names** of **R** and **S** in the diagram, the name of an organelle found in each **structure** and the **function** of these organelles. (6)

	Name	Name of organelle found in the structure	Function of organelle
R			
S			

  ii Use the information in the diagram to explain the role of **T** in the process. (2)

  c **i** Name the part of a plant which has the same function as the structure labelled **X** in the diagram. (1)

  ii State **two** similarities between the process visible in the diagram and the process by which the male nucleus enters a plant ovule. (2)

  Welsh, Paper B3, January 1997

4 The diagram shows some stages in the formation of a mammalian egg cell.

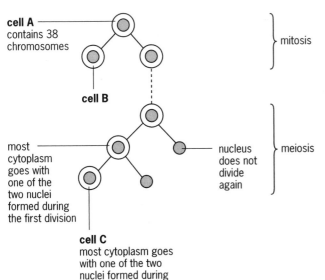

cell A contains 38 chromosomes

mitosis

cell B

most cytoplasm goes with one of the two nuclei formed during the first division

nucleus does not divide again

meiosis

cell C most cytoplasm goes with one of the two nuclei formed during the second division

a How many chromosomes will there be in:
  i cell B (1)
  ii cell C? (1)
b Suggest one advantage in the way the cytoplasm divides during meiosis. (1)
c Describe and explain two ways in which the events of meiosis cause the egg cells to be genetically different from one another. (4)

NEAB, BY02, June 1998

5 a Describe the similarities and differences between male and female gametes. (3)
b The diagram shows a pollen tube entering the ovule of a flowering plant.

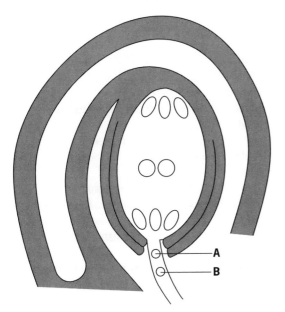

A

B

Explain why gametes **A** and **B** are genetically identical to each other but differ from other male gametes produced by this plant. (6)
c i Explain how a developing plant embryo gains its nutrients from the food reserve in the seed. (4)
  ii Explain two ways in which the placenta is adapted to provide a developing mammalian fetus with its nutrients. (4)

(+3 for quality of language)
AEB, Module 4, Summer 1998

6 The diagram below shows a germinating pollen grain and a mature ovule from a flower of the Papilionaceae. Some nuclei have been labelled.

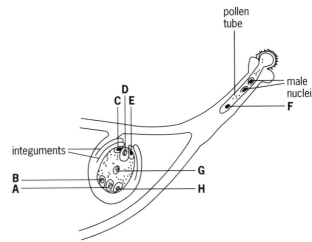

pollen tube

male nuclei

D
C E
F

integuments

B
A

G

H

a Describe how pollination usually occurs in the Papilionaceae. (2)
b Give the letter of the nucleus which fuses with a male nucleus to form each of the following:
  i the zygote
  ii the endosperm (2)
c Describe one mechanism which prevents self fertilisation in flowering plants. (2)

London, Paper B3, June 1998

7 Write an account describing the structure and formation of pollen grains. (10)

Welsh, Paper B3, June 1997

1 a i *Mitosis* results in an increase in numbers;
producing primary oogonia/oocytes and spermatogonia/
spermatocytes.

ii Two of:
- *Meiosis* is a reduction division/produces haploid cells;
- producing secondary ovum/oocytes/spermatids/
spermatocytes;
- which vary from one another.

b

Feature	Insect-pollinated flower	Wind-pollinated flower
stigma	feathery/hangs outside the flower	enclosed inside the flower
pollen	smoother surface/ large amounts/ lighter	spiky/sticky surface/smaller amounts/heavier
presence of nectaries	present (usually)	absent
anthers	versatile/hinged/ hang outside the flower	not hinged/ enclosed inside the flower

2 a A – encloses/protects the flower bud;
B – gives rise to the seed/contains/produces the
female gamete

b large petals/description of petals
knob-like stigma
enclosed stamens/stamens inside the flower

c i Both are haploid.
ii The male gamete of the plant is within the pollen
grain/the human sperm is motile/the human sperm has
a flagellum/refer to structural differences.

3 a oviduct/fallopian tube
b i

	Name	Name of organelle found in the structure	Function of organelle
R	middle piece	mitochondrion	ATP/energy (to maintain sperm/for movement of flagellum)
S	nucleus	chromosomes	carries (paternal) genome/genes

ii Acrosome/**T** contains enzymes;
which hydrolyse/dissolve/break down the zona pellucida/
jelly coat (allowing the sperm nucleus to enter the
oocyte).

c i pollen (grain)
ii Two of:
- Formation/growth of tube.
- Nucleus travels along tube into the egg.
- Enzymes are produced which allow a tube to grow
(and penetrate the egg).
- Only the male nucleus enters the egg/extine and the
rest of the sperm remains outside.

4 a i 38
ii 19

b More food/nutrients for the developing embryo.

c Crossing over/formation of chiasmata;
allows exchange of genetic material/parts of
chromatids.
Independent segregation;
allows different combination of maternal and paternal
chromosomes.

5 a Similarities – any two from:
- haploid
- single cell/single nucleus
- able to fuse with another gamete.

Differences – any two from:
- female is (usually) larger
- male is motile/is moved
- male gametes (usually) formed in larger numbers.

b Six of:
- **A** and **B** are formed by mitosis.
- From same cell/nucleus.
- Meiosis occurred prior to this mitosis.
- Meiosis results in variation.
- Homologous chromosomes are non-identical;
- due to crossing over;
- occurring between different loci/different number of
occasions.
- Independent assortment of chromosomes
(meiosis 1).
- Independent assortment of chromatids (meiosis 2).
- Mutation can occur.

c i *Note: marks are obtained for specific detail rather
than generalisations. Use words such as 'hydrolysis'
rather than 'breakdown'. Give details of specific
enzymes and their substrates and products.*

Four of:
- Water uptake.
- Involvement of gibberellin.
- Enzyme synthesis/correctly named example –
amylase/lipase/protease.
- Hydrolysis/digestion of food reserve.
- Protein to amino acids.
- Lipids.

ii Two pairs (feature plus explanation) from:
- Thin surface/large area/villi/microvilli; allows
diffusion.
- Villi/microvilli (once only for mark); large surface
area.
- Many mitochondria; energy for active transport.
- Rich blood supply/countercurrent blood supply;
maintains diffusion gradient.
- Channel proteins in membrane; for active
transport/facilitated diffusion.

*(3 marks for quality of language – acceptable grammar,
punctuation, spelling; scientific style, correct technical
terms; argument clearly and logically presented)*

6 a *Note: the requirements of exam boards differ so it is important to check your syllabus. This question however, contains many general points.*

Two of:
- The insect is attracted to the flower by scent/colour/nectar.
- Lands on the wing petals/depresses the keel petals.
- Anthers/stamens dust pollen onto the abdomen of the insect.
- The pollen is carried to another stigma/carpel.

b i The zygote – **D**

ii The endosperm – **G**

c *Any two paired marking points for 2 marks.*
- Dichogamy/male and female parts mature at different times.
- Protandry/pollen shed before stigma is mature.
- Dichogamy/male and female parts mature at different times.
- Protogyny/stigma ripe before pollen shed.
- Structure of flower prevents pollen landing on stigma.
- Explanation of heterostyly.
- Dioecious plants; self pollination impossible.
- Self incompatibility/sterility; no germination of pollen.

7 *Note: not all exam boards require all this detail. Many marks can be gained in this essay from information included on a well annotated diagram.*

Ten of:
- Pollen grains consist of an outer exine;
- which is sculptured;
- an inner intine;
- a pollen tube nucleus;
- and a generative nucleus.
- The diploid pollen mother cells divide;
- once by meiosis;
- in the pollen sacs of the anther;
- to produce four (a tetrad);
- haploid microspores.
- The nucleus of each microspore divides once by mitosis;
- to form two nuclei/the generative and pollen nucleus.
- Each pollen grain secretes a resistant coat/exine around itself.
- During their formation pollen grains are nourished and protected by the outer layer of the pollen sac/tapetum.

37 The continuity of life: Meiosis

> **Meiosis is a reduction division which occurs in sexually reproducing organisms to produce gametes. There are two divisions of the nucleus and *four* haploid cells are produced.**

Students should already have studied mitosis which was dealt with in **8** How cells divide – mitosis. In meiosis the main phases of cell division appear to be similar but with the exception of Interphase, which is the same in both mitosis and meiosis, there are marked and important differences.

Meiosis occurs within sex organs during the formation of **gametes** (germ cells) in sexually reproducing organisms. It is a **reduction division** – the number of chromosomes being halved in each gamete.

Sex cells (gametes) are produced in the **gonads**.
In larger animals the gonads are:

- male – **testes** (gametes are spermatozoa)
- female – **ovaries** (gametes are ova or eggs).

In plants the gonads are:

- male – **anthers** (gametes are male nuclei which migrate down the pollen tube)
- female – **ovaries** (gametes are egg cells).

Those cells in an organism which have not divided to form gametes have the **diploid** (*2n*) number of chromosomes in their nuclei. That is, there is one pair of each type of homologous chromosome present: one chromosome from each of the two parents.

Gametes have the **haploid** (*n*) number of chromosomes in their nuclei. There is only one of each homologous pair.

During fertilisation a male and a female gamete fuse together to form a **zygote** (fertilised egg or egg cell) in which the diploid (*2n*) number is restored.

There are differences between male and female gametes in their size, motility and the numbers produced. The facts below refer to human gametes.

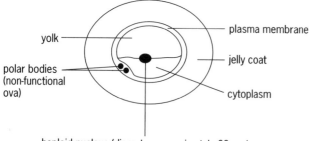

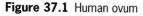

Figure 37.1 Human ovum

In the human female, ovulation takes place on about the 14th day of a 28 day menstrual cycle which happens continuously between the onset of puberty and the menopause. The ova are usually released singly (see **24** The endocrine system).

In the human male, sperm are produced continuously and in vast numbers from puberty to old age. During ejaculation as many as 400–500 million sperm may be released in approximately 3 cm³ of semen (see Figure 36.4 for a diagram of a human spermatozoon).

Meiosis

The process of cell division is a continuous one, but for our own convenience it is divided into stages. These stages have the same names as in mitosis.

During meiosis there are two cell divisions with marked differences between them. The net result is the formation of *four haploid cells* rather than two diploid cells as is the case in mitosis.

The first cell division in meiosis is a **reduction division**. The number of chromosomes is halved, although at this stage each chromosome consists of a pair of 'sister' **chromatids**. The pairs of chromatids, which are joined together at the **centromere**, are still one chromosome and the new cells formed are, therefore, haploid.

In the second division the centromere splits and the two chromatids separate to form two chromosomes. New **genomes** (combinations of genes) are produced. This is partly due to **crossing over**. Crossing over occurs when **bivalents** (pairs of homologous chromosomes) appear in prophase 1. Two non-sister chromatids lying alongside each other cross over at a **chiasma**. The chromatids break at this point and then the two pieces that have broken off rejoin with the non-sister chromatids and new combinations of genes (genomes) are formed. Crossing over can take place at any point along the chromosome.

a

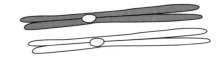

A pair of homologous chromosomes (each consisting of a pair of chromatids) come to lie alongside each other in prophase 1.

b

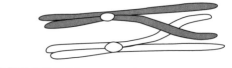

The adjacent non-sister chromatids cross over, break and exchange part of their genetic material.

c

In anaphase 1 the pairs of chromatids pull apart to opposite poles of the cell.

Figure 37.2 Crossing over

It is a matter of random chance as to which chromatid of a chromosome comes to lie alongside which other chromatid of the homologous pair. There are four possibilities (see Figure 37.3). The same is true for all 23 pairs of homologous chromosomes and so there are many possible combinations.

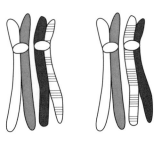
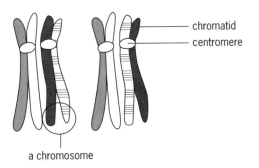
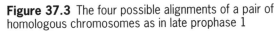

chromatid
centromere

a chromosome

Figure 37.3 The four possible alignments of a pair of homologous chromosomes as in late prophase 1

The reassortment of genes is random. During anaphase 1, when the homologous chromosomes (pairs of chromatids) migrate to opposite poles of the cell, their orientation is purely random. Each chromosome of a pair (irrespective of whether it originated from the male or female parent) moves to one of the poles independently of the migration of all the other chromosomes.

This random reassortment gives rise to a vast permutation of genes and explains why each individual is different from its siblings (brothers and sisters) and has its own genome, with the exception of identical twins.

Figure 37.4 represents the stages in the first meiotic division. Early prophase and details of centrioles and aster formation are the same as in mitosis. Refer back to **8** (How cells divide – mitosis) to check details and to compare meiosis with mitosis.

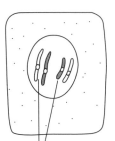

By **mid prophase** the pairs of homologous chromosomes have come together to form bivalents. Compare with mitosis where there is no association of homologous chromosomes.

bivalents

At **late prophase** the chromosomes have formed a pair of chromatids and crossing over with an exchange of genetic material has taken place.

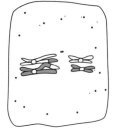

By **metaphase** the nuclear membrane has broken down and the bivalents have arranged themselves parallel to the equator of the cell. In particular compare metaphase 1 with metaphase in mitosis.

During **anaphase** the chromosomes (as pairs of chromatids) are pulled to the opposite poles of the cell where they arrive by early **telophase**. Asters are not shown in the diagram.

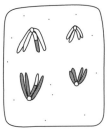

Figure 37.4 Meiosis 1

During **telophase** the two groups of chromosomes take on a granular appearance and the nuclear membranes reform. The cell may divide into two haploid cells or proceed directly into the second division. The second meiotic division is similar to mitosis except that in mitosis there are twice the number of chromosomes. The second meiotic division takes place at right angles to the first meiotic division and, because of this, a **tetrad** of haploid gametes are formed. During the formation of ova (oogenesis) only one of the gametes matures; the other three form small polar bodies with little cytoplasm and no known function (see oogenesis in **36** Sexual reproduction in plants and animals).

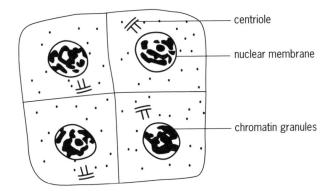

centriole

nuclear membrane

chromatin granules

Figure 37.5 A tetrad of haploid gametes

1 a Give the biological terms for the following:
i A point of genetic crossing over.
ii The stage between cell divisions.
iii The type of cell division which results in a halving of the number of chromosomes.
iv The point which joins together identical chromatids.
v The fertilised egg cell. (5)

b Give two processes which occur in meiosis and are responsible for genetic differences between daughter cells. (2)

2 Give the letter of the correct answer to each of the following:
a Meiosis occurs in:
A – the radicle and plumule of the embryo
B – the pollen mother cell and megaspore mother cell
C – the apical region of cell division of the stem
D – the zone of elongation of the root
b Bivalents (pairs of homologous chromosomes) separate during:
A – anaphase 1
B – prophase 1
C – metaphase 2
D – anaphase 2
c Replication of DNA occurs during:
A – prophase 1
B – telophase 1
C – interphase
D – prophase 2 (3)

3 Models which represent the genetic material in chromosomes can be made using different coloured wires.

The drawings show four possible models made by students, only one of which is a correct representation of one pair of homologous chromosomes at metaphase of meiosis 1, the first meiotic division.

The chromosomes carry an allele which occurred in two forms, D and d, and the model represents an individual heterozygous for this allele. No crossing over has taken place.

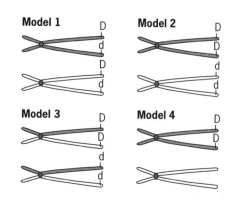

a State which model represents the heterozygous individual. (1)
b Explain why you have rejected the other models. (3)

Welsh, Paper B3, June 1996

4 Complete the boxes in the following table by writing a tick for a correct statement and a cross for an incorrect statement.

	Meiosis, first division	Meiosis, second division	Mitosis
chromosomes show independent assortment			
the two chromatids separate from one another as the centromere splits			
the chromosomes are associated as bivalents on the equatorial plate of the nuclear spindle			
chiasmata formation occurs			

(12)

5 The diagram below shows the chromosomes in the nucleus of a cell.

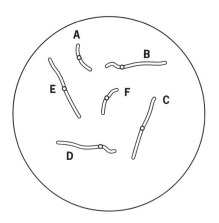

a i Give the letters of one pair of homologous chromosomes. (1)
ii Complete the diagrams below to show the arrangement of these chromosomes at metaphase 1 and metaphase 2 of meiosis. Use the asterisks as the poles of the spindle. (5)

metaphase 1   metaphase 2

*   *

*   *

iii Draw a diagram which will show the chromosome content of a cell which is the product of a meiotic division of the cell shown above (at the start of this question). (2)

A diagrammatic representation of the life cycle of the arctic fox ($2n = 52$) is given below.

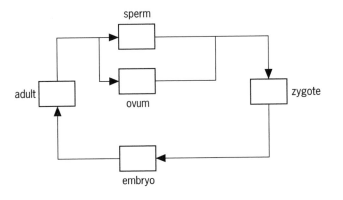

b Complete this diagram by writing in the boxes the number of chromosomes present at each stage. (2)

c In captivity, the arctic fox has been able to breed with the red fox ($2n = 34$). The offspring, however, are $2n = 43$ which makes them infertile.

Suggest why this chromosome number makes the offspring infertile. (2)

Cambridge 4801 Spring 1997 Q3

6 a When a cell divides, the genetic material can divide by mitosis, by meiosis or by neither of these processes. Complete the table with a tick to show the process by which you would expect the genetic material to divide in each of these examples.

	Mitosis	Meiosis	Neither
the division of plasmids in bacterial reproduction			
the stage in the formation of male gametes in a plant in which the haploid daughter cells are formed from a haploid parent cell			
cell division which takes place in the growth of a human testis between birth and 5 years of age			
the stage in the life cycle of a protoctist in which a large number of genetically different spores are produced			

(2)

b The diagram shows a cell during the first division of meiosis.

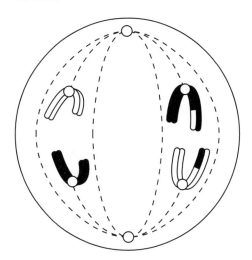

Copy the diagram below and show the appearance of the chromosomes in each of the four daughter cells formed at the end of the second division of meiosis. (2)

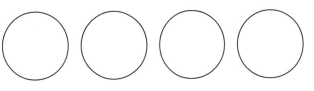

c In an insect, 16 chromatids were visible in a cell at the start of the first division of meiosis. How many chromosomes would there be in a normal body cell from this insect? (1)

AQA (AEB), Paper 2, January 1999

7 There are 64 chromosomes in each body cell of a horse and 62 chromosomes in each body cell of a donkey.

a Complete the table to show the number of chromosomes in the nuclei of these animals at the end of the various stages in cell division. (2)

	Number of chromosomes in one of the nuclei formed at the end of:		
Animal	**Mitosis**	**First division of meiosis**	**Second division of meiosis**
horse			
donkey			

b A mule is the offspring of a cross between a horse and a donkey.

i How many chromosomes would there be in a body cell from a mule? Give a reason for your answer. (1)

ii Suggest why a mule is unable to produce fertile sex cells. (2)

iii Explain why a horse and a donkey are regarded as different species. (1)

NEAB, BY02, March 1998

8 The photomicrograph below shows plant cells which have undergone division during the formation of pollen grains.

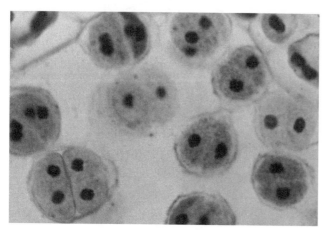

a Name the type of division shown in the photomicrograph. (1)

b i Name one location in a flowering plant in which this type of division occurs. (1)

ii Name one location in mammals in which this type of division occurs. (1)

c Give two reasons why this type of cell division is important in living organisms. (2)

London, Paper B1, June 1997

1 a **i** chiasma **ii** interphase **iii** meiosis **iv** centromere
      v zygote
   b (1) Chiasmata formation leads to swapping of genetic
      material.
      (2) Independent assortment of homologous pairs of
      chromosomes produces random segregation to
      opposite poles (Mendel's second law – each of a pair of
      alleles can be combined with either of another pair from
      a different bivalent).
      *Note: when answering questions of this type it is
      important to place one answer under (1) and the other
      point under (2). If two answers are written in one
      section, only the first will be marked.*

2 **a** B **b** A **c** C

3 a Model 2
   b Model 1 – Different alleles are shown on one pair of
      sister chromatids.
      Model 3 – Different genetic material shown on one pair
      of sister chromatids/chromatids on each chromosome
      differ/equivalent.
      Model 4 – No 'd' alleles shown on sister chromatids/
      sister chromatid on one chromosome of pair blank/not
      homologous/are sex chromosomes.

4

Meiosis, first division	Meiosis, second division	Mitosis
✓	✗	✗
✗	✓	✓
✓	✗	✗
✓	✗	✗

5 a **i** One of:
      • **A** and **F**
      • **B** and **D**
      • **C** and **E**
      ii

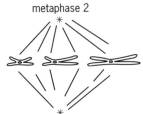

metaphase 1   metaphase 2

      Marks awarded for the general points given below.
      • Spindle drawn and chromosomes aligned along
        equator in both.
      • Chromosomes shown as two chromatids in both.
      Metaphase 1
      • 2/6, chromosomes drawn.
      • Homologous chromosomes paired.
      Metaphase 2
      • 1/3 chromosome(s) drawn.
      iii Three single-stranded chromosomes drawn;
      one of each type.

   b Adult and embryo and zygote boxes showing 52.
      Sperm and ovum boxes showing 26.
   c Two of:
      • Meiosis cannot take place/gametes cannot form.
      • Odd numbers of chromosomes.
      • Few/no homologous pairs/pairing not possible.

6 a *½ mark given for each correct answer.*

Mitosis	Meiosis	Neither
		✓
✓		
✓		
	✓	

   b Cells drawn to show four haploid cells, each with two
      chromosomes.
      Correct combinations of chromosomes (see diagram).

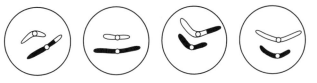

   c eight chromosomes/four pairs

7 a *Table completed with all three columns correct – 2
      marks; two correct – 1 mark.*

| | Number of chromosomes in one of the nuclei formed at the end of: | | |
Animal	Mitosis	First division of meiosis	Second division of meiosis
horse	64	32	32
donkey	62	31	31

   b **i** There would be 63 chromosomes in a body cell of a
      mule, 32 in the gamete from the horse plus 31 from the
      gamete of a donkey.
      ii At meiosis;
      there are an odd number of chromosomes/
      chromosomes in the mule cannot pair.
      iii The mule is infertile/cannot produce young.

8 a meiosis/reduction division
   b **i** anther/ovule/pollen sac/microsporangium/
      megasporangium
      ii ovary/testis/seminiferous tubule

   c *Note: with questions like this the answers must be
      arranged with one answer under (1) and the other
      under (2).*

      (1) It halves the chromosome number/produces haploid
      gametes.
      (2) Meiosis is an important source of genetic variation in
      living things (detail could be included about crossing
      over and independent assortment).

38 The continuity of life: Variation

> **Variation is the difference in characters between members of the same species.**

37 (The continuity of life: Meiosis) dealt with the ways in which genes are reshuffled during gamete formation and sexual reproduction. Namely:

- The mixing of characters during cross fertilisation of gametes.
- Crossing over between chromatids of homologous chromosomes during Prophase 1 of meiosis.
- The random distribution of paired chromosomes at Metaphase 1 of meiosis.

> **A gene is that part of a DNA molecule which codes for one polypeptide (character).**
> **An allele is a particular gene coding for a specific character. In diploid cells there are pairs of alleles, one in each of a pair of homologous chromosomes.**

Such reshuffling of genes does not bring about any major changes but new combinations of existing characters are formed. Each of a pair of alleles is either *dominant* or *recessive*.

Dominant and recessive genes

Chromosomes are present in homologous pairs in somatic cells and therefore there are at least two alleles for each character. These alleles may be either dominant or recessive. A dominant allele masks the expression of a recessive allele. Therefore a recessive character will only be evident if both alleles are recessive. An example is coat colour in mice. Black is dominant to white. The allele for black coat is shown as B whilst that for white coat, being recessive, is shown as b. The results for various combinations of these two alleles are as follows:

- BB = black coat. The mouse is pure-bred dominant (both genes are dominant) for coat colour.
- Bb = black coat. The mouse is hybrid for coat colour and is black because white is recessive.
- bb = white coat. The mouse is pure-bred recessive for coat colour.

The **genotype** of an organism is its genetic constitution expressed in terms of its genes. The genotype for coat colour in a white mouse is therefore 'bb'.

The **phenotype** of an organism is the observable features that arise from its genetic constitution. The phenotype of both the 'Bb' and 'BB' mice is 'black coat'.

Variation of a species within a community may be due to the effect of the environment as well as the genotypes of the organisms. Environmental factors themselves are variable and often form gradations (e.g. temperature may partly be responsible for **continuous variation** in a population).

Continuous and discontinuous variation

Continuous variation

Polygenic characters are controlled by the combined effects of two or more genes found on different loci. Each gene has a small effect on the phenotype. If a number of genes favour a person to be tall, then he, or she, may be very tall. Extremes are rare and a Gaussian (normal) distribution curve would be expected.

Environmental factors may also influence characters to produce overall continuous variation in a population.

Discontinuous variation

When a character within a species is clearly distinguishable, such as blood group in humans where there are no intermediate types, it is controlled by a single gene, although there may be several alleles.

Alleles are a pair of genes responsible for the same character (e.g. tallness/shortness in pea plants) which are found on the same locus (position) in both of a pair of homologous chromosomes (inherited one from each of the two parents). In other words, *alleles are genes found on the same locus in homologous chromosomes.*

ABO blood groups – discontinuous variation

Blood group is controlled by alleles located on chromosome 9. There are three alleles, A, B and O. Since there are two chromosomes number 9 in each cell, there are six possible combinations of genes: AA, AB, BB, AO, BO and OO. These pairs of genes will determine one's blood group to be A, B, AB or O.

Alleles A and B are equally *dominant* (**codominant**) but O is *recessive* and will not have an effect if either A or B is present.

Mutations

> **A mutation is any change in the structure of DNA or the amount of DNA in a cell or an organism.**

Most mutations occur in somatic cells, i.e. those body cells which are not concerned with the formation of gametes. Such mutations will not be transferred to succeeding generations. Other mutations occur in the germ cells (spermatozoa and ova) and will produce differences in the offspring which, if not lethal, will be passed on to future generations.

However, not all mutations in the germ cells will have an effect because the genetic code is degenerate (see **6** The genetic code and protein synthesis).

A change in DNA structure means that there is *a change in the sequence of bases,* which alters the code for the synthesis of a particular sequence of amino acids which make up a protein. A change in DNA structure will be caused by an addition, deletion or a change in the order of the bases in a DNA molecule.

Some mutations may be beneficial and have evolutionary significance. Others may be lethal and therefore never be passed on to a future generation. Many mutations result in genetic diseases and in humans over 2000 are known. In nature, genetic diseases are likely to be quickly bred out of a population if they place the organism at a disadvantage (see **41** Natural selection and classification). In our society **genetic screening** is available and potential parents may be screened to avoid producing a child with certain serious genetic disabilities, or at least to assess the risk. In recent years genetic engineering has promised benefits for mankind with possibilities of treating genetic diseases (see **7** Gene expression and control).

The causes of mutations

Mutations occur all the time. Every species has a natural mutation rate which is more frequent in those organisms which have shorter life cycles and in which meiosis is more frequent.

Mutations are also caused by a number of chemicals including formaldehyde and mustard gas, and most forms of high energy radiation such as X-rays, gamma rays and ultra-violet light.

An agent which causes mutations is known as a **mutagen**.

Types of mutation

Mutations may be either a change in the structure of a locus (single gene) on a chromosome, or a change in the structure of whole, or part of, a chromosome.

Changes in the structure of genes

Remember, a gene is that length of a chromosome which codes for one polypeptide consisting of specific amino acids in a definite sequence (see **6** The genetic code and protein synthesis). Any change in the sequence will code for the wrong series of amino acids and a protein with the wrong shape will be produced. The changed shape will alter the active site, if it is an enzyme, and this will cause a metabolic block, preventing the protein from catalysing its reaction. This may have a serious effect on an organism and result in, for example, an inability to produce normal haemoglobin, causing sickle-cell anaemia (see **41** Natural selection and classification). Fortunately most, but not all, harmful mutations are recessive and will be masked if there is a dominant allele present.

There are a number of types of gene mutation:

- **Deletion** – when a length of nucleotide chain is removed.
- **Duplication** – when a length of nucleotide chain is repeated.
- **Inversion** – when a length of nucleotide chain becomes detached and then rejoins but the 'wrong way round'.
- **Addition** – when an extra length of nucleotide bases becomes inserted into the chain.

Changes in the number of chromosomes – polyploidy

Triploidy and tetraploidy ($3n$ and $4n$ respectively) arise during fertilisation involving diploid gametes.

- $2n + n = 3n$ (triploid). Infertile as there can be no homologous pairings of chromosomes during Meiosis 1. Triploid plants can be propagated asexually by cuttings, etc.
- $2n + 2n$ (self fertilisation) $= 4n$ (tetraploid). Fertile.
- In allopolyploidy the number of chromosomes is doubled when a hybrid is formed between two species.

Polyploidy is rare in animals but common in plants. Polyploid varieties of plants usually have an advantage such as increased size of fruits and seeds.

Additional chromosomes

It sometimes happens that when homologous chromosomes separate during anaphase in meiosis one pair do not separate. Two gametes are formed, one with ($n + 1$) and the other with ($n - 1$) chromosomes. This is known as **non-disjunction** and is usually lethal.

In humans, for example, chromosomes number 21 sometimes fail to separate; one of the gametes produced therefore has both chromosomes and so has ($23 + 1$) chromosomes instead of the usual 23. When this gamete fuses with a normal gamete the resultant fertilised ovum will have 47 chromosomes instead of the normal 23 pairs. Non-disjunction is normally lethal but chromosome number 21 is small and the child is able to survive, but has a condition known as Down's syndrome which results in a low IQ and a shorter than usual life expectancy. Non-disjunction in the case of Down's syndrome occurs with the formation of ova rather than sperm and the age of the mother is a factor. The risk with teenage mothers is one in many thousands, but women of 45 have about a one in 33 chance of producing a Down's syndrome baby.

Changes in the structure of chromosomes

These 'errors' occur when chiasma are formed during Prophase 1 of meiosis. More than one gene is involved.

- **Gene deletion** – a piece of chromosome is lost e.g. 1 2 3 4 5 6 7 8 becomes 1 2 5 6 7 8.
- **Inversion** – a piece of deleted chromosome becomes re-attached the other way round. The actual genes present on the chromosome have not changed, and the genotype is the same, but the phenotype may be different. The order in which the genes appear on the chromosome is important. e.g. 1 2 3 4 5 6 7 8 becomes 1 2 3 4 6 5 7 8.
- **Duplication** – a length of chromosome is repeated e.g. 1 2 3 4 5 6 7 8 becomes 1 2 1 2 3 4 5 6 7 8.

1 Explain the meaning of the following terms:
  a variation
  b mutagen
  c non-disjunction of chromosomes
  d genetic screening   (4)

2 Fill in the following table to indicate with a ✓ which factors are responsible for the variation in phenotype. More than one factor may be involved.

Characteristic	Environmental factors	Genetic factors
ABO blood group		
weight in humans		
cystic fibrosis		
length of human lifespan		
development of yellow coloration in plants left in the dark		

(7)

3 The DNA found in cells is a code for the type of amino acids and their sequence in a polypeptide or protein. Each amino acid is coded for by a triplet of DNA nitrogenous bases. Some amino acids have more than one triplet code. The amino acids arginine and leucine are coded for by six different base triplets.
  a What term is used to describe the fact that the genetic code has more than one triplet for some amino acids, e.g. arginine and leucine?   (1)
  b A mutation changing the code from GCA to GCT might occur.
    i Name a specific chemical which might bring about this change.   (1)
    ii Suggest another type of mutagenic agent which might be responsible other than chemicals.   (1)
    iii Describe two types of gene (point) mutation which might result in this change.   (2)

4 The figure shows the length of cobs in two pure breeding varieties of maize plant and the F_2 generations derived from a cross between them.

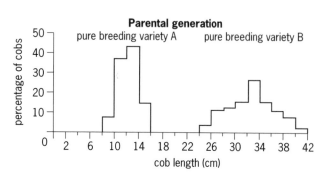

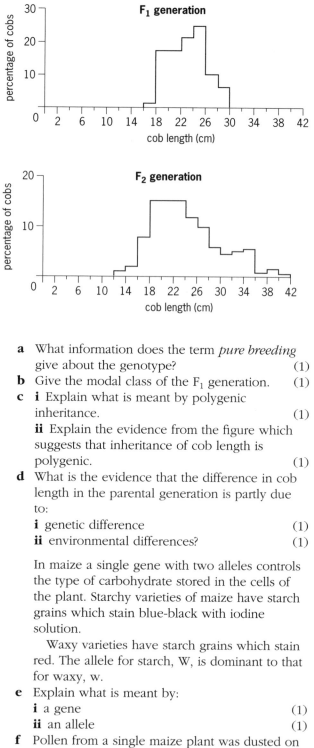

a What information does the term *pure breeding* give about the genotype?   (1)
b Give the modal class of the F_1 generation.   (1)
c **i** Explain what is meant by polygenic inheritance.   (1)
   ii Explain the evidence from the figure which suggests that inheritance of cob length is polygenic.   (1)
d What is the evidence that the difference in cob length in the parental generation is partly due to:
   i genetic difference   (1)
   ii environmental differences?   (1)

In maize a single gene with two alleles controls the type of carbohydrate stored in the cells of the plant. Starchy varieties of maize have starch grains which stain blue-black with iodine solution.

Waxy varieties have starch grains which stain red. The allele for starch, W, is dominant to that for waxy, w.

e Explain what is meant by:
   i a gene   (1)
   ii an allele   (1)
f Pollen from a single maize plant was dusted on a microscope slide and stained red with iodine solution. The results are shown in the table.

Pollen grains stained blue-black with iodine solution	Pollen grains stained red with iodine solution
58	64

What is the genotype of:
   i The pollen grains stained red with iodine solution?   (1)
   ii The parent plant from which these pollen grains were taken?   (1)

AQA, Specification A Specimen Paper 5, for 2002 exam

5 a **i** Explain the importance of genetic mutation in the survival of a species. (2)
ii Explain how gene mutation results in variation. (3)

b Describe three ways that sexual reproduction can bring about new combinations of genes in the offspring. (3)

6 a Explain the meaning of the terms:
i polyploidy (1)
ii autopolyploidy (1)
iii allopolyploidy (1)

b **i** Name a chemical that can be used to induce autopolyploidy. (1)
ii Explain the effect this chemical produces which can bring about polyploidy. (2)

c Describe the advantages that may be conferred in breeding polyploid crops. (2)

7 Karyotype is the chromosome set characteristic of a given species. The following karyotype was obtained from a person suffering a genetic disorder.

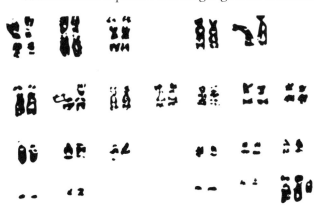

a **i** How many chromosomes are shown in this karyotype? (1)
ii Draw a circle around the Y chromosome in this karyotype. (1)

b Briefly describe the chromosomal events which may have led to this disorder (2)

c Give two ways in which the karyotype of a person with Down's syndrome would differ from the karyotype shown above. (2)

London, Specimen Paper B1, 1966

8 Give an account of the contributions made to genetic variation by:
a point mutations
b gain of extra chromosome(s) (20)

9 The diagram below shows how modern bread wheat is thought to have evolved.

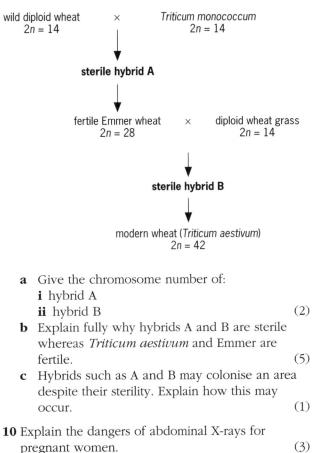

a Give the chromosome number of:
i hybrid A
ii hybrid B (2)

b Explain fully why hybrids A and B are sterile whereas *Triticum aestivum* and Emmer are fertile. (5)

c Hybrids such as A and B may colonise an area despite their sterility. Explain how this may occur. (1)

10 Explain the dangers of abdominal X-rays for pregnant women. (3)

1 a Variation is the differences in characteristics between one organism and another (often within a population or species).

b A mutagen is an agent which brings about mutation.

c Non-disjunction of chromosomes is when a pair of chromosomes fails to segregate at meiosis.

d Genetic screening means identifying the harmful genes carried by potential parents.

2

Characteristic	Environmental factors	Genetic factors
ABO blood group		✓
weight in humans	✓	✓
cystic fibrosis		✓
length of human lifespan	✓	✓
development of yellow colouration in plants left in the dark	✓	

3 a degenerate

b i formaldehyde/mustard gas/nitrous acid, etc.

ii ionising radiation/high energy radiation/X-rays/ gamma rays/UV light/high energy particles such as neutrons/alpha particles/beta particles

iii Two of:
- Insertion of T.
- Deletion of A if the next nitrogenous base is T.
- Substitution of the base A by T.
- Inversion of bases where the following one is T.

4 a homozygous (for the alleles concerned)

b 24–26 cm

c i Polygenic inheritance is where a single characteristic is controlled by many genes/more than one gene.

ii There are many lengths/a wide range of lengths/ phenotypes in the F_2 generation.

d i There are two different groups/they produce intermediate offspring.

ii Each parent shows variation although the genotype is the same.

e i A piece of DNA which codes for a single protein/ polypeptide.

ii A particular form of a gene.

f i w **ii** Ww

5 a i Mutation will result in an increase in the amount of variation within a population or species.
The variation will be selected for or against by natural selection leading to better adaptation of the species/survival advantage.

ii Gene mutations bring about changes in the order of the nitrogenous bases in the DNA;
resulting in changes in the amino acid sequence/a different protein is produced.
The protein may have different properties (variation)/it may result in a changed enzyme affecting which reactions can occur (variation).

b Chiasmata formation between linked genes during meiosis.
Random/independent assortment of unlinked genes as chromosomes segregate at meiosis.
Random fertilisation of one female gamete by one (of millions) male gamete.

6 a i Polyploidy is the condition where the nucleus contains more than two haploid sets of chromosomes/multiple sets of chromosomes.

ii The additional set(s) of chromosomes are derived from the same species.

iii The additional set(s) of chromosomes are derived from a different (but closely related) species.

b i colchicine

ii It inhibits spindle formation;
so that the resulting cell has twice as many chromosomes.

c Two of:
- Polyploidy is often associated with an increase in vegetative size/size of fruit.
- The polyploid may be more hardy/resistant to disease.
- Allopolyploidy may result in new hybrids with different characteristics from the parents.

7 a i 47

ii Circle around the bottom right chromosome (to include one chromosome only).

b It is important that you mention this occurs during *meiosis*;
due to the non-disjunction of a pair of homologous chromosomes/one pair of chromosomes fails to separate and both members of the pair go to one pole of the cell.

c A person with Down's syndrome would have only one X chromosome.
They would have three copies (trisomy) of chromosome 21 instead of chromosome 23.

8 *In general* refer to the causes of mutation/chemical mutagens/ionising radiations.
A mutation is any change in the structure or amount of genetic material/DNA of a cell.
Only changes to the gametes will be passed on to future generations.

a
- Point mutations are a change in the nucleotide sequence of a single gene.
- Many mutations are deleterious/harmful.
- Many mutations are recessive, remaining hidden in the phenotype until a homozygous individual is produced by fusion of two gametes, each carrying the mutated gene.
- Dominant mutations will be selected for or against by the environment immediately since they are expressed in the phenotype.
- Addition.
- Substitution.
- Deletion.
- Inversion.

b • Chromosome mutations may be due to the loss or gain of a single chromosome by the gamete;
- due to failure of a single pair to separate during meiosis/non-disjunction.
- Loss of a whole chromosome usually has a lethal effect.
- Polysomy means the gain of an extra chromosome and this can have marked effects – refer to Down's syndrome/Klinefelter's syndrome and their effects.
- Chromosome mutations may bring about the gain of whole sets of chromosomes/polyploidy;
- due to failure of separation of the chromosome sets during cytokinesis.
- Autopolyploidy is a result of doubling within the species.
- Allopolyploidy occurs as a result of two related species forming a hybrid (often sterile) which then may double its chromosomes.
- Polyploidy can bring about more vigorous growth/increase in fruit yield/increased hardiness.
- This can lead to advantages in competition with parental types, e.g. *Spartina townsendii*.
- This can lead to improved plant crops, e.g. modern wheat ($6n$).

9 a i 14

   ii 21

b The hybrids A and B are both formed by fusion of gametes from different species.
Their chromosomes will differ (shape/size) and be unable to form homologous pairs during meiosis.
Gamete formation will not occur and they will not reproduce sexually.
Chromosome doubling has occurred due to the failure of mitotic spindle formation (to produce Emmer and *T. aestivum*).
This will then ensure that all the chromosomes can pair up into bivalents and produce gametes.

c The sterile plants may be able to reproduce asexually/vegetatively.

10 The fetus and the ovaries are located in the abdomen.
X-rays may cause mutation of the DNA of the developing ova or fetus.
Most mutations are deleterious/harmful.

The continuity of life: Monohybrid and dihybrid inheritance

Monohybrid crosses

Monohybrid crosses are those in which one pair of contrasting characters is considered. As an example we can take a cross between a pure-bred normal-winged fruit fly with a pure-bred vestigial-winged fly. Normal wing is dominant. The pure-bred (or **homozygous**) form can be shown by two letters – NN or nn.

Vestigial-wing is recessive and therefore all vestigial-winged forms must be **homozygous recessive**. If 'N' is used to denote the dominant allele, then a small 'n' *must* be used for the recessive allele. (If you use a 'v', for example, you will not be able to interpret genetic diagrams.)

The first cross in our example is between pure-bred homozygous parents; one is double dominant and the other double recessive.

parents   NN   ×   nn

gametes   N   N   and   n   n

Which male gamete fertilises which female gamete is purely random and, by the law of averages, all combinations are equally possible. The possibilities are often plotted using a Punnett square as follows:

	N	N
n	Nn	Nn
n	Nn	Nn

This example is very simple as only **heterozygous (hybrid)** offspring are produced and a Punnett square was not really necessary in this case. This first generation of hybrids is known as the **F$_1$ generation**.

Hybrids are the offspring resulting from a cross between two genetically dissimilar individuals.

All the F$_1$ will have the same genotype – Nn and the same phenotype   – normal wing

There may be more than two alleles of a gene (**multiple allelism**) although only two can be present at any one time. An example is the ABO blood groups in humans which has three possible alleles, A, B and O. A and B are **co-dominant** and O is **recessive**. Because the loci must be the same on the two homologous chromosomes the genes are referred to as **I^A**, **I^B** and **I^O**.

blood group	possible genotypes
A	I^{A}I^A or I^{A}I^O
B	I^{B}I^B or I^{B}I^O
AB	I^{A}I^B
O	I^{O}I^O

Next is a 'self cross' between the offspring. That is, Nn × Nn.

parents   Nn   ×   Nn

gametes   N   n   and   N   n

	N	n
N	NN	Nn
n	Nn	nn

F$_2$ generation

- Genotypic ratio =   1 : 2 : 1
                          NN : 2Nn : nn
  That is, one homozygous dominant to two heterozygotes (hybrids) to one homozygous recessive.
- Phenotypic ratio = 3 : 1
  That is, three normal-winged flies to one vestigial-winged.

The test cross or back cross

Organisms which express the phenotype of a recessive character must be homozygous (or pure bred) for that character. Organisms which have the phenotype of a dominant character may be either pure bred or hybrid. Only a knowledge of the genotype will give us that information. For example, black coat is dominant to white coat in rats. A black rat can be either Bb or BB and only a particular genetic cross will give us the information. This is the **test cross** (or back cross).

In a test cross the phenotype to be tested is crossed with the double recessive. We can use a black rat from the example above. There are then, two possibilities:

- *possibility* 1 (in which the genotype of the rat is BB)

  parents   BB   ×   bb

  gametes   B   B   and   b   b

  F$_1$: All will be Bb (black).
  This means that the parent can only be pure-bred black. Several matings need to take place before one can be absolutely sure, but if the unknown parent is BB then white rats will never appear.
- *possibility* 2 (in which the genotype of the rat is Bb)

	b	b
B	Bb	Bb
b	bb	bb

F$_1$: In this case both the genotypic and phenotypic ratios are 2:2 which is expressed as 1:1 – 50% of the offspring will be Bb (heterozygous black) and the other 50% will be bb (homozygous recessive, white). Approximately half of the offspring will be white. A phenotypic ratio 1:1 indicates a back (test) cross between a hybrid and the homozygous recessive.

Dihybrid crosses

Dihybrid crosses are those in which two pairs of contrasting characters are considered.

In his classic experiments with pea plants, Mendel crossed homozygous dominant plants that produced round, yellow peas with homozygous recessive plants that produced wrinkled, green seeds. The characters are:

R = round seeds, and r = wrinkled seeds
Y = yellow seeds, and y = green seeds

Note once again that the recessive character takes the same letter as its dominant allele, but for recessive characters the letter is small (lower case).

parents	RRYY	×	rryy
	homozygous		homozygous
	round, yellow		wrinkled, green

gametes	RY	and	ry

F_1 hybrids: All will be heterozygous round, and yellow:

RrYy

The F_1 are then 'selfed'.

parents	RrYy	×	RrYy

gametes all possible combinations for each of the parents are:
RY Ry rY ry

A Punnett square is drawn to calculate all combinations of the pairs of genes in each gamete. (Note that the genes are present in the same gametes but NOT on the same chromosomes.)

	RY	Ry	rY	ry
RY	RRYY	RRYy	RrYY	RrYy
Ry	RRyY	RRyy	RryY	Rryy
rY	rRYY	rRYy	rrYY	rrYy
ry	rRyY	rRyy	rryY	rryy

The various genotypes can be seen at a glance and it would be meaningless to work out a ratio. The phenotypes, on the other hand, fall into a readily discernible pattern. Check out the Punnett above and you will see that there are:

9 round, yellow; 3 round, green; 3 wrinkled, yellow; and 1 wrinkled, green pea seed.

The 9:3:3:1 F_2 phenotypic ratio is always obtained in dihybrid inheritance when F_1 heterozygotes are crossed. Note that there is always just one double recessive for both characters – rryy in the above example. If the genotypes of the parents are known, and provided that a correct Punnett square is drawn, the result of any dihybrid cross can be predicted. The 9:3:3:1 ratio is only obtained when F_1 hybrids are selfed. A statistical routine such as the **chi-squared**

test, which is a test of *significance*, can be applied. Such a test will show if there is a significant difference between observed and theoretical results.

The dihybrid test cross

Crossing with the double homozygous recessive will give similar results as in the monohybrid test cross:

- If the organism is homozygous for the dominant genes (double dominant for both characters) then only one type of offspring will be produced. Using pea plants:

RRYY × rryy

will only produce RrYy
All the offspring will be round yellow hybrids.

- If the test organism is heterozygous, four types of offspring will appear in equal numbers when crossed with the homozygous recessive.

RrYy × rryy

will produce RrYy, round and yellow
rrYy, wrinkled and yellow
Rryy, round and green
rryy. wrinkled and green

The phenotypic ratio will be 1:1:1:1.

Linked genes

If genes are located on the same chromosome they will assort independently (because they are physically linked together) and they will not give a 1:1:1:1 ratio in a dihybrid test cross. This is an example of non-Mendelian inheritance. In such a cross, the proportion of parental types will exceed 50% and the genes are said to be linked.

Sickle cell anaemia

This is an example of an *autosomal recessive* genetic disease in humans. The autosomes are those chromosomes which are *not* concerned with the inheritance of sex. The disease is caused by a recessive gene – a mutation which causes glutamic acid to be replaced by valine in a molecule of haemoglobin. It is a serious disease as it causes as many as 100 000 deaths a year. In some parts of Africa 1 in 50 children may have the disease. The sickle cell gene is shown as Hb^S and the normal, dominant gene as Hb^A. A person with the disease has both recessive genes leading to distortion and destruction of the red blood cells in low oxygen concentrations. This in turn leads to severe bouts of anaemia and even heart failure.

The heterozygotes ($Hb^S Hb^A$) have only half as much normal haemoglobin as the homozygote with two Hb^A genes, and are said to have the 'sickle-cell trait'. They only show symptoms of the disease in conditions of low oxygen concentration. In this case the gene for normal haemoglobin is not completely dominant over the Hb^S gene. It is an example of **incomplete dominance**.

1 Wild turkeys have black feathers and this allele for black feathers is dominant to the white feathered allele common in poultry farms here.
  a Using the symbols B and b:
    i Give the genotype of a homozygous black turkey. (1)
    ii Give the genotype of a white turkey. (1)
  b i What colour feathers would the heterozygous turkey have? (1)
    ii Draw a genetic diagram of and describe the feather colour of the offspring of a cross between two turkeys both heterozygous for feather colour. (3)

2 Fred and Molly appear to be normal phenotypes but their daughter Sally was born with cystic fibrosis. The allele for cystic fibrosis is autosomal. Draw a genetic diagram and explain fully how this can occur. (4)

3 The diagram shows the inheritance of cystic fibrosis in a human family. The allele for cystic fibrosis is recessive to the normal allele.

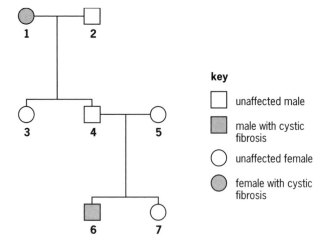

key

☐ unaffected male

▨ male with cystic fibrosis

○ unaffected female

⬤ female with cystic fibrosis

  a What is the probability that the next child born to individuals 4 and 5 will:
    i be a male with cystic fibrosis (1)
    ii have at least one allele for cystic fibrosis? (1)
  b In Britain, the frequency of the cystic fibrosis allele is 0.02. Calculate the proportion of people you would expect to be born with cystic fibrosis. (1)
  c In the past, people born with cystic fibrosis usually died before reaching adulthood. Suggest an explanation for the fact that the cystic fibrosis allele remained at a very high frequency in the population in spite of this. (2)
    AEB, Module 2, January 1997

4 Some cats have white patches on their coats. This effect is produced by the action of the spotting gene, S. This gene has two co-dominant alleles, S^1 and S^2. The coats can have large patches of white, small patches of white or no patches at all.
  a Explain the meaning of the term *co-dominance.* (2)

b A cat with no white patches, homozygous for S^1, was mated with a cat that had small white patches. Some of the offspring produced had coats with small white patches and the rest had no white patches.
   Draw a genetic diagram to show this cross and include the expected ratio of phenotypes on your diagram. (4)
   Edexcel (London), Paper B1, January 1999

5 The diagram below shows a family tree with the blood group phenotypes filled in for some members of the family. A square is used to represent a male and circle is used to represent a female.

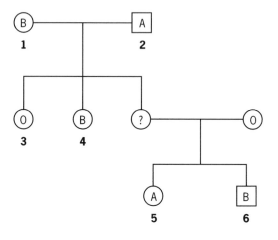

  a There are three alleles represented by the letters I^A, I^B and I^O. Using these letters give the genotypes of the following individuals:
    1 2 3 4 5 6 (6)
  b i What is the missing blood group phenotype of the individual marked? (1)
    ii Explain your answer to part i. (3)

6 Marfan's syndrome is a rare inherited disease which affects the eyes, heart and skeleton. The family tree shows the way it can be inherited.
   The grandfather (individual 1 in the diagram) is a homozygous normal male.

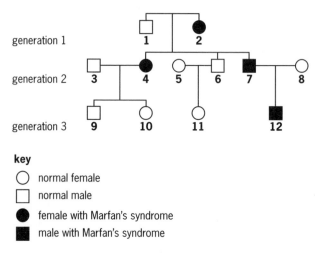

key

○ normal female

☐ normal male

⬤ female with Marfan's syndrome

■ male with Marfan's syndrome

a Which of the following terms can be used to describe the way in which the disease is inherited:
autosomal sex-linked dominant recessive (2)

b The symbols Dd represent an individual heterozygous for Marfan's syndrome. What are the genotypes of the individuals in the diagram labelled **10** and **12**? (2)

c Since individuals **7** and **8** have one affected child they decided to seek the advice of a genetic counsellor before enlarging their family. What is the chance of the next child being affected? (1)

Welsh, Paper B2, January 1997

7 There are four alleles of a gene that determines coat colour in rabbits. The alleles are not sex linked.

a i State the term used to describe this situation. (1)

ii The allele for normal colour, R, is dominant to all the other alleles.
The allele for albino, r, is recessive to all the other alleles.
The allele for chinchilla, r^{ch}, is dominant to himalayan, r^h.
State the phenotypes of the following:
Rr^{ch} $r^{ch}r^{ch}$ $r^{ch}r$. (3)

iii Draw a genetic diagram to show the expected results of a cross between a homozygous chinchilla male and a heterozygous himalayan female. (4)

b In the wild, rabbits have a high reproductive rate.
Explain why such overproduction is necessary. (4)

Cambridge, Module CC, June 1997

8 The inheritance of colour in sweet peas is controlled by two pairs of alleles (A and a, and B and b) at two different gene loci.

The diagram shows how the dominant alleles at these two gene loci determine flower colour by controlling the synthesis of a purple pigment.

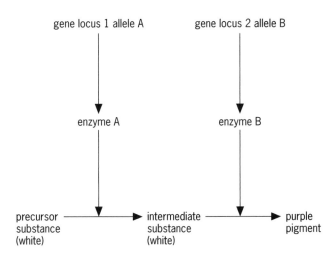

a Plants with the genotypes aaBB, aaBb and aabb all have white flowers. Use information in the diagram to explain why plants homozygous for the mutant allele, a, have white flowers. (3)

b Use a genetic diagram to explain the genotypes and the expected ratio of phenotypes when plants with the genotypes Aabb and aaBb are crossed.

parental phenotypes	white	white
parental genotypes	Aabb	aaBb
genotypes of gametes		
genotypes of offspring		
phenotypes of offspring		
ratio of phenotypes of offspring		(4)

NEAB, BY01, June 1997

9 The stems of some tomato plants can be different colours (green or purple) and may be hairy or without hairs (hairless). Each character is controlled by a different gene. Each gene has two alleles.

The following symbols are used to represent the alleles involved:

A – dominant allele for colour
a – recessive allele for colour
B – dominant allele for hairiness of stem
b – recessive allele for hairiness of stem

In an experiment, a homozygous tomato plant with a green, hairless stem was crossed with a homozygous tomato plant with a purple, hairy stem. The F_1 seeds were collected and grown. All the resulting F_1 seeds had purple, hairy stems.

a State the genotypes of each of the parent plants and of the F_1 seedlings in this cross. (3)

b The F_1 plants were self pollinated (interbred). The phenotypes and numbers of offspring are given in the table below.

Phenotype	Number of offspring
purple, hairy stem	293
purple, hairless stem	15
green, hairy stem	12
green, hairless stem	98

i In this dihybrid cross, what would be the expected ratio of the phenotypes in the offspring? (2)

ii Explain the difference between the expected ratio and the numbers shown in the table. (3)

c A tomato grower wants to find out which of the purple, hairy plants are homozygous for both characters.

i State the genotype of the plant which should be crossed with the purple, hairy plant in this test cross. (1)

ii Explain why this genotype should be used. (2)

London, Paper B1, January 1998

Note that in answering family tree questions it is helpful to write on the diagram the symbols for the two recessive alleles, which an autosomal recessive phenotype must possess. Any dominant phenotype can be given one dominant allele and the other allele can be deduced from the phenotypes of their parents or offspring.

It is important to relate genotype to phenotype in answering genetics questions. Make sure that all the steps are shown, so that if you make a mistake in the answer, the examiner is still able to award some marks for your method of working out the problem.

1 a i BB
ii bb
b i The heterozygous turkey would have black feathers.

ii

heterozygote   ×   heterozygote

genotype: Bb   Bb

gametes:

	B	b
B	BB	Bb
b	Bb	bb

F$_1$ offspring genotype   1BB : 2Bb : 1bb
F$_1$ offspring phenotype   1 black : 2 black : 1 white
                              3 black : 1 white

2 Explain that Fred and Molly must both be heterozygotes, carrying but not showing the recessive allele for cystic fibrosis. *(Note that autosomal means the gene is not sex-linked.)*
Let A be the normal allele and a be the cystic fibrosis allele (use any letter).

Fred   ×   Molly

genotypes: Aa   Aa

gametes:

	A	a
A	AA	Aa
a	Aa	aa

F$_1$ offspring genotypes   1AA : 2Aa : 1aa
F$_1$ offspring phenotypes   3 normal : 1 cystic fibrosis (Sally)

3 a i 12.5%/1 in 8 chance/0.125

(heterozygote × heterozygote = 1 cystic fibrosis to three normal/1 in 4. Multiply by the chance of the child being male 1 in 2)

ii 75%/3 in 4 chance.

b 0.0004 *(A frequency of 0.02 means that at that locus 2 out of 100 genes are for cystic fibrosis. If mating occurs at random the chance of obtaining the two recessive cystic fibrosis genes together will be 0.02 × 0.02 = 0.0004.)*

c The heterozygote carriers of cystic fibrosis will have a normal phenotype and live to adulthood, passing on the allele when they breed.
The heterozygote may have an advantage.
or
Mutation of the normal gene to the cystic fibrosis allele. A high rate of mutation would equal the loss of alleles from the population.

4 a Both alleles have an effect/both alleles are expressed; in heterozygotes/when both alleles are present.
b (S^{1}S^1 × S^{1}S^2)

	S^1
S^1	S^{1}S^1
S^2	S^{1}S^2

Gametes shown correctly.
Genotype of offspring shown correctly.
Ratio of genotypes is   1 S^{1}S^1 : 1 S^{1}S^2
Phenotype ratio is 1 with no white patches : 1 with patches.

5 a 1 = I^{B}I^O; **2** = I^{A}I^O; **3** = I^{O}I^O; **4** = I^{B}I^O; **5** = I^{A}I^O; **6** = I^{B}I^O
b i The individual marked **?** must be blood group AB.
ii The father (individual **7**), since he is blood group O, can only contribute I^O gametes to his children (individuals **5** and **6**).
Individual **5** must have obtained I^A from the mother, individual **6** must have received I^B from the mother.
Alleles I^A and I^B show incomplete dominance/co-dominance.

6 a autosomal; dominant
b 10 = dd; **12** = Dd
c 50%/1 : 1/1 in 2

7 a i multiple alleles
ii Rr^{ch} is normal.
r^{ch}r^{ch} is chinchilla.
r^{ch}r is chinchilla.
iii

homozygous chinchilla (male)   ×   heterozygous himalayan (female)

genotype: r^{ch} r^{ch} r^h r

gametes:

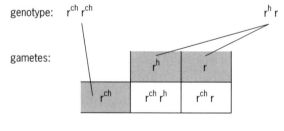

	r^h	r
r^{ch}	r^{ch} r^h	r^{ch} r

Genotypes of offspring   1 r^{ch}r^h : 1 r^{ch}r
Phenotype of offspring   All chinchilla

(The only possibility for the himalayan parent is as shown. It must possess r^h to be himalayan. Since it is heterozygous, the other gene must be a different allele and r is the only allele recessive to r^h and would not therefore show up in the phenotype.)

b Many rabbits do not survive long enough to reproduce and pass on their genes;
due to predators/disease/shortage of food.
Those best adapted (best camouflage/fastest runners/disease resistance) survive/survival of the fittest.
Pass on their genes to their offspring, i.e. natural selection.

8 a Plants homozygous for the mutant allele cannot produce enzyme A.
Give a reason for this, e.g. alteration to genetic code produces a 'different' enzyme.
No intermediate substance (needed to produce purple) is produced.
The flowers produce only a precursor/enzyme B has no substrate to act on, therefore white flowers are produced.

b *Both answers for each part of the question to be correct to get 1 mark.*

genotypes of gametes	Ab or ab	aB or ab
genotypes of offspring	AaBb Aabb	aaBb aabb
phenotypes of offspring	purple white	white white
ratio of phenotypes of offspring	1 purple : 3 white	

9 a Parent with green, hairless stem – aabb
Parent with purple hairy stem – AABB
F_1 seedlings – AaBb

b i 9 purple, hairy : 3 purple, hairless : 3 green, hairy : 1 green, hairless

ii Three of:
- The genes are linked.
- The genes are on the same chromosome/inherited together.
- AB/ab will be inherited together.
- Recombinants/other phenotypes are due to crossing over.

c i aabb

ii Two of:
- This genotype allows the expression of all the alleles.
- The phenotypes of the offspring will show which genotype the purple, hairy parent possessed.
- If there are any green offspring then the parent must be Aa/if there are any hairless offspring the parent must be Bb.
- If there are no green or hairy offspring, the parent is AABB.

40 Sex-linked characters

The inheritance of sex

Many sexually-reproducing organisms exhibit a morphological (structural) difference between the chromosomes of males and females. In male mammals the sex chromosomes are different in both size and shape and are termed the X and Y chromosomes. In female mammals both chromosomes are morphologically the same, and are X chromosomes, being the same in size and structure as the male X chromosome.

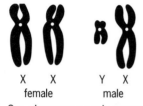

X X   Y X
female   male

Figure 40.1 Sex chromosomes in mammals

The sex that has the two similar chromosomes is said to be **homogametic** and the other is **heterogametic**.

During meiosis four haploid gametes are formed. In mammals all of the female gametes (ova) will have a single X chromosome, while 50% of the male gametes (spermatozoa) will have an X chromosome and the other 50% a Y chromosome.

It follows that during fertilisation there will be a 50% chance of either type of spermatozoon fusing with an egg. An X sperm will produce an XX zygote (female) and a Y sperm an XY zygote (male).

Sex-linked inheritance

Sex-linked inheritance is due to genes being located on the non-homologous portion of the X chromosome. A pattern of inheritance is produced which is different from that shown by genes located on autosomes.

If, in the male, a gene is carried on that portion of the X chromosome which is 'missing' in the Y chromosome, and the gene is recessive, then the male cannot carry a dominant gene to offset the effect of the recessive gene. Because males can carry only one allele of such a gene (on the X chromosome) dominance operates in females only. And since males only require the one recessive gene on the X chromosome to show such a recessive character they are far more likely to exhibit it than women who would need to be homozygous recessive. Women are normally only heterozygous for sex-linked characters (i.e. not homozygous) and therefore show no symptoms themselves.

A classic example of sex-linked inheritance is the disease known as **haemophilia** in which the blood fails to clot normally. The disease has been known for centuries, but there were no cases in the royal family prior to Queen Victoria, who introduced the disease into the royal families of several European countries after a mutation occurred in the earliest stages of her embryonic development.

Figure 40.2 shows a relevant portion of part of the family tree of Queen Victoria and Prince Albert.

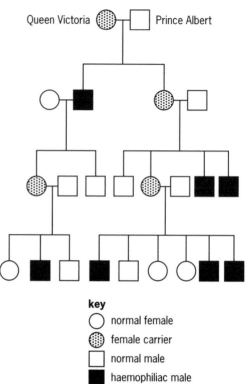

key
- ○ normal female
- ⊛ female carrier
- □ normal male
- ■ haemophiliac male

Figure 40.2 Part of the family tree of Queen Victoria showing the incidence of haemophilia

- Females are usually *carriers* and do not show the symptoms of the disease. Females homologous for the disease are rare and are unlikely to survive the onset of menstruation at puberty.
- If a haemophiliac female carrier pairs with a normal male, 50% of the daughters will be carriers and 50% of the sons will be haemophiliacs.
- If a female of normal phenotype pairs with a haemophiliac male, 50% of the daughters would be carriers and the other 50% would have the disease. Of the males, 50% would be normal and the rest would have the disease.

Another example of a sex-linked disease in humans is **red–green colour blindness**. Many more men than women suffer from some type of colour blindness. The most common colour blindness is an inability to distinguish between red and green. The disease is caused by a recessive gene that is carried on the X chromosome and is far less common in women because they must inherit the recessive gene from both parents. The father must be colour blind and the mother must at least be heterozygous for the condition; that is, a carrier.

A woman who is a carrier (i.e. heterozygous) will produce gametes with or without the colour blindness allele in equal numbers. This means that whatever the genotype of the father, any sons will have a 50:50 chance of being colour blind.

Sex-linked inheritance of dominant mutant genes

This is a rare occurrence. Examples of sex-linked mutations which are dominant include a congenital condition in which the incisor teeth are absent and a rare form of rickets. In both cases the dominant allele is only to be found on the X chromosome. Figure 40.3 shows a family tree in which this form of rickets occurs. Compare this with the inheritance of haemophilia shown in Figure 40.2.

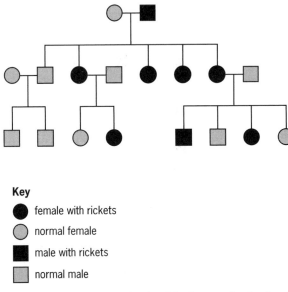

Key

● female with rickets

○ normal female

■ male with rickets

▨ normal male

Figure 40.3 Family tree showing inheritance of a dominant sex-linked gene

Females are *not carriers* as is the case in haemophilia. Since the gene is dominant, if a female has one X chromosome with the rickets gene, she will have the disease and not be a carrier without the symptoms.

- All the daughters of an infected male will have the disease because they must have their father's only X chromosome.
- The sons of an infected male will not have the disease provided that their mother does not carry the rickets gene.

The final example demonstrates the inheritance of the sex-linked character 'white-eye' in the fruit fly (*Drosophila*). 'Red-eye' is the normal gene which controls eye pigmentation in the fruit fly. 'White-eye' is a recessive character which is only carried on the X chromosome and is, therefore, sex-linked.

In this example, a white-eyed homozygous female is crossed with a normal male.

'R' is the wild-type red-eye gene
'r' is the recessive white-eye gene

	white-eyed female		*red-eyed male*	
parents	X^rX^r	×	X^RY	
gametes	X^r	X^r and	X^R	Y

It can be seen that there is one type of ovum and two types of spermatozoa. Therefore the following genotypes and phenotypes will be produced in equal numbers in the F_1:

genotypes	phenotypes
X^rX^R	normal red-eyed female
X^rY	white-eyed male

If the F_1 offspring are selfed we get the following results:

	female		*male*	
parents	X^rX^R		X^rY	
gametes	X^r X^R	and	X^r	Y

	male X^r	male Y
female X^r	X^rX^r	X^rY
female X^R	X^RX^r	X^RY

The phenotypic ratio is $1:1:1:1$ which is (from top left to bottom right):

One female with white eyes
One male with white eyes
One female with red eyes
One male with red eyes

A useful experiment would be to make a **reciprocal cross**, starting with the offspring of the F_1 generation.

A reciprocal cross is one in which the *phenotypes* of the mating partners are reversed. In this case a red-eyed male would be crossed with a white-eyed female. If the results of the two crosses are different it indicates that the gene controlling the character is sex-linked.

The points listed using bullets do not have to be learned by heart. If you do not understand any of them prove the points to yourself by drawing genetic diagrams using the facts given.

1 Explain the meaning of the following terms:
a heterogametic sex
b reciprocal cross
c morphological (3)

2 Red–green colour blindness is a recessive allele carried on the X chromosome.
a Using the symbols 'A' and 'a' draw a genetic diagram to show a cross between a woman carrying but not showing red–green colour blindness and a normal man. Show the genotypes and phenotypes of the expected offspring. (4)
b What percentage of the male children would you expect to be colour blind? (1)

3 Night blindness is a condition in which affected people have difficulty in seeing in dim light. The allele for night blindness, N, is dominant to the allele for normal vision, n. (These alleles are *not* on the sex chromosomes.)

The diagram shows part of the family tree showing the inheritance of night blindness.

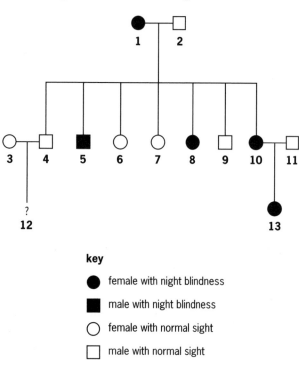

key
● female with night blindness
■ male with night blindness
○ female with normal sight
□ male with normal sight

a Individual **12** is a boy. What is his phenotype? (1)
b What is the genotype of individual **1**? Explain the evidence for your answer. (2)
c What is the probability that the next child born to individuals **10** and **11** will be a girl with night blindness? Show your working. (2)

AEB, Paper 2, June 1998

4 a Doctors in some clinics claim that they can separate individual sperm cells so that the sex of a human child can be predetermined. Explain the genetic principles which enable the sex of a child to be predetermined. (2)

b The diagram shows the inheritance of a rare hereditary form of rickets in a human family.

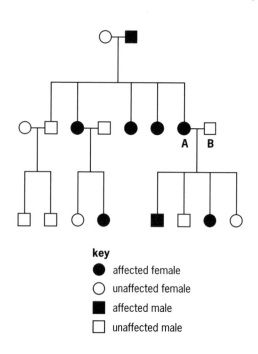

key
● affected female
○ unaffected female
■ affected male
□ unaffected male

The condition is caused by a dominant allele, R. This allele may be present on the X chromosome, X^R, but not on the Y chromosome. Complete the genetic diagram to explain how this type of rickets was inherited by the children of parents **A** and **B**.

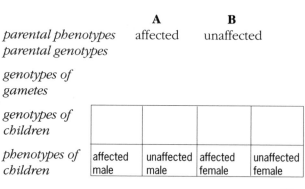

	A	**B**
parental phenotypes	affected	unaffected
parental genotypes		
genotypes of gametes		

genotypes of children				
phenotypes of children	affected male	unaffected male	affected female	unaffected female

(3)

NEAB, BY01, June 1997

5 The sperm of prize-winning bulls can be separated into sperm carrying the X chromosome and sperm carrying the Y chromosome.
a Explain why this could be useful in sperm used for artificial insemination of cows. (2)
b i There are usually approximately equal numbers of 'X sperm' and 'Y sperm'. Explain why this occurs. (2)
ii There is a small difference in the DNA content of the two types of sperm. Suggest a reason for differences between the amount of DNA found in the 'X sperm' and the 'Y sperm'. (1)

6 The way in which sex is determined in birds is different from that in mammals. In birds the male has two X chromosomes while the female has one X and one Y chromosome. In poultry, the gene for chick colour is sex linked and carried on the X chromosome. The allele for light colour is dominant to that for dark colour.

   a Complete the genetic diagram below to show the cross in which all the male chicks will be light coloured and all the female chicks will be dark coloured.

	male (XX)	female (XY)
phenotype of parents		
genotype of parents		
gametes		
genotype of chicks		
phenotype of chicks	light-coloured male : dark coloured female; 1 : 1	

                                                (3)

   b Poultry farmers who keep hens for egg laying usually buy young chicks from a poultry breeder. Explain why sex linkage may be of practical use in poultry farming. (1)

                                  AEB, Paper 2, June 1997

7 Haemophilia is a rare sex-linked disease. It normally occurs at a low frequency in the population; however it is found at a higher frequency in some royal families, descended from Queen Victoria, where inter-marriage of close relatives has been common. The recessive allele for haemophilia was not present in Queen Victoria's ancestors. She, however was a carrier for the haemophilia gene.

   a **i** Explain how Queen Victoria became a carrier for the haemophilia allele. (2)

       ii What is the main symptom of haemophilia? (1)

   b A male haemophiliac married a woman carrying but not showing the haemophilia allele. Use the symbols 'H' and 'h' to fill in the following table. (14)

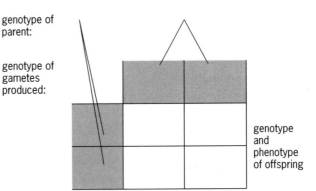

phenotype of parent:	haemophiliac male	normal female (carrier)
genotype of parent:		
genotype of gametes produced:		genotype and phenotype of offspring

1 a The sex which possesses two different sex chromosomes.

  b A reciprocal cross is one in which the genetic characteristics are the same as the initial cross but are associated with the other sex parent.

  c Morphological means related to size and shape.

2 a

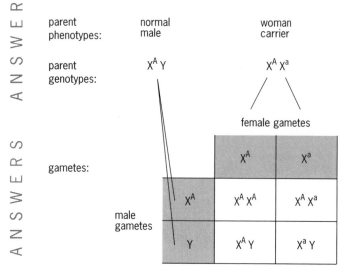

parent phenotypes:	normal male		woman carrier

parent genotypes: $X^A Y$ $X^A X^a$

female gametes

	X^A	X^a
X^A	$X^A X^A$	$X^A X^a$
Y	$X^A Y$	$X^a Y$

gametes:

male gametes

  b 50% of the male children are expected to be colour blind.

3 a normal sight

  b Nn/heterozygous
Shows night blindness and has children who have normal vision.

  c $\frac{1}{4}$/0.25/25%
probability of phenotype $=\frac{1}{2}$

4 a There are two types of sperm.
The selected X sperm will produce a female child/Y sperm produces a male child.

  b *Both answers for each part of the question to be correct to get 1 mark.*

parental genotypes	$X^R X^r$	$X^r Y$
genotypes of gametes	X^R or X^r	X^r or Y (all)
genotypes of children	$X^R Y$ $X^r Y$	$X^R X^r$ $X^r X^r$ (all)

alternative scheme – do not mix and match between the schemes

          $X^R X$ XY
          X^R or X   X or Y (all)
          $X^R Y$ XY $X^R X$ XX

5 a This would allow the sex of the calves to be chosen depending on which type of sperm was used.
X sperm would give female cows for calf and milk production and Y sperm would give males for beef.

  b i The sex chromosomes (XY) of the male;
will segregate during meiosis to produce equal numbers.
    ii The X chromosome has more DNA because it is larger than the Y chromosome.

6 a

	male	female
phenotype of parents	dark coloured	light coloured
genotype of parents	$X^d X^d$	$X^D Y$
gametes	X^d X^d	X^D Y
genotype of offspring	$X^D X^d$	$X^d Y$
	(male/light)	(female/dark)

  b This enables the farmer to identify the sex of the chick from birth and not waste resources rearing unproductive males since only hens are wanted. (These can be identified by colour.)

7 a i It must be caused by **mutation** from the normal allele since it was not present in the ancestors.
It must be a mutation of the **reproductive cells** since it is passed on to the offspring.
    ii The blood fails to clot normally.

  b

phenotype of parent:	haemophiliac male		normal female (carrier)

genotype of parent: $X^h Y$ $X^H X^h$

female gametes

genotype of gametes produced:

	X^H	X^h
X^h	$X^H X^h$ female normal	$X^h X^h$ female haemophiliac
Y	$X^H Y$ male normal	$X^h Y$ male haemophiliac

male gametes

genotype and phenotype of offspring

Natural selection and ecology

41 Natural selection and classification

Natural selection is the mechanism, first proposed by Charles Darwin and Alfred Russell Wallace in 1859, by which members of the same species struggle for existence because more offspring are produced than actually survive. Any small change appearing in an individual, due to a mutation in the gamete of one of the parents, will be perpetuated and passed on to future generations if it is advantageous in the struggle for existence. These advantageous small changes are known as **adaptations** to the environment. Those individuals which possess characters which put them at a disadvantage in this struggle for survival, or which do not change sufficiently to compete with members of the same species, will not reproduce themselves and will die out. Darwin called this process of *survival of the fittest* in which characteristics change slowly due to random mutations over a long period of time, **natural selection**. This process of natural selection is not just a matter of chance; there are *selection pressures* that drive evolution in a variety of strengths and directions. Presumably it was pressures from predators which drove the evolution of the horse from the four-toed form which existed 60 million years ago to the fleet-of-foot single-toed modern horse which appeared about one million years ago. Slower animals would be more likely to end up as prey than those who could outrun their predators.

Natural selection in action

Darwin was unable to demonstrate natural selection in action but today we have numerous examples.

Directional selection

Directional selection is that type in which one phenotype changes into another which has selective advantages. Many species of bacteria have developed a resistance to penicillin. Some grasses, such as *Festuca ovina*, have acquired a tolerance to the presence of heavy metals such as copper and lead in waste tips of mines. These grasses have become so adapted to the presence of high levels of heavy metals that they are less competitive with other species when the heavy metal concentrations are low.

Both predators and prey may develop or inherit features which make them more efficient at either catching prey or avoiding being caught. Such characters will give the organisms a greater chance of reproducing themselves and therefore passing on their advantageous mutations. A classic example is that of the peppered moth (*Biston betularia*) which underwent changes due to a pre-existing mutation which became advantageous in the middle of the nineteenth century when trees in industrial areas, such as Birmingham and Manchester, were blackened by

soot deposits and the lichen on the tree trunks died. Previously, only a light-coloured form of the moth was normal but then a dark variety, which appears as a mutation, became dominant in industrial areas. Few mutations offer an advantage but in this example, where the environment was changing, the mutant form now blended better with the black sooty background and had a selective advantage over the lighter form which was liable to greater predation from insect-eating birds. In Manchester, by 1895, 98% of the population of the peppered moth was of the melanic (dark pigmented) type. By way of contrast, in rural Dorset there were no melanic forms recorded.

Note in Figure 41.1 how inconspicuous are the light form of the peppered moth on the lichen-covered trunk and the melanic form on the sooty trunk. Several distinct forms of a particular species are known as **polymorphism**. Examples of polymorphism include the melanic and light-coloured forms of *B. betularia* and the A, B and O blood groups in humans.

Figure 41.1 Light and melanic forms of the peppered moth at rest on **a** a lichen-covered tree trunk; and **b** a soot-covered tree trunk; **c** and **d** show the forms on contrasting backgrounds

38 (The continuity of life: Variation) dealt with sickle-cell anaemia, a genetic disease in which the red cells become distorted (sickle-shaped) in low oxygen concentrations. Whereas the homozygous condition is usually lethal, the heterozygote inherits a selective advantage because the heterozygous form gives protection against malaria. The malarial parasite is less able to enter the sickle-shaped red cells. Therefore the recessive allele remains at an unusually high frequency within the population. When Africans move to other parts of the world where malaria is not endemic, they lose this selective advantage and there is a selection pressure against the recessive gene.

Stabilising selection

It would be wrong to assume that all selection leads to new and diverse forms in a population. Sometimes natural selection works to preserve characters in a stable environment. For this reason some organisms have changed very little over long periods of time. Horsetails (*Equisetum* sp.), present in the Carboniferous swamps of 33 million years ago, are still represented by a single genus which is found commonly growing in water and damp conditions. Crocodiles were well-developed by the end of the Jurassic period about 135 million years ago.

Evolution

The changes which occur in organisms as a result of natural selection result in **evolution**. It is important to realise that evolution has had virtually immeasurable time at its disposal and that the process of evolution has been very slow. The origin of life on earth is estimated to have been about 2500 million years ago, whereas man has only been on the earth for 0.3 million years – a minute fraction of the total time!

The evidence for evolution comes from a variety of sources:

- palaeontology (the study of fossils)
- comparative anatomy
- cell biology and biochemistry
- embryology
- the distribution of living organisms.

Evidence gained from the above sources enables biologists to place living organisms into groups which *reflect their evolutionary relationships*. The basic unit is the **species** which consists of an interbreeding, morphologically distinct group which cannot successfully interbreed with other organisms. Each population (group) has its own unique collection of genes, represented by their gametes, which constitute the **gene pool**. If a species becomes extinct this gene pool is lost for all time.

The modern process of classification was begun by Linnaeus who, in 1758, was the first to use the **binomial system of nomenclature**. Closely related species are placed together in a genus (plural *genera*) followed by the name given to the species. Humans are placed in the genus *Homo*. The only member of this genus is man who is given the name *sapiens*. The binomial scientific name for man is therefore *Homo sapiens*. The name for neanderthal man, who is believed to have been a separate breeding species, is *Homo neanderthalensis*. Note that only the genus is given a capital letter.

Thus, classification is based on the natural evolutionary relationships of living organisms which have similar structures or common features. The groupings are **hierarchical** with large groups containing smaller groups without any overlap. The names of the groups are known as **taxa** (singular **taxon**). In order of decreasing size the taxa are: **kingdom**, **phylum**, **class**, **order**, **family**, **genus** and **species**.

As knowledge increases and new techniques are discovered the names of some organisms are changed to reflect relationships more accurately.

Examples of the classification of organisms

At present most biologists accept five main **kingdoms**:

- **Prokaryotae** – single-celled organisms – the cyanobacteria (previously called the blue–green algae) and the bacteria. They lack both a membrane-bound nucleus and envelope-bound organelles. There is no mitosis or meiosis.
- **Protoctista** – eukaryotes – (sometimes single-celled or groups of similar cells) – algae and protozoa. They have a membrane-bound nucleus but are not fungi, plants or animals.
- **Fungi** – eukaryotes with cell walls **not** made of cellulose. Non-photosynthetic. Nutrition is by extracellular digestion and absorption.
- **Plantae** – photosynthetic organisms (autotrophic except for some insectivorous plants). Eukaryotic and multicellular. Contain cellulose cell walls.
- **Animalia** – eukaryotes – multicellular, non-photosynthetic organisms with a nervous system. Heterotrophic.

The following is an example of how the arthropods are classified. Orders and families are omitted:

- **Kingdom Animalia** (characteristics given above).

  - **Phylum Arthropoda** – hard exoskeleton; jointed limbs; segmented; bilaterally symmetrical.

    Class Crustacea – mainly aquatic; head not clearly defined; two pairs of antennae, e.g. *Astacus* (crayfish).
    Class Chilopoda – terrestrial; distinct head with one pair of jaws; many segments, each with one pair of similar legs.
      Centipedes (2500 living species).
    Class Diplopoda – terrestrial; many segments, each with two pairs of legs; one pair of jaws.
      Millipedes (10 000 living species).
    Class Insecta – body composed of head, thorax and abdomen; head bears simple eye and usually a pair of lateral compound eyes; thorax of three segments each with a pair of legs (751 000 described species).
    Class Arachnida – mostly terrestrial; body of two parts; four pairs of legs.
      Scorpions, spiders, ticks and mites (70 750 species).

Note: in this topic you are likely to encounter many names of groups of organisms and their characteristics that are unfamiliar. It is important that students should be able to use keys to identify organisms in the field. Questions often include Latin names which are unknown to the majority of students. The information given in the questions should be sufficient to allow students to identify organisms and work out relationships between groups.

1 a Read through the following passage about natural selection and then write a list of the most appropriate word(s) to complete the gaps.

The raw material of evolution is the wide range of variation which exists in living organisms. Most organisms produce _____ offspring than are needed to replace existing members of the population. Those which are best _____ to their habitat will survive, passing on their advantageous _____ to their offspring. Selective forces are climatic factors and _____ between individuals for food and shelter and mates.

When an environmental factor changes slowly but in a particular way over a long period of time this will result in _____ selection. An example of this can be seen in mice populations: as the environment gradually becomes warmer over a period of years, the mean fur length will become _____ . (6)

b Complete the table with the name of the **kingdom** to which the named organism belongs. State a feature which the named organism possesses and which is a characteristic of the kingdom.

Kingdom	Organism	One characteristic feature of the kingdom
	snail	
	daisy	
	mushroom	

(6)

c Write down the correct order of the following classificatory groupings starting with the largest: genus family species kingdom phylum order class (2)

d Explain the principles which are used to classify living things. (4)

2 a Identify the following kingdoms:
i No nuclear membrane or membrane-bound mitochondria.
ii Organelles surrounded by membranes found inside thread-like hyphae.
iii The organisms are either unicellular or multicellular and contain organelles with membranes.
iv Multicellular organisms which have autotrophic nutrition.
v Multicellular organisms which have heterotrophic nutrition. (5)

b The table below shows part of the classification of the honey bee *Apis mellifera*. Complete the spaces in the table.

kingdom	
phylum	Arthropoda
	Insecta
	Hymenoptera
	Apidea
genus	
species	

(6)

3 The diagram below is a simple dichotomous key which could be used to separate three groups of terrestrial plants: mosses, ferns and angiosperms.

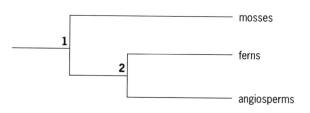

a i State two external features shown by both ferns and angiosperms but not by mosses which could be used to make the separation at **1** on the diagram. (2)
ii State two external features of angiosperms which could be used to separate them from ferns. (2)
b State one feature which distinguishes fungi from mosses, ferns and angiosperms. (1)
London, Paper B2, January 1997

4 The diagram shows how four species of pig are classified.

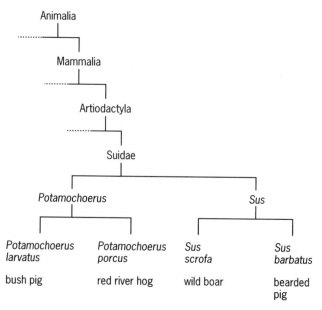

a i To which family does the red river hog belong? (1)

ii To which genus does the bearded pig belong? (1)

b Some biologists think bush pigs and red river hogs belong to the same species. The list below summarises some features of the biology of bush pigs and red river hogs.

- The bush pig has a body length of 100–175 cm and a mass of 45–150 kg.
- The red river hog has a body length of 100–145 cm and a mass of 45–115 kg.
- The red river hog is found in west Africa. The bush pig is found in east Africa.
- Both animals are omnivorous but feed mainly on a variety of underground roots and tubers.
- The ranges of these animals overlap in Uganda. In this area populations of animals which have characteristics intermediate between those of bush pigs and red river hogs have existed for many years.

Do you think that bush pigs and red river hogs belong to the same or different species? Explain how the information above supports your answer. (3)

AQA, Specification A Specimen Paper 5, June 1999 for 2001 exam

5 The diagram below shows the general features of a typical red seaweed.

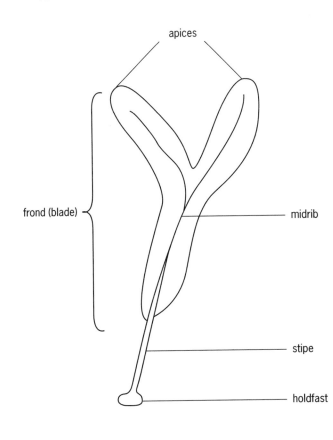

The key below can be used to identify some species of red seaweed. Use the key to identify specimens **A**, **B**, **C** and **D** in the figure below.

A _____

B _____

C _____

D _____

1. Midrib present .... go to 2
   Midrib indistinct or absent .... go to 4

2. Branches arise directly from the midrib
   .... *Hypoglossum woodwardii*
   Branches arise from the stipe not the midrib ......
   .... go to 3

3. Margins of fronds smooth and slightly
   undulating .... *Delesseria sanguinea*
   Margins of frond ragged or toothed ....
   Phycodrys rubens

4. Frond broad, not more than twice as long as
   wide .... go to 5
   Frond narrow, more than four times as long as
   wide .... go to 6

5. Small pointed bladelets on the margin of the
   frond .... *Polyneura gmellinii*
   Small bladelets absent .... *Polyneura billiae*

6. Regular, alternate branches; apices pointed
   .... *Odonthalia dentata*
   Irregular branching; apices rounded
   .... *Acrosorium uncinatum* (4)

Edexcel (London), Paper B2, January 1999

6 The diagram shows the way in which four species of monkey are classified.

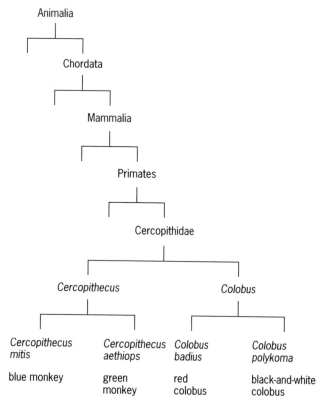

Animalia

Chordata

Mammalia

Primates

Cercopithidae

Cercopithecus   Colobus

Cercopithecus mitis   Cercopithecus aethiops   Colobus badius   Colobus polykoma

blue monkey   green monkey   red colobus   black-and-white colobus

a This system of classification is described as hierarchical. Explain what is meant by a *hierarchical* classification. (1)

b **i** To which genus does the green monkey belong? (1)
ii To which family does the red colobus belong? (1)

c What does the information in the diagram suggest about the similarities and differences between these four species of monkey? (2)

NEAB, BY02, March 1998

7 *Cepaea nemoralis* is a snail which varies in its appearance. The shell can be yellow, brown or pink, and may be banded with dark lines or unbanded. Predation is by thrushes which break open the shells on a large stone.

a Give the term used to describe a species containing several distinct morphological types. (1)

b **i** The majority of *C. nemoralis* found in woodland are the brown, unbanded types. Explain why this occurs. (4)
ii Explain what colour and pattern you would expect to be most common in open grassland. (2)

1 a more; adapted; alleles/genes; competition; directional; shorter

b

Kingdom	Organism	One characteristic feature of the kingdom
Animal(ia)	snail	one of: organisms with nervous co-ordination non-photosynthetic multicellular
Plant(ae)	daisy	one of: multicellular eukaryotic and photosynthetic cell walls containing cellulose
Fungi	mushroom	one of: eukaryotic, non-photosynthetic protective wall not of cellulose absorptive method of nutrition usually organised into multi-nucleate hyphae spores without flagella

c kingdom phylum class order family genus species

d Classification is based on evolutionary relationships between organisms which have similar structures or common features.

The groupings are hierarchical with large groups containing smaller groups without any overlap.

Each organism is given two Latin names (binomial) for the genus and species.

A species is a morphologically distinct group, the members of which are able to interbreed.

2 a i Prokaryotae
ii Fungi
iii Protoctista
iv Plantae
v Animalia

b

kingdom	Animalia
phylum	Arthropoda
class	Insecta
order	Hymenoptera
family	Apidae
genus	Apis
species	mellifera

3 a i Two of:
- (true) roots
- (true) stem
- leaves with vein/true leaves

ii Two of:
- flowers/flowering plants/reference to visible part of the flower
- fruits/seeds
- (axillary) buds

b fungi have hyphae/mycelium/no chloroplasts/no chlorophyll/non-photosynthetic/no photosynthetic pigment/reference to chitin/glycogen

4 a i Suidae
ii *Sus*

b Three of:
- There is considerable variation in body length/mass among members of a species.
- There are differences in appearance across the range.
- Many different species have broadly similar diets.
- The last point provides critical evidence that they are the same species.
- They can interbreed.
- The hybrids are fertile as all the pigs have been involved over many years.

5 A – *D. sanguinea*
B – *A. uncinatum*
C – *P. gmellinii*
D – *H. woodwardii*

6 a This means progressively divided into smaller groups/larger groups divided into smaller groups.

b i *Cercopithecus*
ii Cercopthidae

c Two of:
- All show some similarities as in the same family/Cercopithidae.
- *Cercopithecus*/blue and green monkeys are more closely related and more similar/than to *Colobus*/red and black-and-white.
- All show differences as separate species/have different genes for different colours.

7 a polymorphism

b i The plain brown shells will be well camouflaged against the dark, woodland soil.
Thrushes will more easily find and eat other colour variations.
The plain, brown snails will survive and breed.
The brown and unbanded alleles will be passed on to the offspring/there will be more brown, unbanded offspring.

ii Yellow, banded types would be well camouflaged and survive;
against the striped background pattern of light through the grass.

42 Ecosystems – energy flow and the nitrogen cycle

Energy flow in ecosystems

Energy enters the ecosystem via photosynthesis in green photosynthesising organisms (see **11** Photosynthesis: Cyclic and non-cyclic photophosphorylation). However, about 98% of the light energy reaching the earth is reflected from the leaves or absorbed by other surfaces and converted into heat energy. Only 2% is captured by chlorophyll. The rate at which light energy is assimilated in an ecosystem is measured by an increase in the **biomass**, which is the **fresh or dry mass of vegetation or animal matter**. The productivity of different environments varies and will depend on the photosynthetic rate. This, in turn, will depend on the amount of light and the temperature, as well as the availability of water and nutrients such as phosphorus and nitrogen. In terrestrial environments the highest productivity occurs in tropical rainforests and marshes, while in aquatic environments it is estuaries and reefs that are the most productive.

The energy in a biological community is passed from one **trophic** (feeding) level to the next along the food chain. The shape of the trophic levels forms a pyramid because at each trophic level as much as 90% of the energy may be 'lost' during metabolic processes such as excretion, keeping warm, movement, elimination of faeces and digestion.

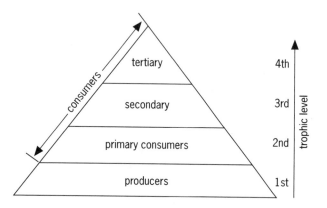

Figure 42.1 Pyramid of numbers or biomass

In Figure 42.1 the producers are **autotrophs** which are able to synthesise their own food material. Most autotrophs are photosynthetic but some are **chemoautotrophic** and get their energy from inorganic oxidation (e.g. sulphur bacteria which oxidise hydrogen sulphide to sulphur). Primary and secondary consumers are herbivores, omnivores or decomposers and they have less available energy than producers. Tertiary consumers (or top carnivores) are carnivores that eat other carnivores and are not preyed upon themselves.

Each level may be measured in terms of numbers or biomass. A pyramid of numbers reflects the numbers of organisms at each level in the food chain. In Figure 42.1 the food chain could be: algae – frog tadpoles – water beetles – pike.

Sometimes the pyramid of numbers may look more like the 'pyramid' shown in Figure 42.2. In this example it is more helpful to show the trophic levels as a pyramid of biomass as shown in Figure 42.3. Neither are drawn to scale.

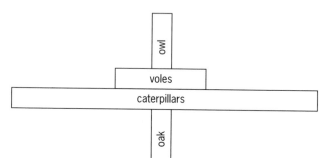

Figure 42.2 Pyramid of numbers

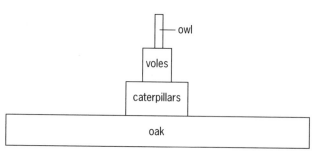
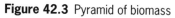

Figure 42.3 Pyramid of biomass

In Figure 42.3 the lengths of the bars in the pyramids are proportional to the biomass. There are still factors which are not taken into account by this method. The biomass of a deciduous tree varies with the seasons as it grows in the spring and sheds its leaves in the autumn. Neither is the *rate* of production taken into account. The population of voles, for example, will remain fairly constant but the turnover is high because of the mortality rate due to predation. Any sample only measures the amount of biomass present at that particular instant. This is called the **standing crop**.

In practice, simple food chains as shown in Figures 42.2 and 42.3 do not exist and the complex food relationships in an environment are better illustrated by **food webs**.

All the above pyramids are illustrating, in slightly different ways, the flow of **energy** through the ecosystem. Pyramids of energy (Figure 42.4) attempt to show the energy flow from one trophic level to the next. The length of the producer bar is proportional to the annual amount of solar energy which has been used by the producer in photosynthesis.

These transfers of energy are very difficult to measure accurately. The following features of energy pyramids are more important to note than the actual amount of energy transferred at each level:

- The transfer of energy from producers to primary consumers is less efficient than the transfer of energy at other trophic levels.
- Not all of the energy is transferred from one level to the next, and therefore they never bulge in the middle or become inverted as in Figure 42.2.

Pyramids of energy are difficult to produce but they give the most accurate picture of the flow of energy through an ecosystem.

In the diagram of an energy pyramid (Figure 42.4), NPP = net primary production. NPP is the amount of dry biomass in grams produced per square metre per year.

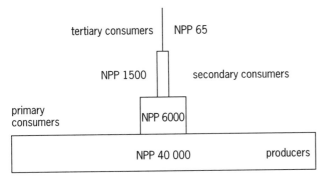

Figure 42.4 Pyramid of energy

Net primary production

The actual amount of biomass originally produced is much greater than the NPP because so much of the energy is 'lost' in respiration and metabolism. Therefore, NPP = gross primary production minus respiration and other metabolic processes.

The NPP of different biomes varies considerably. The most productive biomes are coral reefs with an NPP of about 2500 and the rainforests with an NPP of more than 2000. Unfortunately both of these biomes are threatened by man's intervention and once destroyed cannot be replaced. By 1998 about half of the world's rainforests were destroyed.

The cycling of fixed nitrogen

Proteins are essential compounds in the formation of living matter. (For details of the structure of proteins see **2** Biological molecules.) All proteins contain nitrogen but although 79% of the atmosphere consists of nitrogen gas it cannot be used until it is **'fixed'**. Atmospheric nitrogen is fixed in several ways:

- *Azotobacter* and *Clostridium* are two examples of free-living nitrogen-fixing bacteria which absorb nitrogen from air spaces in the soil and reduce it to ammonia. As much as 90% of nitrogen fixation is brought about by such free-living micro-organisms.

- Aerobic **nitrifying bacteria** present in the soil (e.g. *Nitrosomonas*) convert ammonia into nitrites. Then, other nitrifying bacteria (e.g. *Nitrobacter*) oxidise the nitrites into nitrates – a process known as **nitrification**. The ammonia comes from decomposers breaking down dead organic matter.
- **Nitrogen-fixing bacteria** (e.g. *Rhizobium*) present in the root nodules of **leguminous** plants convert atmospheric nitrogen into ammonia which the plants can then synthesise into amino acids.
- During electric storms small amounts of nitric acid are formed by lightning. The nitric acid then reacts with rock to form nitrates.

Animals can synthesise some amino acids from other amino acids by **transamination**, but there are also a number of **essential amino acids** which must be taken in with the diet (see **6** The genetic code and protein synthesis). For animals, all amino acids are obtained from the food chain either directly, or indirectly, from amino acids synthesised by green plants (see **45** Populations).

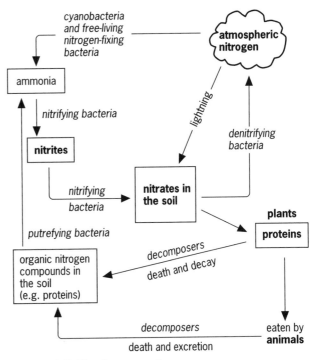

Figure 42.5 The nitrogen cycle

The supply of fixed nitrogen in the soil is variable because all nitrates are very soluble and easily leached. The gain from the nitrogen-fixing bacteria of leguminous plants (e.g. *Rhizobium*) is offset to some extent by anaerobic denitrifying soil bacteria which obtain their energy by converting nitrates in the soil back into atmospheric nitrogen. A low level of fixed nitrogen in the soil is often, therefore, a limiting factor in plant growth. It is for this reason that farmers either grow a leguminous crop, such as clover, which is later ploughed back into the soil, or they add compost, usually animal manure, or artificial fertilisers such as ammonium nitrate (NH_4NO_3) which contain a high percentage of fixed nitrogen.

1 Give the biological terms for the following:
   a A feeding level in a food chain.
   b The biomass of a feeding level at a particular moment in time.
   c A partnership between two organisms where both benefit from the association.
   d Several inter-related food chains. (4)

2 a Which element is found in all proteins but is not present in carbohydrates and fats?
   b What effect does destruction of the rainforest have on the concentration of carbon dioxide in the atmosphere?
   c What happens to most of the sunlight energy reaching Earth?
   d What type of biome is most productive? (4)

3 An investigation was carried out into the microbial decomposition of leaf litter, using bags made of netting with a fine mesh (0.3 mm^2). A large number of leaves were placed into these bags and buried in the leaf litter. The bags were then examined at monthly intervals.
   a i The investigation showed that leaf decomposition occurred more rapidly during months with higher than average temperatures. Give one reason why this occurred. (1)
   ii Suggest and explain how any other environmental factor could influence the rate of leaf decomposition. (2)
   b Proteins in leaves may be converted by a series of reactions, into nitrates which can be absorbed by plants. Describe the role of micro-organisms in these processes. (4)
   NEAB, BY01, February 1997

4 A simplified diagram of the nitrogen cycle is shown below.

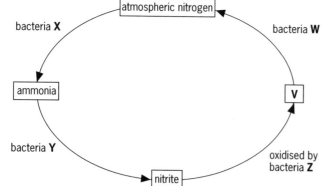

   a Name the ion **V**. (1)
   b What type of conditions would promote the growth of bacteria **W**? (1)
   c Name the bacteria **X**, **Y** and **Z** which could bring about reactions marked in the diagram. (3)

5 lupins → aphids → ladybirds

Organism	Mean mass of organism (g)	Number of organisms (m^{-2})
lupin	52	16
aphid	0.002	5000
ladybird	0.03	19

The diagram represents a food chain and data collected showing mean population densities and masses of the organisms concerned. More information could be shown on a pyramid of numbers, a pyramid of biomass or a pyramid of energy of this food chain.
   a i Use the data from the table to draw the pyramid of numbers for this food chain. (1)
   ii Draw the pyramid of biomass for this food chain. (1)
   iii State one way in which a pyramid of energy would be more useful than a pyramid of numbers. (1)

6 A pyramid of numbers for a food chain is shown in the diagram.

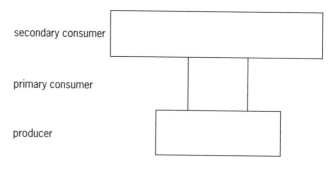

   a Suggest why there are more secondary consumers than primary consumers in this pyramid. (1)
   b The diagram below shows a pyramid of energy for an ecosystem.

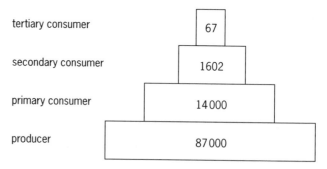

   i Suggest units that might have been used to record the energy flow between trophic levels. (2)
   ii Calculate the percentage energy transfer from secondary consumer to tertiary consumer. (1)
   iii Give two reasons why the percentage of energy transferred between consumers is generally low. (2)
   NEAB, BY01, February 1997

7 The table shows estimates of the production of plant material and of the biomass of all the organisms in various world ecosystems.

	Area (km² ×10⁶)	Primary production dry tons (×10⁹)	Biomass dry tons (×10⁹)
tropical forest	20	40.0	900
temperate forest	18	23.4	540
tropical grass	15	10.5	60
temperate grass	9	4.5	14
open ocean	332	41.5	1

a Which ecosystem shows the greatest productivity per unit area? (1)

b The open ocean biomass is low, although the area of open ocean is very high. Suggest an explanation. (2)

c **i** Compare the primary production figures for the open ocean and tropical forest. (1)
ii Compare the biomass of these two ecosystems. (1)
iii Suggest an explanation for these differences between the two ecosystems. (1)

d Tropical grassland has less than twice the area of temperate grassland but more than four times the biomass. Explain. (2)

e This table was produced 25 years ago when it was estimated that human consumption was a 200th of the world's biomass and that this consumption was increasing exponentially.

For each column in the table describe one different activity which could explain why the figures in the column might be different today. (3)

f Give one example of a way in which this increasing human demand on biomass might be slowed down. (1)

Welsh, Paper B1, June 1997

8 The diagram below shows a simple food web.

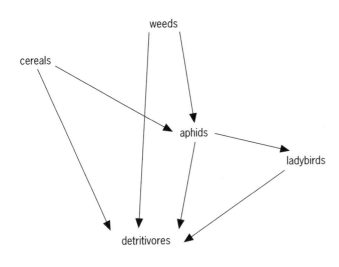

The table below compares the effects of an insecticide, a selective herbicide (specific for weeds) and manure application on the biomass of each trophic level.

	Cereals	Weeds	Aphids	Ladybirds	Detritivores
insecticide					
selective herbicide					
manure					

a Complete the table using one of the following symbols to describe the change in each trophic level. One of these symbols should be placed in each box. (5)
+ = large increase
− = large decrease
0 = a small, or no change

b **i** Name the organisms which represent the trophic level which shows the lowest energy content. (1)
ii Explain fully why this trophic level has the lowest energy content. (2)

Welsh, Paper B1, January 1997

9 a Describe the main features of tropical rainforests. (9)

b Explain why it is important to conserve rainforests. (7)

(content 16, quality of expression 4)
Cambridge, Module 4803, June 1998

10 The diagram below shows the quantity of energy flowing through a food chain in a terrestrial ecosystem. The figures given are in kJ m⁻² yr⁻¹.

incident sunlight 3×10^6

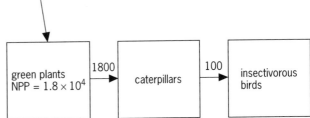

a Calculate the percentage of the incident energy which becomes available as the net primary production (NPP) of green plants. Show your working. (2)

b Give two reasons why not all the energy of the incident sunlight is incorporated into biomass of green plants. (2)

c Using information shown in the diagram, explain why the biomass of insectivorous birds is usually very much less than the biomass of caterpillars. (2)

London, Paper B2, June 1997

11 Write an essay on the cycling of nitrogen through the ecosystem. (10)

1 a trophic level
b standing crop
c symbiosis/mutualism
d food web

2 a nitrogen
b Carbon dioxide concentration is increased (less photosynthesis occurs).
c It is lost as heat.
d coral reef

3 a i Higher temperature will increase the rate of one of: metabolism/respiration/enzyme activity/reproduction.
ii *The marks in this question are related.*
One mark is for a named factor e.g. oxygen/air/pH/moisture, etc.
One mark is for the effect of your named factor, e.g. oxygen is required for respiration/pH is related to the growth of microbes or enzyme activity/moisture is related to hydrolysis (digestion).
b Protein is converted into ammonium compounds;
by bacteria and fungi/saprophytes (saprobionts).
Ammonium compounds are converted to nitrite.
Nitrite is converted to nitrate by nitrifying bacteria/suitable named example.

4 a nitrate
b anaerobic/swampy conditions
c **X** = *Azotobacter/Clostridium/Bacillus/Rhizobium*, etc.
Y = *Nitrococcus/Nitrosomonas*; etc.
Z = *Nitrobacter*

5 a i

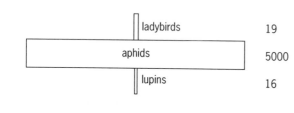

ladybirds	19
aphids	5000
lupins	16

(give credit for correct shape)

ii

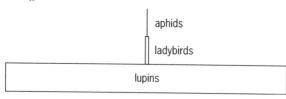

aphids
ladybirds
lupins

(give credit for correct shape)

iii A pyramid of energy takes into account the total energy at each trophic level over a given time period and so it will not be distorted by the size of the organism or be an unrepresentative snapshot in time.

6 a The secondary consumers are parasites/have a small size/have a rapid reproductive cycle.
b i The units must be energy units per unit area per unit time period e.g. kJ/m^2/year

ii $\dfrac{67}{1602} \times 100 = 4.18\%$ (or 4.2%)

iii Energy 'loss' via any two of:
• heat/respiration
• movement/muscle contraction
• faeces/indigestible material
• excretion

7 a tropical forest
b The biomass figure shows an instantaneous picture of the standing crop, which consists largely of small organisms which are short-lived (high turnover). Phytoplankton and zooplankton are thinly distributed at the surface in a medium of low nutrient status.
c i They are very similar/only a small difference.
ii The tropical forest biomass is 900 times greater.
iii Production in the forest forms more permanent biomass/very woody plants whereas the ocean has short-lived small plant forms.
d Tropical grassland occurs in regions with higher mean temperatures and greater light.
The productivity of grass per unit area is much greater.
e *There are many possible answers for this section, e.g.*
• Area of tropical forest is less today due to deforestation for timber or growth of arable crops/area of temperate grass is more due to increased growth of cereals, etc.
• Primary production of oceans has increased due to eutrophication/fertiliser run off, etc.
• Biomass of the ocean has decreased due to fishing with fine mesh/biomass of tropical forest decreased due to clearing of trees and using for grazing, etc.
f Eating more plant foods and less animal foods (which have energy losses, being at the consumer level in the food chain).

8 a

	Cereals	Weeds	Aphids	Ladybirds	Detritivores
insecticide	+	+	–	–	0
selective herbicide	+	–	0	0	+
manure	+	+	+	+	+

(1 mark for each correct column)

b i ladybirds
ii Any two of:
• Energy is lost by death/decay other than being eaten by the next trophic level.
• Energy is lost by excretion of complex molecules.
• Energy is consumed in metabolism/respiration.

9 a Nine of:
- hot climate/20–28 °C all year round
- high rainfall/1500–4000 mm per year
- equatorial
- high productivity
- net primary productivity in the range of 1.0–3.5 kg per metre per year dry mass
- green canopy present all year round
- high diversity of life/greatest diversity of any biome (many, unknown species)
- made up of three to five layers of foliage/stratified vegetation
- each layer has its own range of animal species (detail of an example)
- the canopy is high/30–50 m up
- traps much sunlight/shades the lower parts
- canopy contains many epiphytes
- any valid detail of epiphytes
- credit one correctly named epiphyte
- nutrient cycling within the canopy
- fruit/flowers/leaves available all year as food supply
- soils nutrient deficient/contain little organic matter
- decomposition very rapid
- most of the available nutrients are in plant biomass/biomass 45 kg per m².

b Seven of:
- Rainforests are species rich/have high biodiversity.
- This richness benefits other species.
- Deforestation can result in extinction/reduced diversity.
- Moral responsibility to later generations to conserve diversity.
- Rainforests can be used to supply sustainable crops.
- Example of crop/nuts/rubber/fruits/plant oils (*not plantations*).
- Drugs/useful compounds may await discovery in species that only occur in rainforests.
- Rainforests act as carbon reservoirs/sinks.
- Destruction of rainforest contributes to global warming;
- for example by release of carbon dioxide as a result of burning.
- Less photosynthesis also means less oxygen production.
- Transpiration contributes to atmospheric water content.
- Destruction of rainforest will lead to a reduction in rainfall/disruption of water cycle.
- Soils are nutrient deficient and cannot sustain agriculture for any length of time.

10 a $\dfrac{1.8 \times 10^4}{3 \times 10^6} \times 100 = 0.6\%$

b Two of:
- Light is reflected/not absorbed by the plant leaf.
- Energy is used to evaporate water/heat the plant.
- Light is transmitted/of the wrong wavelength.
- Photosynthesis/biochemical processes are inefficient.
- Energy is released by the respiration of the plant.

c 1800 kJ is transferred to caterpillars but only 100 kJ is transferred to birds/there is 5.6% of the energy transferred to the birds.
There is a loss of energy/biomass due to respiration/excretion/movement.

11 The atmosphere contains over 79% nitrogen but this is relatively inert and must be converted to nitrate so it can be absorbed by plant roots.
Lightning combines atmospheric oxygen and nitrogen to produce nitrates.
Nitrate is needed for the production of amino acids/proteins in plants.
Plants are consumed by heterotrophs which then use the fixed nitrogen for their amino acids/protein/growth.
Dead organisms/faeces/urine are converted into ammonia by bacteria/fungi/decomposers.
Nitrifying bacteria (e.g. *Nitrosomonas*) oxidise ammonia to nitrite.
Nitrifying bacteria (e.g. *Nitrobacter*) oxidise nitrite to nitrate.
Denitrifying bacteria (e.g. *Pseudomonas denitrificans*) reverse these processes changing nitrate to atmospheric nitrogen.
Denitrifiers are anaerobes flourishing in swampy conditions whereas nitrifiers are encouraged by good farming practice/draining etc.
Fixation of atmospheric nitrogen can be brought about by free-living bacteria (e.g. *Clostridium*, *Klebsiella*, *Bacillus*, *Azotobacter*).
Papillionaceae/legumes have symbiotic associations with nitrogen fixing bacteria, e.g. *Rhizobium* which live in root nodules.
Such plants can live in soils deficient in nitrogen/increase soil fertility when their roots decay.

43 Distribution and diversity of species

Distribution is the total area in which a species occurs.

Diversity is how rich the variety of different species is within a community.

The distribution of a species is largely determined by the climate and other **abiotic factors** (i.e. contributions in the environment which are non-living).

Ecologists consider weather differently from meteorologists because local differences occurring within a few metres, or even centimetres, may create microclimates which affect survival. Different rocks near the surface may have an effect on drainage which will alter the moisture content and temperature of the soil. Depressions in the ground may form pools of cold air at night killing species which are not frost-hardy.

Edaphic factors (those resulting from the chemical, physical or biotic nature of the soil) will affect the distribution of land plants and hence the animals which depend on them for their food and shelter.

Abiotic factors

Temperature

- **Poikilotherms** (also called ectotherms) are animals whose body temperature fluctuates with the surrounding environment. The term 'cold blooded' is inaccurate because their body temperature may become higher than that of so-called 'warm blooded' animals due to the sun's warmth.
- **Homoiotherms** (also called endotherms) are the 'warm blooded' birds and mammals that maintain a body temperature within a narrow range.

Normal life exists between 10 and 50 °C. Most poikilotherms cannot live at temperatures that are too high or low unless they are adapted to survive extremes. A few poikilotherms can regulate their temperature slightly. Bees maintain temperatures between 13 and 25 °C within the nest by evaporating drops of water when it is hot and by increasing their metabolic activity when cold. Ants move their larvae within the nest to areas of a suitable temperature.

Deep sea temperatures only fluctuate a few degrees and even in shallow, fresh water the temperature cannot fall much below freezing or rise far above the air temperature. Temperatures on land vary according to the season, daily weather, area and time of day. Many species, especially poikilotherms, grow, mature and reproduce in less than 6 months in temperate or colder latitudes.

Humidity

Too little or too much moisture may be harmful. In damp heat animals may not be able to keep cool by evaporating water. If the air is dry, animals may lose too much water from the body through respiration, excretion or evaporation. The degree of humidity may determine which animals can live in the environment.

Light

Light is a controlling factor in the growth of vegetation and therefore determines which animals are present.

Many insects fly at very definite light intensities. Day fliers, like butterflies, only respond to bright light whereas one of the fruit flies (*Drosophila suboscura*) only flies for 30 minutes at sunrise and sunset. Yet others, like the crane fly, are active from dawn to dusk.

Diversity is affected by pollution

Pollution is the contamination of ecosystems by substances or energy which affect the environment adversely. The richness of the world's fauna and flora is being depleted by pollution in its many forms.

Carbon dioxide

Carbon dioxide, produced from burning fossil fuels, increases the greenhouse effect which reduces the amount of heat energy escaping into space and thereby raises the earth's temperature. Present trends are causing global warming which may result in the melting of the polar ice caps which, in turn, could raise the level of the oceans causing massive flooding. It could also cause extremes of drought and rainfall with consequent problems for food crops.

Acid rain

Acid rain is caused by contamination with sulphuric and nitric acids formed from gases released by petrol and diesel engines, power stations that burn fossil fuels and oil refineries. Acid rain has a pH of 5.6 or lower. It damages agricultural crops and trees. The damage to plants is mostly caused by the inhibition of nitrogen fixation. Acid rain also erodes the external surfaces of buildings, especially those made of limestone.

Heavy metals

Heavy metals, particularly lead and mercury, get into the food chain and accumulate in the body. They may reach concentrations which inhibit enzyme activity. Leaded petrol contained tetraethyl lead and emitted poisonous lead compounds in car exhaust fumes. It is believed that the mental development of children living in town areas have been affected. Mercury may be used in plant sprays and seed dressings to control fungal growth. It is still used extensively in industry and probably as much as 3000 tonnes a year are discharged into the environment. Mercury poisoning can be fatal.

Herbicides

Herbicides are used to control or eliminate weeds. Most herbicides are similar to auxins and are biodegradable, being quickly broken down by bacteria in the soil into harmless substances. Nevertheless, the plants that are killed are removed from the food chain and many of our native plants, and the animals that depend on them, have been severely depleted. Some herbicides are made from trinitrophenol, and during manufacture small amounts of dioxin are formed as an impurity. Dioxin is one of the most deadly poisons known.

Insecticides

Insecticides are used to kill insects. They are indiscriminate and not only destroy those insects such as the mosquito and tsetse fly which transmit diseases, but also kill beneficial pollinators such as bees. Most organic gardeners agree that the best way of dealing with insect pests in the garden is to remove them without the use of chemicals or to let nature produce the predators.

DDT, which is now banned in most developed countries, does not degrade easily and remains in the food chain. In the 1960s this caused a severe decline in the numbers of predatory birds (e.g. the peregrine falcon) in the British Isles.

Many insects have developed a resistance to DDT and related insecticides. This is not a pollution problem but it does mean that there is at present little hope of eradicating insect borne diseases such as malaria.

Nitrates and phosphates

Nitrates and phosphates are used as chemical fertilisers to help farmers get maximum yields from their crops. Crops of clover, or other leguminous plants, are seldom ploughed back into the soil to replace nitrates and other minerals. Instead, artificial fertilisers are applied.

Nitrates and phosphates are very soluble and are easily leached from the soil by rain water into streams and waterways. This causes the rapid growth of algae and other photosynthetic micro-organisms which make the water turbid. When these organisms die they are used as a food supply by aerobic decomposers (mainly bacteria). The **biochemical oxygen demand (BOD)** is the amount of oxygen used by such micro-organisms to break down the organic matter. The BOD is a reliable measure of the degree of pollution of water. The consequent lack of oxygen causes fish and other aerobic organisms to die. Similarly, if untreated sewage enters a river or lake, the BOD increases. This process in which water becomes over-rich in nutrients is known as **eutrophication**.

Thermal pollution

Thermal pollution is caused when hot water is emptied into rivers or the sea by power stations which use natural supplies of water for cooling. Protoctists then multiply rapidly, use up the oxygen and begin to die. The aerobic decomposers then use up even more oxygen.

Diversity is high in less hostile environments

Where the environment is less hostile, diversity is high and it is biotic factors which dominate the abundance of species and succession in an **ecosystem**. An ecosystem is a life supporting environment consisting of a network of **habitats** and the **communities** that live in them. Examples of ecosystems include marine, fresh-water, tropical forests, grasslands and deserts. Each habitat supports different populations of plants and animals. Within the tropical forest, for example, habitats include streams, leaf-litter on the forest floor, the tree canopy and rotting logs.

The structure of communities is always changing. A forest fire, a volcanic eruption or man-made structures such as mining waste tips all give opportunities for new species to colonise a region. The first species to arrive are not usually the ones to survive because they alter the character of the habitat, creating a new environment which is suitable for colonisation by new species. The structure evolves over a period of time and the process is known as ecological **succession**. A succession of plant communities over a period of time are known as a **sere**. The succession is not random. Each new species evolves in such a way as to be able to exploit the particular conditions at that time. There are two different types of succession: primary and secondary.

Primary succession

This occurs in lifeless areas such as lava flows, newly-formed volcanic islands and sand dunes where there is no soil capable of sustaining life.

At first, only a few *pioneer* species from the surrounding area are capable of colonising the new habitat. These will include lichen and algae which can get a grip on the rock and trap particles of organic material and begin to form humus. Grass seeds and the spores of ferns will later germinate in the pockets of humus and the formation of soil will continue.

As new plants arrive and become established the habitat becomes modified by increasing amounts of shade on the ground as well as by the changing mineral composition of the soil. These changes allow new species, which are better adapted to the changed conditions, to become established. A similar succession of animals occurs and there is an interaction between the organisms and the environment which influences the rate and pattern of succession.

Secondary succession

Secondary succession is the evolution of an ecosystem from bare soil which had been previously formed before life was destroyed by fire, flood, or maybe human intervention. The succession is similar to primary succession but, because there is soil, there will be a much greater diversity and abundance of plants and animals from the beginning. Eventually the community will reach a **climax** when it has achieved stability. A good example of a climax community in Britain is an oak woodland.

1 List the most appropriate word(s) to complete the following sentences.

The study of the way organisms interact with each other and their environment is called _____. The non-living factors in the environment are referred to as _____ factors.

The place in which an organism lives is called its _____. In this area there will be populations of different species living together and this is called a _____. Sand dunes are inhospitable areas which are gradually _____ by pioneer species. These gradually improve the fertility of the soil so that new species can replace them. This process of change of species with time is known as _____. The _____ is the final stage or sere. (7)

2 a Explain how the temperature in the environment affects the distribution of organisms. (5)
b Explain how the light in the environment affects the distribution of organisms. (5)

3 The diagram below shows part of a river system.

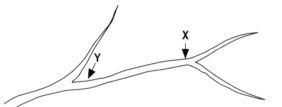

A power failure at a sewage plant has resulted in an excessive release of untreated sewage at point **X**.
a Explain the series of events this may cause in the river downstream from **X**. (4)
b What immediate measures could be taken to prevent serious damage to the ecosystem. (1)
c Further down the river at **Y**, the nitrate concentration is much higher than it was upstream from **X**. Explain the effect this may have on the river flora (vegetation). (1)

4 The table shows the results of investigations of several ponds in mid-Wales.

Pond	pH of pond water	Number of plant species	Number of invertebrate animal species
Mawn Pool	4.4	8	4
Rhulen Hill	4.8	11	5
Llandaban	5.7	16	9
Mere Pool	6.6	23	19
Beilibedw	8.1	21	14

a Describe the relationship between the pH of these pools and the numbers of invertebrate animal species. (1)

b Mere Pool has the greatest species diversity. Use the data to suggest why this is so. (2)
c Which of these pond ecosystems would you expect to be the least stable? Explain your answer. (2)

NEAB, BY01, June 1997

5 The graph below shows the effect of an organic effluent (sewage) on a river system.

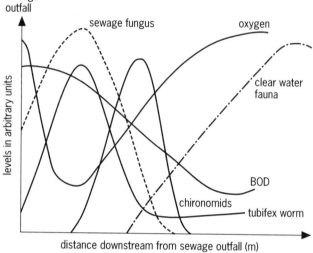

a i Describe the conditions in which the sewage fungus flourishes. (1)
ii Name another organism which can live in similar conditions. (1)
b Explain why the biochemical oxygen demand is high near the sewage outfall. (3)
c Farmers using chemical fertilisers can cause an increase in the concentrations of inorganic ions in nearby waterways.
i Name this process. (1)
ii Explain the sequence of processes, starting with increased river mineral concentrations, that may result in the depletion of some species of fresh water animals. (4)
iii How could the farmer fertilise his land with less damage to the river ecosystems? (1)
iv What type of crop could be planted by the farmer which would not require addition of inorganic nitrates? (1)

6 Rivers and estuaries may become polluted by hot water from the cooling towers of power stations.
a i State two ways in which thermal pollution may cause the death of aquatic organisms. (2)
ii Suggest two ways in which this type of thermal pollution may be reduced. (2)
b The sandhopper *Urothoe* is a small crustacean, which lives in estuaries. An investigation was carried out into the effects of thermal pollution on its growth. Two areas in an estuary were sampled regularly over a period of 16 months (in 1967 and 1968) and the head lengths of the sandhoppers collected in each area were recorded. In one area, hot waste water from a nearby power station entered the estuary, but the other area was unaffected.

The results are shown in the graph below.

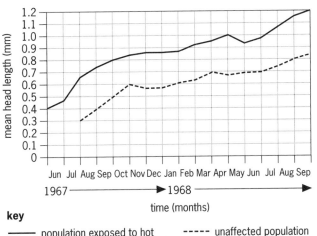

key

—— population exposed to hot waste water from power station

---- unaffected population

(Adapted from: Barnes R.S.K., Estuarine Biology, 1994.)

i Calculate the percentage difference between the mean head lengths of the two populations at the beginning of January 1968. Show your working. (3)

ii Suggest an explanation for this difference. (2)

c The graph below shows the effect of thermal pollution on the percentage of female sandhoppers carrying eggs from the beginning of March to the end of August 1968. Females carry eggs only during the breeding season.

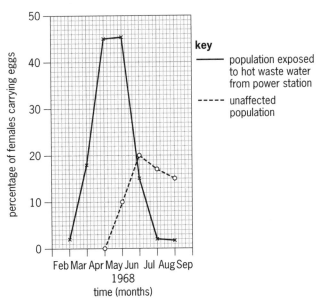

key

—— population exposed to hot waste water from power station

---- unaffected population

(Adapted from: Barnes R.S.K., Estuarine Biology, 1994.)

i Comment on the difference between the two curves. (3)

ii Suggest why this thermal pollution effect may be a disadvantage to the sandhoppers. (2)

London, Paper B6, January 1997

7 Write an essay on atmospheric pollution. (20)

(You may illustrate your answer where appropriate with diagrams but they should be included only if they are relevant and complement the rest of your answer)

London, Paper B6, June 1997

8 The volcanic island of Krakatoa is 40 km away from two large tropical islands, Java and Sumatra. In 1883 volcanic eruptions blew away two thirds of Krakatoa and 100 m depth of volcanic ash accumulated on the remaining land surface. Scientists visiting the island shortly after the eruption found no signs of life. By 1886 cyanobacteria (blue–green algae), ferns and grasses had begun to colonise and, on the beaches, seedlings of coastal plants were becoming established. This was rapid and by 1920 around 50 species were thriving.

Recolonisation of the interior rainforest was considerably slower, only wind-dispersed species being found in 1886. By 1898 the scattered ferns and grasses had given way to tall grassland with trees and shrubs and 8 years later discontinuous patches of woodland, including figs, had developed, most being dispersed by birds and fruit bats.

By 1921 there were 21 species of forest trees and 10 years later the forest canopy was complete.

A survey in 1983, a hundred years after the eruption, found 27 species of tree in an area of 30 km², compared to 211 species in one tenth of this area on the adjacent islands. The table below shows the number of tree species according to dispersal mechanism on Krakatoa in the years when surveys were carried out.

Year	Number of tree species		
	Sea dispersed	Wind dispersed	Animal dispersed
1883	0	0	0
1886	8	2	0
1898	24	16	1
1906	46	20	16
1921	54	46	46
1931	53	55	64
1983	55	77	115

(Adapted from: Asquith N. and Whittaker R., Rain Forest Disturbance and Recovery. Geography Review, Vol.5 No.5. Philip Allan Publishers, 1992.)

a Plot these data on a single pair of axes on graph paper. (5)

b From the graph estimate the number of animal-dispersed tree species in 1958, 75 years after the eruption. (1)

c What type of succession is described in the passage? (1)

d Construct a 'flow' diagram to show the key stages in the succession on Krakatoa. (3)

e Suggest reasons to explain the different rates of colonisation by sea-, wind- and animal-dispersed tree species. (6)

f Suggest why recolonisation was not complete in 1983. (1)

OCR (Oxford), Environmental Biology Paper 6916, March 1999

1 ecology; abiotic; habitat; community; colonised; succession; climax

2 a Environments with extremes of temperature will have less species diversity. High temperatures result in increased water losses due to evaporation. Organisms with xeromorphic or behavioural adaptations will survive better. The distribution of homoiothermic/endothermic organisms is less affected by temperature than poikilothermic/ectotherms. In very cold conditions plants are dormant much of the time, with seeds germinating only with an increase in temperature and then completing the life cycle quickly/ephemeral.

Surface area to volume ratio is high in small birds and mammals (ectotherms) which would use up vast amounts of energy to maintain their body temperatures in very cold conditions. Hibernation will conserve food reserves. Migration to a warmer climate will help avoid extremes of cold and alter the distribution in different seasons. Ectotherms/poikilotherms will be inactive in cold climates since this will more directly affect the rate of enzyme-controlled reactions.

b Light is essential for the photosynthesis of plants and so this will have a major effect on distribution. Some plants need full sun and cannot compete if shaded. Shade plants can compete in dim light such as the undergrowth of woods. The length of the daylight affects flowering in many plants except those which are day-length neutral, and since day-length varies at different latitudes this will affect distribution.

In water, blue light penetrates further thus affecting the distribution of algae, e.g. red algae are found deeper in the water since they have accessory pigments which utilise blue light.

In animals light may be required to see food and therefore affect the period of their activity, but others may be adapted to dim light, e.g. tapetal layer in eyes of cats so they can hunt at night. Others use sonic devices to locate food, e.g. bats. Other animals may be restricted to totally dark areas where they can thrive, e.g. deep in the soil or in caves. Many fungi and bacteria are independent of light for their distribution, although light may be required for the alignment of the gills of mushrooms, etc.

3 a The organic material (sewage) is food for decomposers.
The numbers of decomposers will increase.
They use up much of the dissolved oxygen in the water/the biochemical oxygen demand increases.
This can result in death of many species of fresh water fauna (depending on their tolerance of low oxygen concentrations).
b Use a machine to bubble air/oxygen into the river.
c An increase in growth of algae and other aquatic plants.

4 a As the pH increases (from pH 4.4) the number of invertebrate animal species increases up to pH 6.6 after which the number of species decreases to pH 8.1.

b The pH of Mere Pool is the least extreme environment in terms of pH/nearest to neutral 7;
so has the least effect on distribution of organisms/has the greater number of species.
c Mawn Pool (no mark for the name alone).
Two of:
- This consists of fewer species;
- forming a simpler food web/has the lowest diversity;
- so that slight changes affecting one or more of these species will have a proportionally larger effect on the whole ecosystem/low diversity makes the ecosystem more susceptible to change.

5 a i anaerobic/low oxygen concentrations
ii tubifex worms
b Saprophytic bacteria and fungi/decomposers;
grow/reproduce using the organic material as an energy source;
use up oxygen for their respiration.
c i eutrophication
ii Algae and photosynthetic bacteria increase exponentially.
The algal bloom on the surface prevents lower vegetation from receiving light.
Decomposers increase.
Oxygen used for respiration/oxygen becomes depleted (resulting in death of fresh water animals).
iii Use compost/farmyard manure/kelp, etc.
iv Legumes/Papillionaceae/clover/peas/beans, etc.

6 a i Two of:
- Lack of oxygen due to lower solubility.
- Inactivation/denaturation of enzymes or proteins.
- Loss of food species.
- Increased decomposition leading to a lack of oxygen/increased pH/release of toxins.
ii Two of:
- Allow the water to cool in ponds.
- Describe a method of transferring the waste heat to air.
- Use to heat homes/glasshouses/fish farms.
- Refer to use of an alternative source of energy.
b i (Read the figures correctly from the graph)
0.86 to 0.88; 0.56 to 0.58
The answer from your use of these figures is allowed e.g.

$0.28/0.86 \times 100 = 32.6\%$ or
$0.28/0.58 \times 100 = 48.2\%$ etc.
ii Two of:
- Increase in temperature increases rate of growth.
- There is more abundant food material/energy for growth.
- The higher temperature increases the enzyme activity/rate of metabolism.
c i The population subjected to hot water starts its breeding season earlier in the year.
The peak is maintained longer.
The maximum percentage of females carrying eggs is higher (or give the figures).

ii Two of:
- The hatching time of the larvae may not coincide with the maximum food availability.
- There may be increased numbers of predators.
- The increased numbers may increase intra-specific competition.

7 *In general essays such as this, care needs to be taken to give a broad, balanced account, whilst including scientific detail. Careful planning of your time is essential. Different exam boards include different amounts of detail on this topic so it is important to consult your syllabus.*

A **general introduction** to the essay should start with a definition of atmospheric pollution – fouling of the **air** with **materials harmful to living organisms**. These materials can be solids, liquids or gases. Reasons for an increase in pollution and sources of pollution such as industry, motor vehicles and domestic fires need to be mentioned. The use of fossil fuels and the contribution made by natural causes such as volcanoes should be included.

Ozone depletion – refer to the use of CFCs and how they get into the atmosphere (aerosols/fridges, etc). CFCs decompose to form chlorine atoms which cause the breakdown of ozone. This depletion of the protective ozone layer allows the passage of more UV down to the earth's surface etc/increases mutation/skin cancer etc/affect on survival of species. Preventative measures (give details – less use of aerosols etc).

Greenhouse gases include water vapour, carbon dioxide, methane/nitrous oxide/ozone/CFCs. Comment on the sources of these gases, e.g. methane from the digestive tract of cattle and from swampy land. These gases are able to absorb and emit heat, some of which returns down to the Earth's surface. This may lead to global warming. Climatic effects such as the melting of the ice caps and the rise in sea level will affect species distribution and may result in changes in flora and fauna and some extinctions. Preventative measures need to be discussed – less use of private transport, less burning of trees and fossil fuels, etc.

Acid rain is formed when sulphur dioxide and oxides of nitrogen from industrial processes are oxidised in the air to form sulphuric and nitric acids. These processes may be catalysed by ozone/hydrogen peroxide/ammonia. The pollutants may be blown hundreds of miles by winds and then deposited on vegetation, affecting particularly conifers (denaturing proteins, decreasing resistance to fungal disease, loss of leaves, etc). Soils are affected because ammonium sulphate liberates toxic aluminium ions – may increase senile dementia/Alzheimer's disease/Parkinson's disease. The pH of lakes is lowered which causes changes in the distribution of algae and affects fish stocks (eggs do not hatch/gills affected).

Other references: particular pollutants (smoke/quarrying/cement); radioactive dust; use of catalytic converters to remove toxic gases from car exhausts; lead-free petrol; flue-gas and fuel desulphurisation as preventative measures.
(Scientific content = 17, balance = 3, coherence/clarity = 3; max. 20 marks total)

8 a Graph: time scale correctly scaled (to use most of the graph paper) and labelled 'time in years' (it must start at 1880);
number of species axis correctly scaled and labelled;
all points plotted accurately;
all points joined with a ruler;
plots suitably keyed or labelled.

b 90

c primary

d cyanobacteria, ferns, grasses→coastal plants→tall grassland→woodland patches→forest canopy cover

(4 links = 3 marks, 3 links = 2 marks if correct sequence, 2 links = 1 mark if correct sequence)

e *Note: with 'suggest' questions, read through the text for relevant facts and then apply your general knowledge and biological principles. Always attempt to answer such questions, even with a 'best guess'!*

Sea – two of:
- Land only 40 km away.
- Seeds washed up on coastline.
- Take root easily in thick layer of ash.
- No competitors.

Wind – two of:
- Land only 40 km away (allow only once).
- Wind speed and direction may vary.
- Wind not blowing towards Krakatoa much of the time.
- Wind not strong enough to blow seeds all that way.
- Only light or specially adapted seeds reach the island.

Animal – two of:
- There is nothing to attract animals or birds to the island.
- Time lag behind other mechanisms.
- Further time lag before permanent animal life on the island.

f Not complete – by comparison with the number of species per area on the main islands.

44 Adaptation and competition

Adaptation is a process of natural selection through which a living organism becomes suited to its environment (see **41** Natural selection and classification).

Body size and shape in relation to the environment

During evolution there have been trends towards an increase in the size of both animals and plants. Large aquatic animals have comparatively small skeletons and depend on the buoyancy of the water in which they live for support. Water offers more resistance to movement than does air because it is more viscous and more dense. Those animals which move at speed through water (especially fishes) have acquired a streamlined shape which offers less resistance to movement. Many other animals, such as frogs and otters attain a streamlined shape by keeping their limbs pressed closely to the sides of the body when gliding through the water.

Large organisms have a much smaller surface–volume ratio than smaller organisms of similar shape. Elephants, which live in hot climates, have such a small surface–volume ratio that they cannot lose heat quickly enough. Their surface area is increased by their large ears and folds of skin. In contrast, the extremities of animals living in cold climates tend to be smaller.

On the other hand, there is a limit to how small a homoiothermic animal can be before it is unable to maintain its constant body temperature. Small mammals, such as mice, eat their own weight in food every day and maintain their body temperature by constant activity with only short spells for sleep. In winter the body temperature cannot be maintained and the animal goes into hibernation.

Reducing water loss in plants

A xerophyte is a plant which has become adapted to grow in areas where there is either a permanent or a seasonal shortage of water. Modifications result in less water being lost from the leaves by **transpiration**. The leaves of terrestrial plants typically have a waxy cuticle which is much thicker on the upper surface while most of the stomata are on the underside of the leaves. Both of these characteristics are important in conserving the water balance in plants. Xerophytes have further modifications which may include one or more of the following.

- Leaves which are reduced to spines (e.g. cacti). The stem stores water and also acts as the main photosynthetic organ. Such plants are known as **succulents**.
- Very small leaves (e.g. heather).
- Stomata in sunken pits (e.g. pine trees). These pits create a microenvironment of moist air over their surface and lower the diffusion gradient between

the intercellular spaces of the leaf and the air immediately outside the stomata. The loss of water from the leaves by diffusion (the transpiration rate) is therefore reduced.

- Curled leaves (e.g. marram grass). The stomata are confined to the inner surface and moist air is trapped within the curled leaf. The curled shape also gives the leaf rigidity which helps to prevent wilting. The leaves of the marram grass, which colonises sand dunes, have large epidermal cells, called hinge cells, at the bases of the furrows on the inner epidermis. These hinge cells have thin walls and shrink rapidly when there is a loss of water. This makes the leaf curl into a tubular shape (see Figure 44.1).
- Hairs on the underside of leaves, which slow down the movement of air over the stomata and create a more humid layer of air which reduces transpiration.

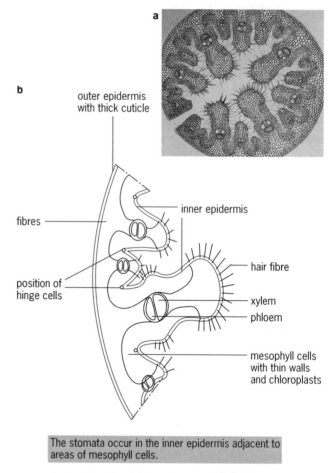

The stomata occur in the inner epidermis adjacent to areas of mesophyll cells.

Figure 44.1 TS of a rolled leaf of marram grass.
a Photomicrograph. **b** Tissue map of part of the leaf.
(Adapted from: Clegg C.J. and Cox G., Anatomy and Activities of Plants. John Murray, 1978.)

Within any **habitat** there is a fine balance involving both abiotic and biotic factors. The biotic factors include those of predation and competition between species (**inter-specific competition**) and between members of the same species (**intra-specific competition**).

Predation

Fluctuations, which are slightly out of phase, occur in the populations of prey and predator. An increase in the population density of prey is followed by an increase in the population density of its predators. If the prey is a herbivore, a decrease in the density of plant food, probably due to unfavourable abiotic factors such as colder weather, will result in a decrease in the prey population which feeds upon the particular vegetation. This, in turn, will lead to a fall in the number of predators. Figure 44.2 is a graph showing the relationship between the population densities of a predator and its prey.

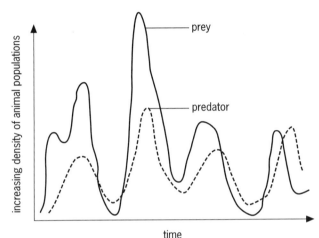

Figure 44.2 Relationship between the densities of prey and predator

Man is a predator. Although much of our meat is farmed, wild species are still killed in vast numbers. This has led to the extinction of some species.

Deep sea fishing is now highly mechanised and organised. Factory ships stay out at sea for months at a time and such intensive fishing may lead to stocks being overfished. Fishing boats should neither use nets with a mesh smaller than regulation size nor fish in the breeding season. If the number of individuals in a species within a population falls too low, the species may not be able to survive.

Some of the world's largest mammals, and thousands of smaller species of animals and plants, are in danger of extinction because they cannot compete with man for the land and its resources. Large mammals sometimes pose a threat to small village communities or may be illegally hunted as trophies or for unproven 'medical' properties. The sizes of populations fluctuate naturally but sometimes a change in abiotic and/or biotic features results in an organism flourishing in an uncontrolled manner and becoming a pest. The introduction of the rabbit into Australia is a good example of what happens when an organism is introduced into an area where it has no natural predators. In such cases biological control, in which predators are introduced into the environment, has been tried with varying degrees of success.

The biological control of pests

When the European rabbit was introduced into Australia in the 1800s its population grew unchecked because there were no natural predators present. Eventually the myxoma virus, which causes the disease myxomatosis, was released and the population of rabbits was dramatically reduced. However, not all of the rabbits died and there is now a large population of myxomatosis resistant rabbits. The myxoma virus was also released in England in the 1950s with similar results.

Gardeners are now able to purchase from garden centres organisms for the biological control of a number of pests including whitefly and slugs. The predators have to be reintroduced from time to time and it is more expensive than using conventional sprays or slug pellets. The advantage is that the predators are specific. The predator which attacks slugs is a nematode which only attacks slugs and snails, and obviates the use of slug pellets which can be harmful to other organisms such as frogs which may eat the slugs and be killed by the chemical. The predatory wasp grub which feeds on whitefly, and which is used as a method of biological control, only eats the whitefly. Insect sprays are indiscriminate and kill beneficial insects such as bees. Therefore, plants infected with insect pests need to be sprayed under very carefully controlled conditions such as in a greenhouse.

Weeds can also be controlled by using biological methods although the control tends to be effective only for a period of time. The classic example is the control of the prickly pear cactus (*Opuntia*) in Australia in the 1920s. Nearly 25 million hectares of cactus scrubland were converted to arable land and pastures by the introduction of a moth (*Cactoblastis*) from Argentina whose larvae bore into and destroyed the cactus. The cactus was well under control by 1933. However, many other similar attempts at biological control have failed.

Competition in the community is intense. There is both inter-specific and intra-specific competition for every imaginable type of resource. The same two species may compete for more than one resource, for example food and shelter. Often there is one resource, known as the limiting resource, which limits the growth of populations more than any other. The availability of sunlight could be the focus of competition for green plants in a particular ecological habitat such as a forest floor or the entrance to a cave. Light would be a limiting resource in this example.

1 a Use your knowledge of adaptation to decide whether the following are likely to be true or false.
i Foxes in arctic regions have smaller outer ears than those from warmer climes.
ii Plants on windy moors will have larger leaves than related species in more sheltered regions.
iii Marram grass leaves will be less tightly rolled on calm, humid days than dry, windy days.
iv The upper cuticle of dicotyledonous leaves tends to be thicker than the lower cuticle (4)
b lupins → aphids → ladybirds
Using the information in the food chain shown above:
i Suggest why using a chemical insecticide spray may eventually result in more insect damage to the lupins. (2)
ii Give two advantages in using the ladybirds as a means of biological control of aphids rather than using a chemical insecticide. (2)

2 a Write a list of the most suitable words to complete the following passage.

Data from the Hudson Bay Company have been used over the years to calculate and plot graphs of the sizes of populations of hares and the lynx which feed on them. The lynx is the _____ and the hare is the _____. As the population density of the hares rises the population density of lynx will _____, however there will be a _____ between the peaks of the two graphs. Such relationships can be made use of in the _____ control of pests. (5)
b Give the correct biological term for the following:
i One species of plant A grows quickly to obtain light at the expense of the slower growing species plant B.
ii A predator or parasite is introduced into a community to curb the growth of a pest.
iii A plant which is adapted to grow in dry conditions.
iv Plants with water-storage tissue.
v The evaporation of water from the surfaces (mainly the leaves) of the plant. (5)

3 *Ammophila arenaria* (marram grass) is a xerophyte which lives on sand dunes. In dry conditions the leaf rolls up so that the hairy inner epidermis is concealed inside.

TS *Ammophila* leaf

Explain how: **a** rolling and **b** hairs, help to reduce transpiration. (4)

4 The table shows the results of an investigation into how the distribution of the roots of three species of grass varied with depth. The figures are given as percentages of the total root dry mass of the species concerned.

Soil depth (m)	Species of grass		
	Panicum maximum	Themedra triandra	Eragrostis superba
0–0.4	64.9	66.5	73.6
0.4–0.8	14.2	25.9	15.5
0.8–1.2	12.1	5.6	7.4
1.2–1.6	4.7	1.4	2.6
1.6–2.0	2.6	0.6	0.8
2.0–2.4	1.2	0	0.1
2.4–2.8	0.3	0	0
total dry mass (g per plant)	114	58	27

a All three of these species grow in hot, dry conditions. Which species would you expect to grow best if the ground was lightly sprinkled with water at regular intervals? Give an explanation for your answer. (2)
b *Panicum maximum* is able to survive better than the other two species during lengthy periods of hot, dry weather. Use the data in the table to suggest an explanation. (2)
c Describe two ways in which the leaves of plants may be adapted for reducing water loss in hot, dry conditions. (2)
AQA (NEAB), BY05, March 1999

5 European carp feed on plants. They were introduced into Australia in the last century. One strain of carp, the Boolarra strain, was used to stock fish farms on the Murray River. From there it spread rapidly through the neighbouring river systems. In some sections of the Murray River, the carp is now almost the only type of fish left. The Boolarra strain can grow to a length of 80 cm and a mass of 10 kg.
a Suggest and explain why the carp is now almost the only type of fish left in the Murray River. (2)
b To solve the carp problem, river authorities are considering releasing a European virus called spring viraemia which is known to kill carp.
i Suggest one test the river authorities should carry out before releasing spring viraemia into Australian rivers.
ii Explain why it would be necessary to carry out this test. (2)
NEAB, BY01, June 1997

6 The whitefly *Trialeurodes* is a pest of glasshouse cucumber and tomato crops and can be controlled biologically using a minute parasitic wasp, *Encarsia*. In an experiment, adult *Encarsia* wasps were introduced to a glasshouse crop infested with *Trialeurodes* and the density of each species was measured over 20 generations. The results are shown in the table below.

Generation number	Density of *Trialeurodes* (insects per unit area)	Density of *Encarsia* (insects per unit area)
0 (start of experiment)	28	10
5	20	8
10	3	26
15	21	5
20	10	22

a Describe the changes in the density of *Trialeurodes* over the course of this experiment. (3)

b Comment on the relationship between the density of *Trialeurodes* and the density of *Encarsia*. (3)

c Suggest two adverse effects on a crop of an infestation by *Trialeurodes*. (2)

d i Give two advantages of biological rather than chemical control. (2)
ii Give one disadvantage of biological rather than chemical control of pests. (1)
London, Paper B2, June 1997

7 Moose are large herbivorous animals. Isle Royale is a large island in Lake Superior. Moose first colonised this island in 1900. At the time they had no predators on the island. Wolves, which are predators of moose, were introduced to the island in 1950. The graph shows the moose population from 1900.

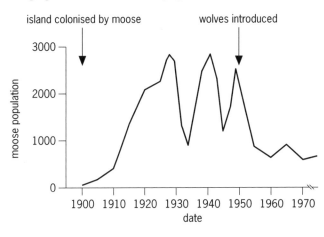

a i Suggest an explanation for the changes in the moose population before the introduction of wolves. (2)
ii Describe the effect of introducing wolves on the moose population. (1)
AQA (AEB), Module 2, June 1999

8 One of the most valuable crops planted by the Forestry Commission is spruce because its yield of timber is very high. Early planting trials showed that spruce trees grew very slowly when planted on land on which heather was also growing.

a Name the type of competition shown between spruce and heather. (1)

b Give two resources for which spruce and heather are likely to be competing. (1)

c Further trial plantings on land dominated by heather showed that the growth of spruce was greatly assisted by planting another tree species at the same time. This use of a 'nurse' crop is now standard practice. The table shows the results of some of these trial plantings.

	Height of spruce after 15 years (metres)
spruce, heather and Japanese larch	4.5
spruce, heather and Scots pine	3.1
spruce, heather and Corsican pine	3.5
spruce and heather	2.0

i Young spruce trees were 50 cm high when first planted. Calculate the difference in the rate of growth when these trees were grown with Japanese larch compared to the control. Show your working. (2)
ii Suggest one way in which a 'nurse' crop may aid the growth of spruce trees. (1)
AQA (NEAB), BY02, March 1999

9 What is meant by biological control? (4)
Using examples to illustrate your answer, explain the advantages and possible limitations of using biological control methods. (11)
OCR (Oxford), Environmental Biology Paper 6916, March 1999

10 Describe ways in which xerophyte plants are adapted to survive. (10)

1 a i true
  ii false
  iii true
  iv true
 b i Some aphids will be missed by (or be resistant to) the insecticide and can reproduce quickly/parthenogenesis. Many ladybirds will also be killed by the insecticide and their numbers will recover more slowly than the aphids.
   ii The biological control often has a longer lasting effect. It does not result in poisonous residues getting into the food chain/is not harmful to other organisms.

2 a predator; prey; increase; delay/time lag; biological
 b i inter-specific competition
   ii biological control
   iii xerophyte
   iv succulent
   v transpiration

3 a Rolling reduces transpiration by reducing the area of the outer leaf surface which is in contact with the air. The evaporation of water will also decrease because the air inside the leaf will become saturated with water vapour which cannot easily escape.
 b Hairs reduce transpiration by reducing air movement over the surface of the leaf.
  A layer of still air will be trapped by the hairs and this will become saturated with water vapour and so reduce further evaporation.

4 a *E. superba*; this plant has the largest proportion of shallow roots.
 b The roots go to the greatest depths.
  Such roots are able to get water when the surface dries out.
  or
  It has the greatest root mass;
  so it is able to store more water.
 c Two of:
  • curled leaves
  • thick cuticle
  • sunken stomata
  • hairs

5 a The carp is a successful competitor for food.
  Because of its large size/there are no other herbivores left for carnivores to feed on/no plants for other fish to feed on.
 b i Test whether spring viraemia virus would infect other species in the Murray River.
   ii In case the food webs are disrupted/'the balance of nature' is upset.

6 a The density decreases for the first 10 generations;
  and then rises again between 10 and 15 generations;
  then it falls for the last 5 generations.
 b Three of:
  • There is an inverse relationship between the density of *Encarsia* and the density of *Trialeurodes*.
  • *Trialeurodes* provides food for *Encarsia*.

• More *Encarsia* means more *Trialeurodes* are parasitised/killed.
• Few *Trialeurodes* will result in less breeding by *Encarsia*.
 c Two of:
  • Decreased yield of the crop.
  • *Trialeurodes* may spread infestation/disease.
  • The crop will be of decreased value due to spoilage/it may be inedible.
 d i One of:
  • There will be no residues on the crop/it can be marketed as organic.
  • Biological control does not harm other organisms/is not toxic.
  • It has a long-lasting effect.
  • The pest is less likely to become resistant.
   ii One of:
  • Biological control does not remove all the pests.
  • Can only control one pest at a time.
  • Takes longer to work.
  • Is more dependent on environmental conditions.
  • Can only be used if the pests are present.
  • May have an effect on other organisms.

7 a i Moose numbers rose in the absence of predators/with plenty of food.
  Depletion of food/spread of disease/climatic change resulted in a fall in numbers.
   ii Introduction of wolves caused the moose numbers to stabilise at a lower level.

8 a inter-specific
 b One mark for two resources, e.g. light; named nutrient e.g. nitrate; water.
 c i Difference in growth
$$= 4.5 - 2.0 \text{ (spruce and heather control)}$$
$$= 2.5\,\text{m}$$
  $2.5\,\text{m}/15\text{ years} = 0.166$
   Answer $= 0.17$ metres per year.

   ii *Specialised knowledge is not required here. Use your biological knowledge and information provided to make a likely suggestion.*
  The nurse crop may produce a substance or nutrient which stimulates the growth of spruce/it may produce a substance which inhibits the growth of heather/it may provide nitrogen-fixing bacteria.

9 Biological control (any four of the following):
 • The use of a living organism.
 • This is usually a predator.
 • It is used to control a pest.
 • It reduces the numbers to below the economic damage threshold.
 • It does not eradicate the pest altogether.
Examples
A maximum of 3 marks for any one correct example with details.
Advantages
 • No use of chemical pesticides.
 • Biological control is specific – only target species are affected.

- There are side effects in using insecticides, for example, indiscriminate killing of beneficial insects e.g. bees.
- It is cheap, resistance does not build up as with pesticides.

One mark for other examples illustrating the effects.

Limitations

- Biological control may take some time to achieve a result.
- If a pest has been introduced (i.e. an alien species) the predator may not adapt well to the new environment.
- There may be no natural predator so the control species may become a pest.
- The pest is never completely eradicated.
- It is necessary to tolerate some damage/loss.

(to a max of 11 marks)

Other marks are available for information which is biologically accurate and relevant e.g. releasing sterile males. Give named examples.

Use of parasites, bacteria, viruses (see text and questions 5 and 6 for examples)

10 *The Plant Kingdom contains many varied examples of xerophytes and xerophytic adaptations. Try to ensure that you mention a wide range in your essay.*

Ten of:

- Land plants need to ensure that their water losses by transpiration do not exceed their water uptake through the roots or this could lead to **permanent cell damage due to plasmolysis**.
- Xerophytes are plants living in conditions **where water losses tend to be higher than water uptake**.
- Diffusion and evaporation of water vapour/transpiration will occur through the **stomata**. These are necessary to allow gaseous exchange of carbon dioxide and oxygen during photosynthesis and respiration. At night the stomata shut which reduces transpiration. Some plants in tropical regions **also close their stomata in the very hot midday period to cut down water loss**.

- Excessive transpiration is likely in high winds as found on upland moors. *Calluna vulgaris* (heather) has **sunken stomata** to trap a pocket of saturated water vapour in the substomatal chamber. This reduces transpiration.
- **Rolling of leaves** to produce a saturated atmosphere next to the stomata occurs in *Ammophila* in dry conditions. This decreases the rate of diffusion of water vapour/transpiration.
- Many evergreen plants have a **thick, shiny** cuticle to reduce water losses through the upper epidermis. This may also reflect the light.
- Leaves may be **reduced in surface area**, e.g. heather and pine. In some cacti they are reduced to spines and the photosynthetic function has to be taken over by the stem.
- Many plants in hot climates have a **white, hairy covering of the leaves** which reflects light and traps a layer of air which becomes saturated with water vapour, decreasing the diffusion gradient.
- **Small cell size** is common in xerophytes since such cells will still provide some support if turgor pressure is reduced.
- Cells may have a **more negative osmotic potential** to reduce water losses.
- The root system may be **extensive and near the surface** to absorb water from light showers.
- Other xerophytes may have **long root systems/rhizomes** to tap into underground water supplies.
- Succulent plants have **water storage tissue** to aid survival in dry periods. This may be located in the stem or leaves.
- Other plants may **align their leaves so that they receive less sunlight** and therefore transpire less.

Populations

A **population** is the number of individuals of the same species living in a particular area at the same time.

A **community** consists of all the organisms that live in a particular area (the **habitat**).

Population growth

If the population of a species consists of only a few individuals, the rate of growth will be very slow at first. This is known as the **lag phase**, which involves adaptation to the environment.

If the numbers of the population increase, and more individuals become available for reproduction, the population will enter the **log (exponential) phase** when it will grow at its fastest rate.

Growth cannot continue at this rate indefinitely because limiting factors such as food supply will determine the number of individuals that can be supported. This is known as the **carrying capacity** of the habitat and at this point the population growth will enter a **stationary phase** in which the number of births equals the number of deaths.

Finally a population may reach a stage in which the carrying capacity has declined due to such factors as lack of food or oxygen, or an increase in toxic waste. The population will decline rapidly as this **death phase** is entered.

An example which illustrates all of the above phases in a relatively short period of time would be a laboratory culture of micro-organisms such as bacteria or protozoa (Figure 45.1).

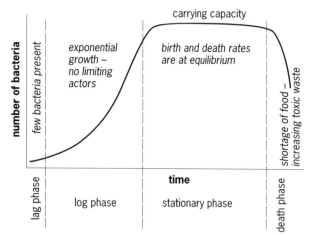

Figure 45.1 Growth curve of a bacterial culture in a nutrient medium

The most usual growth curve for populations (or organisms or even organs) when growth is measured against set intervals of time is an S-shaped **sigmoid curve** (Figure 45.2). The early stage of exponential growth is replaced by the stable state where there is an equilibrium between births and deaths. As long as there are no other influencing factors, the stable phase will also be the carrying capacity of the environment.

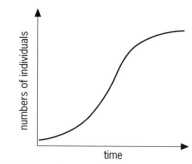

Figure 45.2 Sigmoid growth curve

Such a population growth curve as shown in Figure 45.2 will happen rarely in nature.

Charles Darwin calculated that a pair of breeding elephants, at the minimum breeding rate of six offspring per pair over a 60-year breeding span, would produce a population of 19 million elephants in 750 years. This would be a case of **hyper-exponential growth** and is what has been happening for the last 300 years with human populations, where the interval time required for the population to double has halved as time proceeds instead of remaining constant as is the case with exponential growth.

The factors which maintain the stable growth rates of populations are related to **population density** which is controlled by the number of individuals of a species that the region can support. Periods of increase or decline are usually followed by periods of change in the reverse direction. In some species the population sizes fluctuate very little but in others the fluctuations may be erratic or cyclical where the population reaches a peak and then crashes every few years as happens, for example, with some rodent populations.

Density-dependent growth

Populations may experience exponential growth when a new habitat is being colonised. If the population density becomes too great, the supply of food becomes reduced and the population, having reached a certain size, becomes stable. Examples of density-dependent factors include a shortage of food, the build-up of toxic waste and the availability of nesting sites. In such circumstances mammals will respond by maturing later and by producing smaller and/or fewer litters. Birds will mature later and produce fewer but larger eggs.

Density-independent growth

Many populations fluctuate according to the seasons. A drop in temperature will kill large numbers, or all, of a population regardless of its size. Such seasonal fluctuations occur particularly in species which reproduce once a year, during a particular season, or only live for one year.

Density-independent factors include abiotic factors such as temperature, floods and fires.

Factors affecting population size

- The birth rate is:

$$\frac{\text{the number of births in the population}}{\text{the number of adults}}$$

- The death rate is:

$$\frac{\text{the number of deaths in the population}}{\text{the number of adults}}$$

The size of a population is determined largely by a balance between **fecundity** (the reproductive capacity of individual females) and **mortality** (the death rate from whatever cause). But there are other factors which influence the size of a population:

- **Dispersal**, which extends the range of a species. An example is successful seed and spore dispersal by plants and fungi.
- **Dispersion**, in which breeding adults redistribute themselves within an area.
- **Migration**, in which there is a recurring cycle of movement from one specific area to another in response to seasonal changes. By definition there is always a return trip to where the individuals commenced their journey. The classic examples are the migration of many species of birds. Swallows migrate annually from northern Europe to South Africa. The outward journey is termed **emigration** and the return journey **immigration**.
- **Relocation**, in which animals move to a less predated, more fertile or less densely populated area. There is no immigration at a later date.

Hardy–Weinberg law

This law states that in a large population where there is random mating, the frequency of alleles remains constant. The Hardy–Weinberg law shows why there are not necessarily more dominant phenotypes in a population than recessive ones. Consider two alleles A and a which have frequencies of p and q respectively. There will be a genetic equilibrium between the two alleles which equals $(p + q)^2 = 1$.

If this equation is expanded we get:

AA + Aa + aa = 1
$p^2 + 2pq + q^2 = 1$

AA represents a homozygous dominant
Aa represents a hybrid showing the dominant phenotype
aa represents a homozygous recessive showing the recessive phenotype

Examination questions set on the Hardy–Weinberg law often involve simple calculations which involve using one or other of the above equations.

The above genetic equilibrium only works under the following four conditions:

- no emigration or immigration
- no mutations taking place
- random mating
- large population size.

Human populations

Thomas Malthus published an essay on population in 1798 in which he said that the world's food supply was increasing arithmetically but the human population was increasing geometrically. The series 1, 2, 3, 4, 5, 6 illustrates an arithmetic progression and 1, 2, 4, 8, 16, 32, where the number is *multiplied* each time (in this example by 2) is a geometric progression.

Prior to the agricultural revolution, which took place some 10 000 years ago, the earth could not support more than about 10 million humans (one thousandth of the present population).

By the beginning of the Industrial Revolution in Britain (AD1750) the world population had risen to as much as 800 million. The annual population growth rate had been about 0.1% for the previous 750 years.

The world's human population reached its first billion in about AD1800. This doubled to 2 billion by AD1930 (a gap of 130 years) and doubled again to 4 billion by AD1974 (a gap of only 45 years). The growth peak was reached in the 1960s when it was about 2%.

The population increase in some developing countries in the nineteenth century has been much greater than in industrialised nations. By the late 1960s, some countries in South America and Asia were experiencing an annual growth rate of up to 3%. At this rate a population doubles in only 23 years! Compare this with Belgium, an industrialised country, which has a growth rate of only about 0.4% per year.

There have been improvements in agriculture that Malthus could not have anticipated but there must be a limit to food production levels. The rapid growth of the world's human population has been largely due to increased food supplies brought about by the use of fertilisers and intensive farming techniques, increased medical knowledge, including the discovery of new and effective medicines (especially antibiotics), better supplies of fresh water and disposal of sewage.

The virtual eradication of previously fatal diseases such as malaria, smallpox, diphtheria and cholera has resulted in large increases in population which, in turn, has meant even more mouths to feed.

Family planning in industrialised countries has reduced the birth rate, but most developing countries, where the population is much poorer and many people cannot afford contraceptives, or some are illiterate, have a much higher birth rate.

Other problems at present include:

- **Urbanisation.** It is estimated that more than 50% of the world's population now live in cities. This puts an increasing strain on housing, medical care and sanitation.
- **Increased use of energy.** The supply of fossil fuels will eventually cease. The present increase in the burning of fossil fuels will add to global warming.
- **Global warming.** Changes in temperature and weather patterns will make it more difficult to grow crops as the earth becomes hotter.

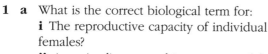

POPULATIONS

1 a What is the correct biological term for:
i The reproductive capacity of individual females?
ii An animal's seasonal journey to and from different areas?
iii The number of individuals of a species that a certain habitat can support?
iv The type of population growth where the numbers double each generation? (4)
b Describe what is meant by the following terms:
i population
ii community (2)
c Human population numbers are thought to have been fairly constant at a low level until 10 000 years ago when they started to rise slowly. Four hundred years ago the numbers began to rise much more rapidly and continued to rise at an ever increasing rate to the present day.
i What term is used to describe the present phase in the growth of the human population? (1)
ii Give two ways in which governments can encourage measures to slow the rate of growth of the human population. (2)
iii Suggest a way in which food supply for humans can be increased without dependence on traditional agricultural methods. (1)

2 The graph below shows changes in the size of a population of yeast cells growing in a glass flask containing nutrient broth with glucose as a source of carbohydrate.

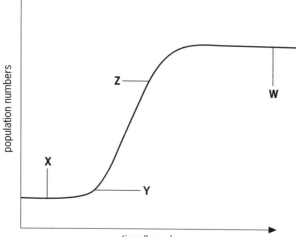

a Give the letter from the diagram which represents the following stages in the growth of the population. You may use a letter more than once.
i The point at which birth rate and death rate are equal.
ii The point where population growth is most rapid.
iii The exponential growth phase.
iv The lag phase (4)

b i Suggest why the rate of growth of the population is less at **W** than at **Z**. (1)
ii Name a factor which should be kept constant during this experiment. (1)
iii The yeast was left in the flask for a further 24 hours and the population size was then found to have decreased substantially. Suggest two factors which may have caused this decrease. (2)
c The population curve obtained when a single female toad lays eggs in a shallow pool and the surviving offspring breed for many generations, would differ from the curve for yeast cells.
i Describe why the population curve for toads would differ from that of yeast. (2)
ii What factors would affect the survival of the larvae to the breeding stage? (2)

3 a Which of the following ecological terms can be used to describe human activity in the fishing industry?
density dependent factor   predator
density independent factor   parasite (2)
b State two reasons why regulating the size of fishing net mesh may not prevent overfishing. (2)
c Arctic cod always lay their eggs in the same place. The adults migrate from their feeding ground **R**, to the spawning area, **S**. The young larvae drift on the ocean currents to the nursery area, **T**, eventually joining the adult stock in their deep water feeding area. The simplified diagram shows the migration circuit used by these fish.

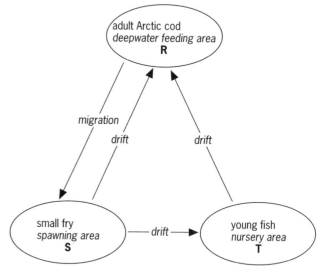

(Adapted from: Cushing D.H., Fisheries Biology: A study in Population Dynamics. *University of Wisconsin Press, 1968*.)

i What would be the overall effect on the cod population, if the area labelled **T** in the diagram was very intensively fished with nets of small mesh size? (1)
ii Use the information in the diagram to help you explain your answer fully. (3)
iii Suggest two reasons why it would be difficult to gain an accurate estimate of the size of the Arctic cod population. (2)

Welsh, Paper B3, June 1997

4 Froghoppers are plant bugs which pierce the phloem of plants and suck the sap, producing the characteristic bubbles often called 'cuckoo spit'.

The graph shows the changes occurring in the population numbers of adult froghoppers during the course of a year in a forest.

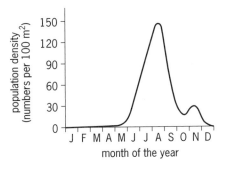

a Name the type of population growth curve shown. (1)

b What term is used to describe the nature of the population growth during June and July? (1)

c On the graph rule a line, clearly labelled K to indicate the likely carrying capacity for froghoppers. (1)

d Give two possible reasons for the decline in numbers from August onwards. (2)

e Suggest a reason for the small peak in October. (1)

f Suggest an explanation for the observed results between November and April. (1)

OCR (Oxford), Environmental Biology Paper 6916, March 1999

5 A type of wild goat was introduced into an island and they bred successfully. The following graph shows the numbers of the wild goat population over the period of 120 years after their first introduction to the island.

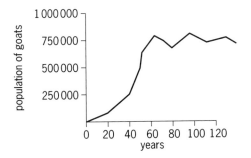

a Describe the pattern of population change shown in the graph. (3)

b i Define what is meant by the term 'carrying capacity'. (1)

ii What is the carrying capacity for the island with respect to goats? (1)

iii Name two factors which might limit the size of the island's carrying capacity for goats. (2)

iv Are the factors named in (iii) density dependent or density independent? (1)

v Explain what is meant by the term density independent factor. (2)

6 The table shows the average number of maize grains per plant produced when plants are grown at different densities under conditions of high and low nitrogen availability.

Density (number of maize plants/unit area)	Average number of maize grains per plant produced when nitrogen availability is	
	High	Low
4	52	43
8	85	64
12	105	72
16	111	62
20	115	60
24	102	45

(Adapted from: Silvertown J., Introduction to Plant Population Ecology. Longman, 1990.)

a Suggest two general conclusions which can be drawn from the data. (2)

b i State the optimum density for maize grain yield when nitrogen availability is:
1) High
2) Low (1)

ii Use these figures to calculate the percentage increase in optimum density when maize is grown under conditions of high nitrogen availability. (1)

iii Use the information in the table and your knowledge to describe one ecological principle underlying the reason for the optimum density being smaller when nitrogen availability is low. (1)

iv When nitrogen availability is high, suggest two other reasons for the change in yield when the density of plants increases beyond the optimum. (2)

v The average number of maize grains per plant has never been observed to exceed 115 under any set of conditions. Suggest one reason for this observation. (1)

Welsh, Paper B3, January 1997

7 a The graph shows four survival curves.

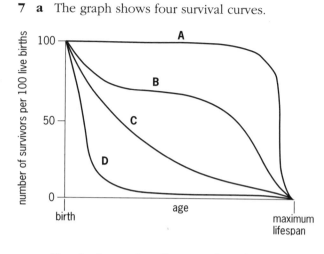

Use the letters **A** to **D** to match each curve with the relevant descriptions below.
i The percentage of the population dying at each age is approximately constant.
ii Most deaths are attributable to old age.
iii Almost all mortality occurs in the very young.
iv Mortality is highest in the very young and the very old. (2)

b The diagrams show population pyramids of two countries.

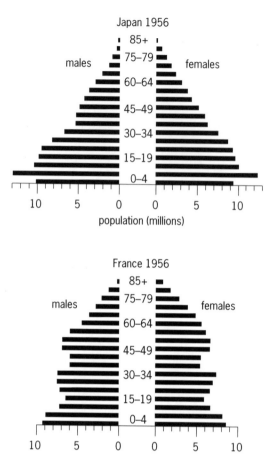

i Suggest two reasons why the ratio of females to males above the age of 60 in the French population differs from the ratio in the younger age groups. (2)
ii Use the information from the Japanese population pyramid to sketch the shape of the pyramid you would expect for Japan in 1996.

Explain the difference in the shape of the two pyramids. (2)

NEAB, BY09, June 1998

1 a i fecundity **ii** migration **iii** carrying capacity
iv geometric progression
b i A population is a localised group of organisms of the
same species.
ii A community consists of localised interacting
populations of different species which also interact with
the abiotic factors of the environment.
c i exponential/log phase
ii education (especially for females); contraception
iii Use of microbial food (unicellular/filamentous
protein) produced in fermenters.

2 a i W ii Z iii Z iv X
b i The rate of growth decreases as the substrate
glucose is used up.
ii temperature
iii shortage of food;
an increase in toxic products in the growth medium.
c i Two of:
• The time for a breeding cycle is longer in toads.
• Only female toads produce eggs/all yeast cells have
the potential for asexual reproduction.
• Many other organisms would be present in the
pool/only yeast cells are present.
• Reproduction only occurs in the toad breeding season.
ii Two of:
• extremes of temperature
• availability of water
• predation
• availability of food (oxygen will not be limiting since
the pool is shallow)

3 a density independent factor; predator
b i Two of:
• The number of fishing boats in an area is unregulated.
• The period of time spent fishing is unregulated.
• The optimum net mesh size for different species will
not be the same.
• It will not stop fishing in nursery areas/exclusion zones.
• It is difficult to enforce/police net mesh size.
c i The cod population would decline (*note the answer –
become extinct is not correct*).
ii Three of:
• The young fish in **T** will be removed.
• These fish are too immature to have reproduced.
• Fewer migrate/drift to **R** to replace the breeding adults.
• These adults are lost from the population by ageing
or predation.
• **R** reduces the number of eggs spawned/young fish in **S**.
iii Two of:
• It is impossible to count the number of females which
produce eggs.
• It is impossible to count the numbers of eggs which
hatch.
• The location of all individuals within a population at
any time is not known.
• The total area occupied by all members is very large
and difficult to sample.
• It is difficult to accurately age fish caught in a sample.
• It is difficult to sample in deep water.

4 a density independent **b** linear (exponential)
c A horizontal line, labelled K, at the 120 population
density number/100 m^2.
d Two of:
• All or most of their food plants die.
• There is insufficient sap available/plants stop growing.
• It is the end of the froghopper's natural life span.
• The weather starts to deteriorate (colder/wetter).
• The effect of insecticide spray.
• The crops are harvested.
e A few early autumn/late summer eggs hatch/these are
late developing young from a spring brood.
f The froghoppers overwinter as eggs/diapause as
larvae/pupae (i.e. no adults).

5 a The population increases slowly with time at first (give
figures from the graph).
The population then increases more quickly.
Population size then becomes more constant
(fluctuating around 750 000).
b i The carrying capacity is the population maximum
which can be sustained on the island (habitat) for an
indefinite period of time.
ii 750 000
iii The numbers of predators; availability of food.
iv density dependent
v A density independent factor is one which is not
affected by the numbers of the population per unit area.
Examples are abiotic factors such as temperature.

6 a (*Values are not needed – just general conclusions.*)
Two of:
• The yield increases to a maximum and then
decreases/reference to the optimum yield.
• Density affects the production of maize grains.
• The yield depends on the nitrogen availability in the soil.
b i *Both answers are required.* 1) 20 plants per unit area
2) 12 plants per unit area
ii $(20 - 12) \times 100/12 = 66.6\%$
iii Nitrogen becomes a limiting factor earlier when
nitrogen availability is low.
iv Pest infestation.
Lack of potassium/lack of phosphate/lack of other
nutrients/competition for light.
v There is genetic control of the number of maize
grains per plant.

7 a i C ii A iii D iv B
(*two correct for one mark*)
b i Two of:
• more males smoke
• males have a lower life expectancy than females
• more males died in world wars
• there are greater occupational risks for males
• males are more susceptible to diseases.

ii *Note that the pyramid has 1956 written above it. All
the people will therefore be 40 years older and will be in
a higher age band (+40) or have died. The diagram is
awarded 1 mark.*
Fewer babies are being born and this will continue to
narrow the base of the pyramid.

Synoptic questions

Synoptic papers are usually set at the end of a modular course to overcome the problems inherent in such a course, since otherwise the student tends to concentrate only on the subject matter of one particular module at a time. The synoptic questions test a student's overall view of the subject and gives them the chance to show their understanding of the connections between one area of biology and another. For example, questions on Ecology which link the distribution of the organisms concerned with a knowledge of their physiological processes to explain their habitat preferences.

Applied skills such as manipulation of numbers from experimental data, applying statistics, drawing graphs and appreciating the nature of three-dimensional biological structures form an integral part of many synoptic questions. Skills of communication and problem solving are also often tested. Frequently students are presented with unfamiliar examples and data. This allows good students to apply biological concepts and analyse data logically.

It is important that students revise the basics of earlier modules even though these have been successfully completed at an earlier stage of the course. Synoptic skills should have been gradually acquired throughout the course, but can be improved by working through the examples given here *before* looking at the mark scheme. Careful comparison of the student's answer with the mark scheme and an analysis of the points which were omitted or incorrect should lead to an awareness of the type of answers which lead to high grades. Synoptic questions make up a substantial percentage of the marks in both A/S and A level exams.

General types of questions

(Note: these overlap in many examples)

Analysis of results

The student is given a table of results to analyse. They are asked to describe patterns shown in the data and to compare one set of data with another, possibly suggesting reasons for the differences.

They may be asked to suggest and explain the reasons for the link between an increase in a factor and the change in the rate of a process. They may be asked to calculate the rate, the mean, or the percentage difference between one result and another.

Species	Percentage frequency	
	1954	1957
Fescue grasses (*Festuca* sp.)	74	93
Spring sedge (*Carex caryophyllea*)	14	32
Wild thyme (*Thymus drucei*)	25	4
Creeping bent grass (*Agrostis stolonifera*)	10	6
Common rock rose (*Helianthemum chamaecistus*)	34	44
Birdsfoot trefoil (*Lotus corniculatus*)	1	4

	1954	1957
mean height of turf (cm)	2	11.5

1 An investigation was carried out into the effect of grazing by rabbits in an area of chalk grassland.

In 1954 and 1957, the percentage frequency of a number of plant species was determined. At the same time, the mean height of the turf (vegetation) was also measured. In 1956, the rabbit population was greatly reduced in the area due to an outbreak of the disease myxomatosis. In 1956, brambles, trees and shrubs began to spread into the area.

The results are shown in the table (right).
a Explain the effect of the reduction in the rabbit population on the height of the turf. (2)
b i Name two species which grew better in the short turf. (1)
ii Suggest three reasons why the species you have named in **b i** grew better in the short turf. (3)

c Suggest why there were no brambles, shrubs and trees in the area before 1956. (2)
d In this area, there was a marked increase in the number of rabbits in 1960. Sampling was continued in order to monitor the changes in the vegetation.

State two precautions which would need to be taken when sampling was carried out in order to make a valid comparison with the earlier results. (2)
e i Suggest two changes that might occur in the vegetation after the rabbit numbers increased. (2)
ii Suggest how the vegetation in this area might have changed if the rabbit population had remained very low. (3)

London, Paper B6, January 1998

Using specification information

Using information from different areas of the specifications to produce a coherent explanation which fits the facts and demonstrates correct use of biological terms, and includes the analysis of tables.

2 a Micro-organisms present in a rabbit's gut are able to digest carbohydrates in the plant material that they eat. The biochemical pathways by which starch and cellulose are digested in the gut of a rabbit are shown below.

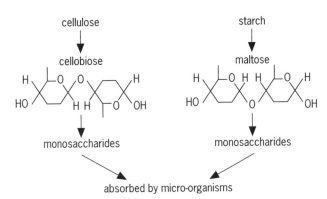

i Describe how a molecule of cellulose differs from a molecule of starch. (1)
ii Draw a diagram to show the molecules produced by the digestion of cellobiose. (2)
iii Cellobiose and maltose are both disaccharides. Explain why amylase enzymes produced by the rabbit are unable to digest cellobiose. (3)

b One way in which rabbits cause considerable damage to agricultural land is by competing for plant material which would normally be eaten by domestic animals.

The table below shows some features of the energy budgets of rabbits and cattle living under the same environmental conditions. All figures are kilojoules per day per kilogram of body mass.

	Rabbits	Cattle
energy consumed in food	1272	424
energy lost as heat	567	311
energy gained in body mass	68	17

i What is the purpose of giving these figures per kilogram of body mass? (1)
ii Explain the difference in the figures for the amount of energy lost as heat. (2)
iii Use the information in the table above to explain why all the energy consumed in food cannot be converted to body mass or lost as heat. (2)

c Rabbits were introduced to Australia in the middle of the last century. Their population grew rapidly and they are now major agricultural pests.

The table below compares some features concerned with heat loss in cattle and rabbits at a temperature of 30°C.

	Cattle	Rabbits
percentage of body heat which is lost by evaporation	81.0	17.0
core temperature of body	38.3	39.3

Use the information given in parts **b** and **c** of this question to explain:
i How evaporation helps cattle maintain a constant body temperature. (2)
ii The main way in which a rabbit would lose heat at an environmental temperature of 30°C. (2)
iii Why rabbits are a major environmental pest in Australia. (2)
iv Why rabbits are better able to survive than cattle in the hot dry conditions found in many parts of Australia. (3)
AQA, Specification A Specimen Unit 8 (terminal synoptic), for 2002 exam

Drawing a graph from data

It is important to ensure that the axes are arranged with the independent variable (the factor being varied by the experimenter) along the horizontal and the dependent variable (experimental results) along the vertical.

The scale should be chosen so that the graph occupies most of the graph area provided. Points should be plotted accurately with a cross or dot enclosed by a circle.

Students should ensure they are familiar with their exam board's preferred way of representing a curve, e.g. using a ruler to join up the dots or drawing a curve of best fit.

Note: the curve should not be extrapolated back to zero if this is not one of the readings.

They may need to use the curve to find the rate of a process.

They may be required to comment on the results and explain these in relation to biological processes.

In other questions, graphs are presented in the text and the student is asked to interpret the information in the light of knowledge which may link several modules.

3 An investigation was carried out into the effects of cytokinin on the glucose content of radish seedlings.

Twelve batches of 20 radish seedlings were used. Six of the batches were treated with a solution of cytokinin. The remaining six batches were used as controls without cytokinin.

All the seedlings were kept in the dark for 5 days and the glucose content of the cotyledons was determined each day.

The results are shown in the graph below.

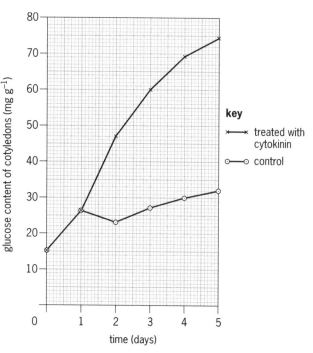

a Compare the glucose content of the cotyledons of the seedlings treated with cytokinin with that of the cotyledons of the control seedlings. (3)

b i Suggest why the seedlings were kept in the dark. (2)

ii Suggest why the glucose content of the cotyledons was determined rather than that of the whole seedling. (2)

c Suggest how cytokinin might contribute to the growth of the seedlings. (3)

Edexcel (London), Paper B6, January 1999

Statistical analysis of experimental data

Students need to check their specification to find which tests are required by their exam board, e.g. chi-square test, Student's *t* test, Mann–Whitney U test. It is also important to know in which circumstances and for which size of sample the tests are applicable.

An understanding of the terms *null hypothesis* and *significant difference* is often required in interpreting the results of these statistical tests and using them to evaluate a hypothesis.

4 An investigation was carried out into the effect of detergent on the rate of growth of germinating radishes.

Sixty radish seeds taken from the same packet, were soaked for 24 hours. Thirty seeds were soaked in distilled water (treatment A) and 30 seeds

were soaked in detergent (treatment B). The seeds were then grown under the same conditions. The lengths of the radicles (roots) of the seedlings were measured after 14 days.

The results are shown in the table below.

Radicle lengths (cm)			
Treatment A		Treatment B	
2.1	4.1	4.0	8.1
4.9	8.4	6.0	9.0
3.2	9.2	7.1	8.1
6.6	6.0	11.1	7.8
6.6	6.4	9.5	5.3
7.7	5.3	8.3	7.7
2.8	4.1	6.0	9.8
9.2	7.3	9.6	8.1
7.2	8.1	8.8	10.7
5.3	5.0	6.5	9.5
5.8	6.4	8.0	8.9
6.5	8.5	10.5	7.1
6.0	7.2	7.8	6.3
8.1	6.7	8.7	8.8
3.5	7.7	5.8	9.0
total A = 185.9		total B = ?	
mean A ($\bar{x}_A$) = 6.20		mean B ($\bar{x}_B$) = ?	

a A *t* test can be used to determine whether there is a significant difference between the two treatments.

i Suggest a null hypothesis for this investigation. (1)

ii The formula used in the *t* test is given below.

$$t = \frac{\bar{x}_A - \bar{x}_B}{\sqrt{\left(\dfrac{s_A^2}{n_A} + \dfrac{s_B^2}{n_B}\right)}}$$

where $\bar{x}_A$ is the mean radicle length of treatment A

$\bar{x}_B$ is the mean radicle length of treatment B

s_A^2 is the variance for treatment A

s_B^2 is the variance for treatment B

n_A and n_B are each 30, the number of radicles measured for each treatment

The value for s_A^2 is 3.4 and the value of s_B^2 is 2.7. Using the values from the table calculate the value of *t*. Show your working. (3)

iii State the number of degrees of freedom for this investigation. (1)

iv A statistical table showed that the critical value at the 5% level with this number of degrees of freedom was 2.00. What does this tell you about the difference between the mean radicle lengths from the treatments A and B? (1)

b Suggest two reasons why the seeds were taken from the same packet. (2)

London Paper, B6 June 1998

5 Warfarin is a pesticide which is used to kill rodents such as rats and mice. Some rodents are resistant to warfarin and such resistance was first discovered in wild rats on farms around Welshpool in 1959. Resistance to warfarin is controlled by a gene with two alleles W^1 and W^2. In 1959, a study was made of the genotypes of rats on farms around Welshpool where warfarin had been used. The genotypes of 74 trapped rats were determined. The results are shown in the table below. The table also shows the expected numbers of each genotype, assuming the frequency of the two alleles, W^1 and W^2, in the population remains constant.

Genotype	Phenotype	Observed number	Expected number
W^1W^1	susceptible to warfarin	28	32
W^1W^2	resistant to warfarin	42	33
W^2W^2	resistant to warfarin	4	9

a i Determine whether the allele for resistance for warfarin, W^2 is dominant or recessive, giving a reason for your answer. (1)
ii The chi-squared test was used to determine whether the difference between observed and expected results is significant. A value of 5.73 was obtained for χ^2.

Using the extract from the table of values, state whether the difference is significant, giving a reason for your answer. (2)

Probability levels P (%)	99	10	5	2	1	0.1
χ^2		0.00	2.71	3.84	5.41	6.64	10.83

b Rats with the genotype W^2W^2 require much more vitamin K in their diet than those with the other genotypes.
i Calculate the observed number of rats of each homozygous genotype as a percentage of the expected number of that genotype. Show your working. (3)
ii Both homozygous genotypes are at a disadvantage compared to the heterozygous genotype. Using your calculated results from **b i**, state which of the homozygous genotypes is at the greater disadvantage. (1)
iii Give a reason why each of these genotypes is at a disadvantage compared with the heterozygous genotype. (2)
c A further study was carried out in a nearby area where warfarin was also used. In 1973, nearly 60% of the rats were resistant to warfarin. The use of warfarin was then discontinued and after 2 years the number of resistant rats had dropped to less than 40%. Explain why the number of resistant rats dropped after the warfarin was discontinued. (3)

d Explain why the allele for resistance to warfarin is likely to remain in the gene pool when the use of warfarin is discontinued. (2)
Edexcel (London), Paper B6, January 1999

Understanding and interpretation of diagrams

> The student may be asked to put a series of diagrams or photographs in sequential order, for example the stages in meiosis. They may be asked to complete flow charts or to redraw a longitudinal view from a diagram of a transverse section in order to demonstrate their understanding of the three-dimensional nature of the cell/tissue/organism.

6 In Europe, the common shrew, a small insectivorous mammal, seems to include two species, one found in western Europe and the other in more northern and eastern parts. The two species are virtually indistinguishable, recognised only by differences in their chromosomes. The basic diploid complement of these species is ten pairs of autosomes plus sex chromosomes.

The diagram below shows chromosomes from the males of each species. One chromosome of each pair of autosomes (numbered **1** to **10**) of the western species is shown alongside the equivalent chromosome of the eastern species.

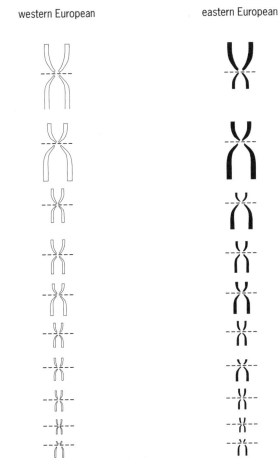

a i State the numbers of all the chromosomes which differ between the two species. (1)
ii In what way do these chromosomes differ? (1)
iii Explain how these differences may have occurred. (2)
b The diagram below shows the sex chromosomes of a male western European common shrew.

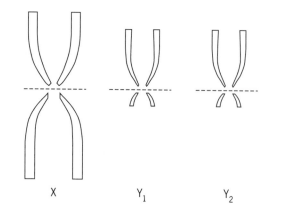

State one way in which the sex chromosome complement of the male common shrew differs from that which is typical of most mammals. (1)
c In areas of Europe where the two species of common shrew overlap in their natural populations they do not interbreed successfully. Suggest two reasons why they do not interbreed successfully. (2)

London, Paper B6, June 1998

7 The diagram below illustrates energy flow diagrams for **A** a deciduous forest and **B** a marine community. The units on the flow charts are kJ m⁻² day⁻¹. The major plants in the marine community are phytoplankton.

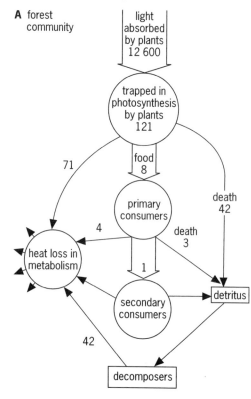

a Calculate the efficiency with which solar energy is trapped by the forest plants. (1)
b With reference to the diagram above, describe and explain the major differences between the energy flow in these two communities. (*For this question, 1 mark is available for the quality of written communication.*) (8)
c Suggest two ways in which the deciduous forest may be managed for timber production. (2)
OCR Sample Synoptic Paper for first teaching in 2000

Comprehension questions

In this type of question a passage from a scientific journal is used as a basis to test the student's knowledge of that particular topic by asking them to explain phrases within the text. Such questions also test their grasp of scientific English.

8 Read the following passage.

Madagascar
The island of Madagascar has been described as the laboratory of evolution. It broke away from mainland Africa at least 120 million years ago and, following this, many new species developed. Estimates of the number of plant species on the island vary from 7370 to 12000, making it botanically one of the richest areas in the world. Of 400 flowering plant families found worldwide, 200 only grow in Madagascar. Among animals, true lemurs are found nowhere else, and 95% of the country's 235 known species of reptiles evolved on the island. Over the past 25 years the human

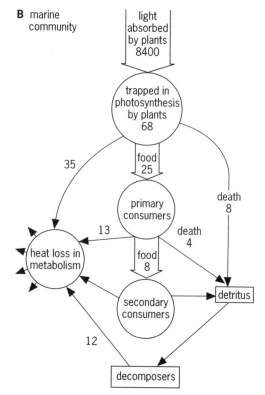

266

population has doubled. Land shortage is leading to clearing of the forest, and Madagascar is now facing deforestation on a massive scale. Scientists have estimated that even if the forest could recover, regeneration could take up to 100 years.

(Adapted from an article in New Scientist*)*

a Explain what is meant by the term *species*. (2)
b i Explain the processes which might have led to the evolution of new species on the island. (4)
   ii Suggest and explain how the number of animal species on the island may have changed as it became 'botanically one of the richest areas in the world'. (2)
c Describe the processes by which a forest is able to regenerate after being cleared. (4)

AQA (NEAB), BY02 (Section B), March 1999

9 Read the passage and answer the questions which follow.

Red and grey squirrels

The red squirrel *Sciurus vulgaris*, is not as common in Britain as it was a century ago. The grey squirrel, *Sciurus carolinensis*, is now extremely common.

One suggestion for the relative success of the grey squirrel in Britain is that it is able to out-compete the red squirrel. The popular view is that the larger grey squirrels attack the red squirrels. In fact, the two species take very little notice of each another.

Grey squirrels spread slowly and where they have been established for more than 15 years, red squirrels were usually missing. Detailed studies have shown that in some areas the two species have co-existed 16 years or more, but in other areas the reds had disappeared before the greys arrived.

Comparisons of the two species suggest ways in which their ecologies differ so that red squirrels probably do better in coniferous woodland and grey squirrels better in deciduous woods. For instance, reds spend much more time in the canopy and less time on the ground than greys; this matches the fact that they are lighter, more nimble and put on less fat for the winter. Most conifers take two years to ripen their cones, so there are always cones available in the canopy for a squirrel which is light enough and nimble enough to reach them. The grey squirrel is something of a specialist, feeding mainly on large seeds, such as acorns and beechnuts, that are abundant on the ground in autumn in broad-leaved woodlands. Both squirrels can produce two litters a year; a female grey squirrel, which has the ability to exploit the rich autumn seed crop, will be better placed to produce a strong litter of young early in the year.

Grey squirrels have a further advantage which has probably been decisive. They can digest acorns more efficiently than red squirrels who do eat acorns but they cannot digest them properly. Conservationists have criticised the extensive use of alien conifers by commercial forestry in Britain and have asked for more native conifers and broad-leaved trees to be planted. This is just what grey squirrels prefer. Serious consideration is now being given to felling oaks in areas which are red squirrel strongholds.

(Adapted from: Yalden D.W., Squirrel Dynamics, Biological Sciences Review, Vol. 6 No. 2. Philip Allan Publishers, 1993.)

a State **three** pieces of evidence that support the idea that grey squirrels have not **actively** displaced the native red ones. (3)
b Explain whether there is any evidence for the idea that the two species of squirrel occupy the same niche. (3)
c Explain why:
   i Red squirrels are better adapted than grey squirrels to live in coniferous woodland. (2)
   ii Grey squirrels are better adapted than red squirrels to live in deciduous woodland. (2)
d Describe the measures that may be taken to encourage the red squirrel to gain an advantage over the grey squirrel. (3)

OCR Sample Synoptic paper for first teaching in 2000

Synoptic essays

A choice of essays may be offered including essays on topics which require information from several modules and are a test of the student's breadth of knowledge and understanding. Other essays may be based on the specialised options offered within a particular module.

Synoptic essays usually include marks for punctuation, spelling and grammar as well as scientific knowledge and a balanced overview. It is important to take care with use of English as this could improve your grade.

Examples of synoptic essay topics:
Chemical co-ordination in plants and animals (sample answer provided for this question only) (20)

London, Paper B6, June1998

The role of lipids in living organisms
Causes and effects of water pollution
Homeostasis
The roles of enzymes in metabolism
Gas exchange in living organisms
Gene technology
Pigments in living organisms
The nitrogen cycle
Cell surface membranes
The structure and functions of proteins
Ectothermy and endothermy
ATP and its role in living organisms

Synoptic answers

1 *Note: this question is deceptively simple but many of the sections require careful wording in order to qualify for marks.*

   *In section **a** a mark is available for manipulating the figures to show the percentage reduction in turf height. It is often a useful idea to manipulate the numbers to give a comparison between different sets of data in questions of this type.*

   Note: the question uses Latin binomial names for the organisms but it is not necessary for the student to have come across these species before. It is important not to be put off by unfamiliar words but to try to analyse the data calmly.

   *Note in **c** the emphasis is on the seedling stages of brambles, etc, since once these are established as large plants/trees, the rabbits would find them more difficult/have less effect on them.*

   *In **d** the question is evaluating the student's knowledge of experimental technique.*

   a Two of:
   - An increase in the height of the turf.
   - The turf height had increased by $9.5/11.5 \times 100\%$ (or some other calculation).
   - The turf was not being eaten/grazed by the rabbits.

   b **i** Wild thyme and creeping bent grass.
   ii Three of:
   - The species are low growing and so are missed by the rabbits.
   - The rabbits find those plants distasteful.
   - Those plants cannot compete with taller plants/they need more light, etc.
   - In low turf there is less competition for water/nutrients.
   - Those species benefit from nutrients in the rabbit dung.

   c Two of:
   - The seedlings could not become established.
   - They were eaten by the rabbits in the young stage/seedling.
   - They could not compete with other plants/slower growing.

   d *This section is testing knowledge of a fair test.*

   Two of:
   - The same area should be sampled as before/same number of quadrats.
   - Same size of quadrat/same method/technique.
   - Sampling at the same time of year.

   e **i** Two of:
   - The turf would be grazed/be shorter.
   - An increase in wild thyme/creeping bent grass.
   - A decrease in a named species, e.g. spring sedge/birdsfoot trefoil.
   - Fewer seedlings of brambles/shrubs/trees.
   ii Three of:
   - An increase in larger plants/trees/shrubs.
   - Explain the changes by reference to succession.
   - Refer to climax vegetation/development of woodland.
   - A reduction in the numbers of ground species/loss of ground species.
   - A further reduction in the percentage frequency in ground species.

2 *This question tests knowledge from many areas of the specifications including the hydrolysis of carbohydrates and the energy changes involved in changes from liquid to gas and methods of transfer of heat energy.*

   a **i** Cellulose is made from β-glucose and starch from α-glucose.
   ii See diagram showing two monosaccharides; correct groups produced as a result of hydrolysis.

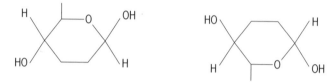

   iii Cellobiose and maltose have different shapes. Cellobiose will not fit/bind to/form an enzyme–substrate complex;
   with the active site of amylase.

   b **i** To take into account the different size of the animals.
   ii Heat is lost from the surface.
   The rabbit is smaller and has a greater surface area–volume ratio.
   iii Some of the energy consumed in food is used by micro-organisms;
   for respiration or microbial growth.

   c **i** Two of:
   - Evaporation requires (latent) heat;
   - to convert water to vapour;
   - this heat is drawn from the animal's body.
   ii The rabbits must lose heat directly to the cooler environment by radiation/convection.
   iii Two of:
   - Rabbits are not kept in check by natural predators/there are no predators of rabbits in Australia.
   - They consume much more food per kilogram than cattle.
   - There is competition for scarce resources.
   iv Three of:
   - Rabbits have a higher body temperature and so do not have to lose so much heat.
   - They can lose more heat to the environment by radiation.
   - They do not have to rely as much on sweating as cattle.
   - They are less dependent on drinking water.

3 *Note in this question the first part is a **comparison** so it is important to read glucose values from the graphs and compare them mathematically.*

   a Three of:
   - The glucose content is the same for both treated and untreated from day 0 to day 1/or converse/or the treated is always higher than the control after day 1.
   - The treated increases faster/at a greater rate than the control/or converse.

- Glucose content rose steadily in the treated whereas the control fluctuates.
- Some mathematical comparison e.g. after 5 days the treated contained $74/34 \times 100\% = 218\%$ more glucose than the controls/more than twice as much.

b **i** Two of:
- Photosynthesis would occur in the light;
- this would change the level of carbohydrate/glucose.
- Variables such as light intensity need to be controlled.

ii Two of:
- Other cells/the rest of the seedling, use more carbohydrate in their respiration/metabolism.
- Glucose is produced in the cotyledons;
- from starch.

c Three of:
- Cytokinin stimulates the synthesis of carbohydrases/amylases.
- Glucose is produced by the hydrolysis of starch.
- Glucose is released for respiration.
- Glucose/energy is available for anabolic reactions/synthesis of named example, e.g. DNA synthesis.
- Cytokinin may stimulate cell division.
- May promote cell expansion.

4 *It is important to ensure that you have a calculator and spare battery.*

a **i** There is no significant difference between the radicle lengths from the two treatments.

ii *Since 3 marks are available, it is important that you show your working out. Start by substituting the relevant figures in the formula. The mean values in the table are given to two decimal places and this can be used as a guide to the number of decimal places in your answer.*

$$t = \sqrt{\frac{6.20 - 8.06}{(3.4/30 + 2.7/30)}} = \frac{1.86}{0.45} = 4.13/4.12$$

iii 58

iv There is a significant difference/the null hypothesis should be rejected/the difference is not due to chance.

b Two of:
- To reduce genetic variation/ensure seeds are of the same variety.
- To ensure seeds are of the same age.
- To ensure they have a similar nutrient content.
- A reference to being kept under the same storage conditions, e.g. temperature.

5 **a** **i** It is dominant, as rats are resistant to warfarin in the heterozygote/W^1W^2 condition.

ii The difference is significant (at the 5% level). The probability of obtaining a value of 5.73 is less than 5%/would expect to find such variation in 1% to 2% of observations by chance.

b **i** W^1W^1 $\frac{28}{32} \times 100 = 87.5\%$

Alternative method
W^2W^2 $(4/9 \times 100) = 44.4\%$

ii Homozygous dominant/W^1W^2

iii W^1W^1 killed by warfarin/not resistant to warfarin. W^2W^2 needs so much more vitamin K in the diet.

c Three of:
- The homozygous recessive will no longer be at a disadvantage (if warfarin is not used)/resistant rats will no longer be at an advantage;
- so the numbers of non-resistant rats will increase.
- The homozygous dominant rat still needs high quantities of vitamin K/those carrying W^1 need less.
- There will be greater competition for food resources in the environment.
- There is no effect on the heterozygotes.
- Correctly refer to a change in selection pressure.

d Two of:
- The resistant allele is still present in the heterozygotes.
- It may occur as a result of mutation.
- There is no selection against heterozygotes/refer to the many generations occurring before the allele frequency changes.

6 **a** **i** 1, 3, 4, 7

ii One of:
- the arms of the chromosomes/chromatids differ in length
- the centromere is in a different position.

iii Two of:
- mutation
- translocation/chromosome mutation
- caused by ionising radiation/e.g. radiation/chemical mutagens.

b The male shrew has two Y chromosomes.

c Two of:
- different habitats/niches
- different behaviour patterns/do not mate with each other
- the breeding cycles do not coincide
- failure of fertilisation
- chromosomes do not pair
- sterile offspring/hybrid inviability.

7 **a** 0.96% ($121/12\,600 \times 100$)

b *Note: it is important to make sure that the text is legible with accurate spelling, punctuation and grammar to gain an extra mark.*

*Make sure that the **differences** are both **described** and **explained**.*

There are many points of comparison which are worthy of marks but remember to support your description by reference to the figures.

Seven of:
- 50% more energy absorbed by leaves of forest plants than phytoplankton.
- Light is a limiting factor of photosynthesis and the lower levels of light available underwater will reduce the amount of food energy at the base of the food chains.
- Carbon dioxide and temperature may also limit photosynthesis – comparison for **A** and **B**.

- There is a higher energy flow through the consumer food chain in **B**.
- There is a higher energy flow through the detritus food chain in **A**.
- There is a larger population of decomposers in the forest.
- There is much inedible material produced in the forest, e.g. wood.
- There is a smaller number/biomass of secondary consumers in **A**.
- Compare the efficiency of the transfer of energy between trophic levels.

c The large trees are felled and replaced by planting young trees.
Coppicing is used – fast growing trees are cut to ground level, every 2–4 years.

8 *This question is from AQA Continuity of Life module but is a synoptic type question. It requires an understanding of a scientific passage and also interlinks knowledge of both ecological and evolutionary processes.*

a Two of:
- A species has similar characteristics/physically similar/DNA similar.
- Breed among themselves;
- to produce fertile offspring.
- Do not share an ecological niche with other species.

b i Four of:
- isolation
- no gene flow between populations
- variation
- different environmental factors
- natural selection/selection for specific alleles/characteristics
- change in allele/phenotype frequency
- changes over a long period of time.

ii Two of:
- more habitats/niches
- more/greater range of food for herbivores
- more/greater range of food for carnivores/predators
- more detritus.

c Four of:
- colonisation/pioneer species
- succession
- alteration of the habitat/more humus/deeper soil
- development of a herbaceous/field layer (followed by a shrub layer).

9 a They have lived together for 16 years.
They ignore each other.
In some areas the reds disappeared before the greys arrived.

b They do not occupy the same niche because:
Three of:
- The reds are found in coniferous woods and greys in deciduous woods.
- The red feeds on small seeds, e.g. conifer seeds; the grey feeds on large seeds, e.g. acorns.

- The red spends more time in the canopy than the grey.
- They can coexist in the same areas.

c i Two of:
- light
- nimble
- they put on less fat in the winter.

ii Two of:
- They can digest acorns.
- Eating lots of acorns in autumn increases their body reserves.
- This increases their breeding success.

d *Note: some of the acceptable answers to this question are not found in the passage.*
Three of:
- cull/kill greys
- feed red squirrels
- plant trees with small seeds
- remove oaks
- use contraceptives on the greys.

Synoptic essay – Chemical co-ordination in plants and animals

This is typical of a synoptic essay with a wide area of subject. It is important to plan the essay so that too much time is not spent on one aspect at the expense of others, since marks are awarded for a well-balanced essay. The rough plan can be written in abbreviated form in pencil and later crossed out. Care should be taken however not to spend more than 3 to 5 minutes on the plan. During the essay writing, any new ideas can be slotted into the plan at the appropriate place. In synoptic essays there should be a brief introduction and a logical order. Care should be taken with spelling, grammar and punctuation. Students will be penalised for incorrect use of terminology and irrelevant or incorrect statements.

Chemical co-ordination in animals – brief summaries of:
- Reference to the need for regulation of the internal environment.
- The concept of negative feedback.
- Hormones and chemical co-ordination – insulin and glucagon in the regulation of blood glucose.
- Action of ADH, the role of osmoregulators.
- Hormonal control of heart rate.
- Hormones in reproduction.
- Roles of luteinising hormone, follicle stimulating hormone, testosterone, oestrogen, progesterone, oxytocin and prolactin.

Chemical co-ordination in plants – reference to the effects of:
Auxins; cytokinins; gibberellins; abscisic acid; ethene.

(scientific content 17 marks, balance 3 marks, clarity 3 marks, total max 20 marks)

AS and A level specification matrix

Board	AQA (AEB) Biology A						AQA (NEAB) Biology B						Edexcel Biology							OCR Biology						WJEC			
Module	1 (B/HB)	2 (B)	3 (HB)	5 (B/HB)	6 (B)	7 (HB)	1	2	3	4	5	6	1	2B	2H	3	4	5B	5H	A	B	C1	D	E1	E5	1	2	4	5
1	✓						✓						✓							✓						✓			
2	✓						✓						✓							✓						✓			
3	✓						✓						✓							✓						✓			
4	✓						✓						✓							✓						✓			
5		✓	✓					✓					✓							✓						✓			
6		✓	✓					✓					✓							✓						✓			
7		✓	✓					✓					✓							✓									✓
8		✓	✓					✓					✓							✓						✓			
9	✓						✓						✓							✓						✓			
10	✓						✓						✓							✓						✓			
11				✓						✓								✓			✓							✓	
12				✓						✓								✓			✓							✓	
13				✓						✓								✓			✓							✓	
14				✓						✓							✓				✓							✓	
15				✓						✓							✓				✓							✓	
16				✓						✓							✓				✓							✓	
17					✓	✓				✓							✓				✓								✓
18					✓	✓				✓							✓				✓								✓
19					✓	✓				✓							✓				✓								✓
20					✓	✓						✓												✓					
21			✓									✓			✓	✓					✓							✓	
22					✓	✓				✓							✓							✓					✓
23					✓	✓				✓							✓					✓							✓
24					✓	✓				✓							✓					✓							✓
25																						✓							✓
26	✓							✓						✓	✓						✓						✓		
27	✓							✓						✓	✓						✓						✓		
28				✓				✓					✓								✓						✓		
29				✓				✓					✓								✓						✓		
30				✓	✓		✓										✓				✓							✓	
31				✓	✓		✓										✓							✓				✓	
32				✓				✓						✓	✓						✓						✓		
33				✓	✓			✓						✓	✓						✓						✓		
34				✓						✓							✓					✓							✓
35				✓						✓							✓					✓							✓
36						✓	✓						✓										✓						✓
37			✓				✓							✓	✓							✓				✓			
38				✓						✓								✓	✓			✓							✓
39				✓						✓								✓	✓			✓							✓
40				✓						✓								✓	✓			✓							✓
41				✓						✓								✓	✓			✓							✓
42				✓					✓							✓				✓							✓		
43				✓					✓									✓				✓					✓		
44									✓					✓								✓					✓		
45									✓									✓				✓					✓		

271

Index